ATLANTIS STUDIES IN MATHEMATICS

VOLUME 3

SERIES EDITOR: J. VAN MILL

AF739402

Atlantis Studies in Mathematics

Series Editor:

J. van Mill

VU University Amsterdam, Amsterdam, the Netherlands

(ISSN: 1875-7634)

Aims and scope of the series

The series 'Atlantis Studies in Mathematics' (ASM) publishes monographs of high quality in all areas of mathematics. Both research monographs and books of an expository nature are welcome.

For more information on this series and our other book series, please visit our website at:

www.atlantis-press.com/publications/books

ATLANTIS PRESS

AMSTERDAM – PARIS

© ATLANTIS PRESS

Analytic Inequalities

Recent Advances

B.G. Pachpatte †

57, Shri Niketen Coloney
Aurangabad
India

ATLANTIS
PRESS

AMSTERDAM – PARIS

Atlantis Press

8, square des Bouleaux
75019 Paris, France

For information on all Atlantis Press publications, visit our website at: *www.atlantis-press.com*

Copyright

This book, or any parts thereof, may not be reproduced for commercial purposes in any form or by any means, electronic or mechanical, including photocopying, recording or any information storage and retrieval system known or to be invented, without prior permission from the Publisher.

Atlantis Studies in Mathematics

Volume 1: Topological Groups and Related Structures, A. Arhangel'skii, M. Tkachenko
Volume 2: Topics in Measure Theory and Real Analysis, A.B. Kharazishvili

ISBNs	
Print:	978-94-91216-43-5
E-Book:	978-94-91216-44-2
ISSN:	1875-7634
ISBN:	978-94-6239-059-1

© 2012 ATLANTIS PRESS
Softcover reprint of the hardcover 1st edition 2012

To the memory of my father.

Preface

Analytic inequalities are widely acknowledged as one of the major driving forces behind the development of various branches of mathematics and many applied sciences. The study of inequalities has increased enormously over recent decades as it has been demonstrated that they have applications in many diverse fields of mathematics. There exists, for example, a very rich literature related to the Čebyšev, Grüss, Trapezoid, Ostrowski, Hadamard and Jensen inequalities. This monograph is an attempt to organize recent progress related to these in the hope that it will further broaden developments and the scope of applications. It does not intend to be comprehensive, but rather it is meant to be a representative overview of the recent research related to the fundamental inequalities noted above.

A large part of the material included in the book can only be found in the research literature although it should be understandable to any reader with a reasonable background in real analysis and its related areas. It will be a valuable source of reference in the field for a long time to come. All results are presented in an elementary way and it could also serve as a textbook for an advanced graduate course.

The author is grateful to Professor Jan van Mill and Arjen Sevenster for the opportunity to publish this book and their invaluable professional cooperation for the work reported here. I am also indebted to the editorial and production staff of the publisher for the care they have taken with this book. I would like to thank my family members for providing their strong support and constant encouragement during the writing of this monograph.

B.G. Pachpatte

Publisher's Note

Unfortunately the author passed away during the last phase of the manuscript preparation and was therefore unable to see the final realization of this topical and interesting work.

Contents

Introduction

Mathematical inequalities have played an important role since the time of A.L. Cauchy, P.L. Čebyšev, C.F. Gauss and many others in establishing the foundations for methods of approximation. Around the end of the nineteenth and the beginning of the twentieth century, numerous inequalities were investigated and used in almost all branches of mathematics as well in other areas of science and engineering. The pioneering work Inequalities by Hardy, Littlewood and Polya [68] appeared in 1934, transformed the field of inequalities from a collection of isolated formulas into a systematic discipline. This work presents fundamental ideas, results and techniques and it has had much influence on research in various branches of analysis. Since 1934, a considerable variety of inequalities have been proposed and studied in the literature. Excellent surveys of the work done up to the years of their publications, together with many references can be found in the books by Beckenbach and Bellman [10] and Mitrinović [78]. These three major books serve as mere stepping-stones to the recent vast literature in the subject.

The study of various types of inequalities has been the focus of great attention for well over a century by many researchers, interested both in theory and applications. Various approaches are developed by different researchers for handling a variety of analytic inequalities. There are several classical and notable books that introduce new researchers to the basic results, methods and applications and at the same time, serve the dual purpose of textbooks for graduate students in many different fields of mathematics.

Over the past two decades or so, the field of inequalities has undergone explosive growth. Concerning numerous analytic inequalities, in particular a great many research papers have been written related to the inequalities associated to the names of Čebyšev, Grüss, Trapezoid, Ostrowski, Hadamard and Jensen. A number of surveys and monographs published during the past few years described much of the progress. However, these expositions are still far from being a complete picture of this fast developing field. The literature related to the above mentioned inequalities is now very extensive, it is scattered in various journals encompassing different subject areas. There is thus an urgent need for a book that brings readers to the forefront of current research in this prosperous field. The rapid development of this area and the variety of applications drawn from various fields motivated and inspired us to write the present monograph.

The subject of inequalities being so vast, most recent books on this subject cover only a certain class of inequalities and either try to be encyclopedic within that class or bring the rapidly expanding field of inequalities up-to-date in some area. The present monograph is an attempt to provide its readers with a representative overview related to the above noted inequalities and it is not the intention to attempt to survey this voluminous literature. We mostly focus on certain advances, particularly not covered in the recent surveys and monographs. Below, we briefly summarize some fundamental inequalities, which grately stimulated the present work. By doing this, we hope to help the reader to prepare for more recent results that will be considered in subsequent chapters.

One of the many fundamental mathematical discoveries of P.L. Čebyšev [13] is the following classical integral inequality:

$$|T(f,g)| \leqslant \frac{1}{12}(b-a)^2\|f'\|_\infty\|g'\|_\infty, \tag{1}$$

where $f,\ g : [a,b] \to \mathbb{R}$ are absolutely continuous functions, with first derivatives of which, $f',\ g'$ are bounded and

$$T(f,g) = \frac{1}{b-a}\int_a^b f(x)g(x)dx - \left(\frac{1}{b-a}\int_a^b f(x)dx\right)\left(\frac{1}{b-a}\int_a^b g(x)dx\right), \tag{2}$$

provided the involved integrals in (2) exist. The representation (2) is known in the literature as the Čebyšev functional. The inequality (1) which first appeared in 1882 is now known in the literature as Čebyšev's inequality. Over the years, this inequality has evoked the interest of numerous researchers and a large number of results related to the inequality (1) have been published, see [79,144] and the references given therein.

In 1935, G. Grüss [61] proved an interesting integral inequality that gives an estimate of the difference between the integral of the product of two functions and the product of their integrals, as follows:

$$|T(f,g)| \leqslant \frac{1}{4}(\Phi-\phi)(\Gamma-\gamma), \tag{3}$$

where $f,\ g : [a,b] \to \mathbb{R}$ are integrable on $[a,b]$ and satisfy the condition

$$\phi \leqslant f(x) \leqslant \Phi, \quad \gamma \leqslant g(x) \leqslant \Gamma,$$

for each $x \in [a,b]$, where ϕ, Φ, γ, Γ are given real constants and $T(f,g)$ is given by (2). For a simple proof of (3) as well as some other integral and discrete inequalities of the Grüss type, see the book [79] by Mitrinović, Pečarić and Fink.

The following inequality is well known in the literature as the Trapezoid inequality:

$$\left| \int_a^b f(x)dx - \frac{b-a}{2}[f(a)+f(b)] \right| \leqslant \frac{1}{12}(b-a)^3 \|f''\|_\infty, \tag{4}$$

where the mapping $f : [a,b] \to \mathbb{R}$ is assumed to be twice differentiable on the interval (a,b), with the second derivative bounded on (a,b), that is,

$$\|f''\|_\infty = \sup_{x\in(a,b)} |f''(x)| < \infty.$$

The inequality (4) has received a considerable attention and many results related to this inequality have appeared in the literature. A detailed discussion related to the inequality (4) can be found in a recent paper [17] by Cerone and Dragomir, see also [3,58,140].

Now if we assume that $I_n : a = x_0 < x_1 < \cdots < x_{n-1} < x_n = b$ is a partition of the interval $[a,b]$ and the function f is as above, then we can approximate the integral $\int_a^b f(x)dx$ by the trapezoidal quadrature formula $A_T(f,I_n)$, having an error given by $R_T(f,I_n)$, where

$$A_T(f,I_n) = \frac{1}{2}\sum_{i=1}^{n-1}[f(x_i)+f(x_{i+1})]h_i, \tag{5}$$

and the remainder satisfies the estimation

$$|R_T(f,I_n)| \leqslant \frac{1}{12}\|f''\|_\infty \sum_{i=0}^{n-1} h_i^3, \tag{6}$$

with $h_i = x_{i+1} - x_i$ for $i = 0,1,\ldots,n-1$. Expression (5) is known as the trapezoidal rule, if $n = 1$, and as the composite trapezoidal rule for $n > 1$. The trapezoidal rule is widely used in practice since it is easy to implement in an efficient fashion, especially if the partitioning is done in a uniform manner.

In 1938, A.M. Ostrowski [81] proved the following useful inequality (see also [80]).

Let $f : [a,b] \to \mathbb{R}$ be continuous on $[a,b]$ and differentiable on (a,b) with derivative $f' : (a,b) \to \mathbb{R}$ being bounded on (a,b), that is, $\|f'\|_\infty = \sup_{x\in(a,b)} |f'(x)| < \infty$. Then

$$\left| f(x) - \frac{1}{b-a}\int_a^b f(t)dt \right| \leqslant \left[\frac{1}{4} + \frac{\left(x - \frac{a+b}{2}\right)^2}{(b-a)^2} \right](b-a)\|f'\|_\infty, \tag{7}$$

for all $x \in [a,b]$.

The inequality (7) gives an upper bound for the approximation of the integral average

$$\frac{1}{b-a}\int_a^b f(t)dt$$

by the value $f(x)$ at the point $x \in [a,b]$. In the last decade a great number of important results on this topic have been appeared in the literature. An excellent survey of the work on Ostrowski type inequalities together with many references are contained in the recent book [50] edited by Dragomir and Rassias.

Among numerous inequalities involving convex functions, the following inequality (see [45,108]):

$$f\left(\frac{a+b}{2}\right) \leqslant \frac{1}{b-a}\int_a^b f(x)dx \leqslant \frac{f(a)+f(b)}{2}, \tag{8}$$

which holds for all convex functions $f : [a,b] \to \mathbb{R}$ is known in the literature as Hadamard's inequality. The inequality (8) is remarkable in terms of its simplicity, the large number of results to which it leads, and the variety of applications which can be related to it. Due to its importance in various applications, this result has attracted a great deal of attention over the years and a number of papers related to it have appeared in the literature.

The following inequality is well known in the literature as Jensen's inequality (see [78,108,144]):

$$f\left(\frac{1}{P_n}\sum_{i=1}^{n} p_i x_i\right) \leqslant \frac{1}{P_n}\sum_{i=1}^{n} p_i f(x_i), \tag{9}$$

where $f : X \to \mathbb{R}$ be a convex mapping defined on the linear space X and $x_i \in X$, $p_i \geqslant 0$ $(i = 1,\ldots,n)$ with $P_n = \sum_{i=1}^{n} p_i > 0$.

There are many well known inequalities which are particular cases of inequality (9), such as the weighted Arithmetic mean-Geometric mean-Harmonic mean inequality, the Ky-Fan inequality, the Hölder inequality etc. For a comprehensive list of recent results on Jensen's inequality, see the book [108] where further references are also given.

A large number of results related to the above inequalities have recently appeared in the literature. Indeed, a particular feature that makes these inequalities so fascinating arises from the numerous fields of applications. The literature related to the above inequalities is vast and rapidly growing vaster. The part of this growth is due to the fact that the subject is genuinely rich and lends itself to many different approaches and applications. Some of the results, recently discovered in the literature provide simple and elegant extensions, generalizations and variants of the above inequalities and thus have a wider scope of applicability.

Taking into account the great variety of results related to the above inequalities, the choice of material for a book is a difficult task. The selection of the material is largely influenced with a view to provide basic tools for researchers working in mathematical analysis and applications. The material it presents is new and never appeared in the book form before, and will be a valuable source to both experts and non-experts in the field.

A brief description of the organization of the book is as follows.

The work is arranged in five chapters and references. Chapters 1 and 2 presents a large number of new basic results related to Grüss, Čebyšev, Trapezoid type inequalities involving functions of one and many independent variables investigated by various researchers. These results offers a representative overview of the major recent advances in the field as well as the diversity of the subject. Chapters 3 and 4 are devoted to present most recent results on Ostrowski type inequalities involving functions of one and several independent variables. These results reflect some of the major recent advances in the field. Chapter 5 contains some basic inequalities involving convex functions investigated by various researchers during the past few years. We hope that these results will provide new directions of thinking besides extremely important inequalities due to Hadamard and Jensen. Each chapter contains sections on applications and miscellaneous inequalities for further study and notes on bibliographies. A list of references does not include titles related to the topics, which we have not covered in this book. Without any intention of being complete, here only those references used in the text are given.

Throughout, we let $\mathbb{R}$, $\mathbb{C}$, $\mathbb{Z}$ and $\mathbb{N}$ denote the set of real, complex, integers and natural numbers respectively. Let $\mathbb{R}_+ = [0,\infty)$, $[a,b] \subset \mathbb{R}$ $(a<b)$, $\mathbb{N}_0 = \{0,1,2,\ldots\}$, $\mathbb{N}_{\alpha,\beta} = \{\alpha, \alpha+1, \ldots, \alpha+n = \beta\}$ for $n \in \mathbb{N}$, α, $\beta \in \mathbb{N}_0$. The derivatives of a function $u(t)$, $t \in \mathbb{R}$ are denoted by $u^{(I)}(t)$ for $i = 1, \ldots, n$. The function $u : [a,b] \to \mathbb{R}$ is said to be bounded on $[a,b]$, if $\|u\|_\infty = \sup\limits_{x\in[a,b]} |u(x)| < \infty$. The notation, definitions and symbols used in the text are standard or otherwise explained.

The book is largely self-contained. It thus should be useful for those who are interested in learning or applying inequalities in their studies regardless of their specific subject focus. It will be an invaluable reading for mathematicians, physicists and engineers and also for graduate students, scientists and scholars wishing to keep abreast of this important area of research. Most of the inequalities included in the book are recent innovations and it is hoped that they will provide motivation for future research work.

Chapter 1

Grüss-and Čebyšev-type inequalities

1.1 Introduction

In 1882, P.L. Čebyšev [13] proved the remarkable inequality given in (1). In a celebrated paper of 1935, G. Grüss [61] proved the well-known inequality given in (3). Applications of these inequalities have been found in statistics, coding theory, numerical analysis and various other branches of mathematics. Over the years, a multitude of papers related to the classical inequalities (1) and (3) have been published, see the books by Mitrinović, Pečarić and Fink [79] and Pećarič, Pochan and Tong [144], where further references are also given. In the past few years, an enormous amount of attention has been given to these inequalities and numerous generalizations, extensions and variants have appeared in the literature. The main goal of this chapter is to present a number of new and basic inequalities related to (1) and (3) involving functions of one independent variables, recently investigated by various researchers. Applications of some of the inequalities are also given.

1.2 Grüss-type inequalities

In this section we present some Grüss-type inequalities established by different investigators in [34,72,96,105]. In what follows we shall make use of the notation set to define Čebyšev functional $T(f,g)$ in (2).

We start with the following Grüss-type inequality proved by Matić, Pećarić and Ujević in [72].

Theorem 1.2.1. Let $f,\ g:[a,b]\to\mathbb{R}$ be two integrable functions and $\gamma\leqslant g(x)\leqslant\Gamma$, for all $x\in[a,b]$, where $\gamma,\ \Gamma\in\mathbb{R}$ are constants. Then

$$|T(f,g)|\leqslant\frac{1}{2}(\Gamma-\gamma)\sqrt{T(f,f)}. \tag{1.2.1}$$

Proof. By direct computation it is easy to observe that the following Korkine's identity holds (see [79, p. 242]):

$$T(f,g)=\frac{1}{2(b-a)^2}\int_a^b\int_a^b(f(t)-f(s))(g(t)-g(s))dtds. \tag{1.2.2}$$

From (1.2.2) we observe that

$$T(f,f)=\frac{1}{b-a}\int_a^b f^2(x)dx-\left(\int_a^b f(x)dx\right)^2. \tag{1.2.3}$$

Furthermore, by using the Cauchy-Schwarz integral inequality, it is easy to observe that $T(f,f)\geqslant 0$. Similarly, $T(g,g)\geqslant 0$. From (1.2.2) and using Cauchy-Schwarz integral inequality for double integrals, we have

$$\begin{aligned}|T(f,g)|^2&=\left\{\frac{1}{2(b-a)^2}\int_a^b\int_a^b(f(t)-f(s))(g(t)-g(s))dtds\right\}^2\\ &\leqslant\left\{\frac{1}{2(b-a)^2}\int_a^b\int_a^b(f(t)-f(s))^2dtds\right\}\left\{\frac{1}{2(b-a)^2}\int_a^b\int_a^b(g(t)-g(s))^2dtds\right\}\\ &=T(f,f)T(g,g).\end{aligned} \tag{1.2.4}$$

It is easy to observe that the following identity also holds:

$$\begin{aligned}T(g,g)&=\left(\Gamma-\frac{1}{b-a}\int_a^b g(x)dx\right)\left(\frac{1}{b-a}\int_a^b g(x)dx-\gamma\right)\\ &\quad-\frac{1}{b-a}\int_a^b(\Gamma-g(x))(g(x)-\gamma)dx.\end{aligned} \tag{1.2.5}$$

Using the fact that $(\Gamma-g(x))(g(x)-\gamma)\geqslant 0$ in (1.2.5) and then the elementary inequality

$$cd\leqslant\left(\frac{c+d}{2}\right)^2;\quad c,\ d\in\mathbb{R},$$

we observe that

$$T(g,g)\leqslant\left(\Gamma-\frac{1}{b-a}\int_a^b g(x)dx\right)\left(\frac{1}{b-a}\int_a^b g(x)dx-\gamma\right)\leqslant\left(\frac{\Gamma-\gamma}{2}\right)^2. \tag{1.2.6}$$

The required inequality in (1.2.1) follows from (1.2.4) and (1.2.6). The proof is complete.

Remark 1.2.1. We note that the inequality (1.2.1) is called a *premature Grüss inequality* (see [72]). The term premature is used to denote the fact that the result is obtained from not completing the proof of the Grüss inequality if one of the functions is known explicitly. In [72], it is observed that (1.2.1) provides a sharper bound than the Grüss inequality (3).

The following Theorem deals with a Grüss-type inequality proved by Dragomir and McAndrew in [34], which can be used in certain applications.

Theorem 1.2.2. Let $f,\ g:[a,b]\to\mathbb{R}$ be two integrable functions. Then

$$|T(f,g)|\leqslant\frac{1}{b-a}\int_a^b\left|\left(f(x)-\frac{1}{b-a}\int_a^b f(y)dy\right)\right.$$
$$\left.\times\left(g(x)-\frac{1}{b-a}\int_a^b g(y)dy\right)\right|dx. \tag{1.2.7}$$

The inequality (1.2.7) is sharp.

Proof. First we observe that

$$\frac{1}{b-a}\int_a^b\left(f(x)-\frac{1}{b-a}\int_a^b f(y)dy\right)\left(g(x)-\frac{1}{b-a}\int_a^b g(y)dy\right)dx$$
$$=\frac{1}{b-a}\int_a^b\left(f(x)g(x)-g(x)\frac{1}{b-a}\int_a^b f(y)dy\right.$$
$$\left.-f(x)\frac{1}{b-a}\int_a^b g(y)dy+\frac{1}{b-a}\int_a^b f(y)dy\frac{1}{b-a}\int_a^b g(y)dy\right)dx$$
$$=\frac{1}{b-a}\int_a^b f(x)g(x)dx-\frac{1}{b-a}\int_a^b g(x)dx\frac{1}{b-a}\int_a^b f(y)dy$$
$$-\frac{1}{b-a}\int_a^b f(x)dx\frac{1}{b-a}\int_a^b g(y)dy+(b-a)\frac{1}{b-a}\int_a^b f(y)dy\frac{1}{b-a}\int_a^b g(y)dy$$
$$=\frac{1}{b-a}\int_a^b f(x)g(x)dx-\left(\frac{1}{b-a}\int_a^b f(x)dx\right)\left(\frac{1}{b-a}\int_a^b g(x)dx\right)$$
$$=T(f,g). \tag{1.2.8}$$

From (1.2.8) and using the properties of modulus, we have

$$|T(f,g)|\leqslant\frac{1}{b-a}\int_a^b\left|\left(f(x)-\frac{1}{b-a}\int_a^b f(y)dy\right)\times\left(g(x)-\frac{1}{b-a}\int_a^b g(y)dy\right)\right|dx,$$

and the inequality in (1.2.7) is proved.

Choosing $f(x)=g(x)=\operatorname{sgn}\left(x-\frac{a+b}{2}\right)$, the equality is satisfied in (1.2.7).

In [96], Pachpatte proved the following Grüss-type inequality, which bounds $T(f,g)$ in terms of integral involving inherent functions and their derivatives.

Theorem 1.2.3. Let $f,\ g:[a,b]\to\mathbb{R}$ be continuous on $[a,b]$ and differentiable on (a,b) with derivatives $f',\ g':(a,b)\to\mathbb{R}$ being bounded on (a,b). Then

$$|T(f,g)|\leqslant\frac{1}{2(b-a)^2}\int_a^b\left[|g(x)|\|f'\|_\infty+|f(x)|\|g'\|_\infty\right]E(x)dx, \tag{1.2.9}$$

where

$$E(x)=\frac{1}{4}(b-a)^2+\left(x-\frac{a+b}{2}\right)^2, \tag{1.2.10}$$

for $x\in[a,b]$.

Proof. Define the mapping

$$p(x,t)=\begin{cases} t-a & \text{if } t\in[a,x] \\ t-b & \text{if } t\in(x,b] \end{cases} \tag{1.2.11}$$

Integrating by parts, we have

$$\int_a^b p(x,t)f'(t)dt=\int_a^x (t-a)f'(t)dt+\int_x^b (t-b)f'(t)dt$$

$$=(t-a)f(t)|_a^x-\int_a^x f(t)dt+(t-b)|_x^b-\int_x^b f(t)dt$$

$$=(x-a)f(x)+(b-x)f(x)-\int_a^b f(t)dt$$

$$=(b-a)f(x)-\int_a^b f(t)dt.$$

From this, we obtain

$$f(x)-\frac{1}{b-a}\int_a^b f(t)dt=\frac{1}{b-a}\int_a^b p(x,t)f'(t)dt, \tag{1.2.12}$$

for $(x,t)\in[a,b]^2$. The representation (1.2.12) is known as Montgomery's identity (see [50,80]). Similarly, we obtain

$$g(x)-\frac{1}{b-a}\int_a^b g(t)dt=\frac{1}{b-a}\int_a^b p(x,t)g'(t)dt, \tag{1.2.13}$$

for $(x,t)\in[a,b]^2$. Multiplying both sides of (1.2.12) and (1.2.13) by $g(x)$ and $f(x)$ respectively, adding and then integrating the resulting identity with respect to x from a to b, we have

$$2\int_a^b f(x)g(x)dx=\frac{2}{b-a}\left(\int_a^b f(x)dx\right)\left(\int_a^b g(x)dx\right)$$

$$+\frac{1}{b-a}\int_a^b\left[g(x)\int_a^b p(x,t)f'(t)dt+f(x)\int_a^b p(x,t)g'(t)dt\right]dx. \tag{1.2.14}$$

From (1.2.14) and using the properties of modulus, we observe that

$$|T(f,g)|\leqslant\frac{1}{2(b-a)^2}\int_a^b\left[|g(x)|\int_a^b |p(x,t)||f'(t)|dt+|f(x)|\int_a^b |p(x,t)||g'(t)|dt\right]dx$$

$$\leqslant\frac{1}{2(b-a)^2}\int_a^b\left[\|f'\|_\infty|g(x)|+\|g'\|_\infty|f(x)|\right]E(x)dx,$$

and the inequality (1.2.9) is proved.

Next, we give the following Grüss-type inequality established by Pachpatte in [105].

Theorem 1.2.4. Let $f,\ g,\ h:[a,b]\to\mathbb{R}$ be continuous functions on $[a,b]$ and differentiable on (a,b) with derivatives $f',\ g',\ h':(a,b)\to\mathbb{R}$ being bounded on (a,b). Then

$$\begin{aligned}&\Bigg|\frac{1}{b-a}\int_a^b f(x)g(x)h(x)dx-\frac{1}{3}\Bigg[\left(\frac{1}{b-a}\int_a^b g(x)h(x)dx\right)\\ &\times\left(\frac{1}{b-a}\int_a^b f(x)dx\right)+\left(\frac{1}{b-a}\int_a^b h(x)f(x)dx\right)\left(\frac{1}{b-a}\int_a^b g(x)dx\right)\\ &+\left(\frac{1}{b-a}\int_a^b f(x)g(x)dx\right)\left(\frac{1}{b-a}\int_a^b h(x)dx\right)\Bigg]\Bigg|\\ &\leqslant\frac{1}{3(b-a)}\int_a^b\big[|g(x)||h(x)|\|f'\|_\infty+|h(x)||f(x)|\|g'\|_\infty\\ &+|f(x)||g(x)|\|h'\|_\infty\big]A(x)dx,\end{aligned}\tag{1.2.15}$$

where

$$A(x)=\left[\frac{1}{4}+\frac{\left(x-\frac{a+b}{2}\right)^2}{(b-a)^2}\right](b-a).\tag{1.2.16}$$

Proof. From the hypotheses, for any $x,\ y\in[a,b]$, we have the following identities:

$$f(x)-f(y)=\int_y^x f'(t)dt,\tag{1.2.17}$$

$$g(x)-g(y)=\int_y^x g'(t)dt,\tag{1.2.18}$$

$$h(x)-h(y)=\int_y^x h'(t)dt.\tag{1.2.19}$$

Multiplying both sides of (1.2.17), (1.2.18) and (1.2.19) by $g(x)h(x)$, $h(x)f(x)$ and $f(x)g(x)$ respectively and adding the resulting identities, we have

$$\begin{aligned}&3f(x)g(x)h(x)-[g(x)h(x)f(y)+h(x)f(x)g(y)+f(x)g(x)h(y)]\\ &=g(x)h(x)\int_y^x f'(t)dt+h(x)f(x)\int_y^x g'(t)dt+f(x)g(x)\int_y^x h'(t)dt.\end{aligned}\tag{1.2.20}$$

Integrating both sides of (1.2.20) with respect to y over $[a,b]$ and rewriting, we have

$$\begin{aligned}&f(x)g(x)h(x)-\frac{1}{3(b-a)}\left[g(x)h(x)\int_a^b f(y)dy+h(x)f(x)\int_a^b g(y)dy+f(x)g(x)\int_a^b h(y)dy\right]\\ &=\frac{1}{3(b-a)}\left[g(x)h(x)\int_a^b\left(\int_y^x f'(t)dt\right)dy\right.\\ &\left.+h(x)f(x)\int_a^b\left(\int_y^x g'(t)dt\right)dy+f(x)g(x)\int_a^b\left(\int_y^x h'(t)dt\right)dy\right].\end{aligned}\tag{1.2.21}$$

Integrating both sides of (1.2.21) with respect to x from a to b and rewriting, we have

$$\frac{1}{b-a}\int_a^b f(x)g(x)h(x)dx - \frac{1}{3}\left[\left(\frac{1}{b-a}\int_a^b g(x)h(x)dx\right)\left(\frac{1}{b-a}\int_a^b f(y)dy\right)\right.$$

$$+\left(\frac{1}{b-a}\int_a^b h(x)f(x)dx\right)\left(\frac{1}{b-a}\int_a^b g(y)dy\right)$$

$$\left.+\left(\frac{1}{b-a}\int_a^b f(x)g(x)dx\right)\left(\frac{1}{b-a}\int_a^b h(y)dy\right)\right]$$

$$=\frac{1}{3(b-a)^2}\int_a^b\left[g(x)h(x)\int_a^b\left(\int_y^x f'(t)dt\right)dy\right.$$

$$\left.+h(x)f(x)\int_a^b\left(\int_y^x g'(t)dt\right)dy + f(x)g(x)\int_a^b\left(\int_y^x h'(t)dt\right)dy\right]dx. \tag{1.2.22}$$

From (1.2.22) and using the properties of modulus, we have

$$\left|\frac{1}{b-a}\int_a^b f(x)g(x)h(x)dx - \frac{1}{3}\left[\left(\frac{1}{b-a}\int_a^b g(x)h(x)dx\right)\left(\frac{1}{b-a}\int_a^b f(x)dx\right)\right.\right.$$

$$+\left(\frac{1}{b-a}\int_a^b h(x)f(x)dx\right)\left(\frac{1}{b-a}\int_a^b g(x)dx\right)$$

$$\left.\left.+\left(\frac{1}{b-a}\int_a^b f(x)g(x)dx\right)\left(\frac{1}{b-a}\int_a^b h(x)dx\right)\right]\right|$$

$$\leqslant \frac{1}{3(b-a)^2}\int_a^b\left[\left[|g(x)||h(x)|\|f'\|_\infty + |h(x)||f(x)|\|g'\|_\infty\right.\right.$$

$$\left.\left.+|f(x)||g(x)|\|h'\|_\infty\right]\int_a^b |x-y|dy\right]dx. \tag{1.2.23}$$

It is easy to observe that

$$\int_a^b |x-y|dy = \frac{(x-a)^2+(b-x)^2}{2}, \tag{1.2.24}$$

and

$$\frac{(x-a)^2+(b-x)^2}{2(b-a)} = \left[\frac{1}{4}+\frac{\left(x-\frac{a+b}{2}\right)^2}{(b-a)^2}\right](b-a) = A(x). \tag{1.2.25}$$

Using (1.2.24) and (1.2.25) in (1.2.23), we get the required inequality in (1.2.15) and the proof is complete.

Remark 1.2.2. In the special case, when $h(x) = 1$ and hence $h'(x) = 0$, it is easy to observe that the inequality (1.2.15) reduces to

$$|T(f,g)| \leqslant \frac{1}{2(b-a)} \int_a^b \left[|g(x)|\|f'\|_\infty + |f(x)|\|g'\|_\infty\right] A(x)dx. \tag{1.2.26}$$

We note that the bound obtained in (1.2.26) is the same as the bound in (1.2.9).

Another Grüss-type inequality established by Pachpatte in [96] is embodied in the following theorem.

Theorem 1.2.5. Let $f,\ g : [a,b] \to \mathbb{R}$ be continuous on $[a,b]$ and twice differentiable on (a,b), with second derivatives $f'',\ g'' : (a,b) \to \mathbb{R}$ being bounded on (a,b). Then

$$\left| T(f,g) - \frac{1}{2(b-a)} \int_a^b \left(x - \frac{a+b}{2}\right)(fg)'(x)dx \right|$$

$$\leqslant \frac{1}{2(b-a)^2} \int_a^b \left[|g(x)|\|f''\|_\infty + |f(x)|\|g''\|_\infty\right] B(x)dx, \tag{1.2.27}$$

where

$$B(x) = \int_a^b |k(x,t)|dt,$$

for $x \in [a,b]$, in which

$$k(x,t) = \begin{cases} \dfrac{(t-a)^2}{2} & \text{if } t \in [a,x] \\ \dfrac{(t-b)^2}{2} & \text{if } t \in (x,b] \end{cases} \tag{1.2.28}$$

Proof. Integrating by parts, we have successively

$$\int_a^b k(x,t)f''(t)dt = \int_a^x \frac{(t-a)^2}{2} f''(t)dt + \int_x^b \frac{(t-b)^2}{2} f''(t)dt$$

$$= \frac{(t-a)^2}{2} f'(t)\Big|_a^x - \int_a^x (t-a)f'(t)dt + \frac{(t-b)^2}{2} f'(t)\Big|_x^b - \int_x^b (t-b)f'(t)dt$$

$$= \frac{(x-a)^2}{2} f'(x) - \left[(t-a)f(t)|_a^x - \int_a^x f(t)dt \right] - \frac{(b-x)^2}{2} f'(x) - \left[(t-b)f(t)|_x^b - \int_x^b f(t)dt \right]$$

$$= \frac{1}{2}\left[(x-a)^2 - (b-x)^2\right] f'(x) - (x-a)f(x) + \int_a^x f(t)dt + (x-b)f(x) + \int_x^b f(t)dt$$

$$= (b-a)\left(x - \frac{a+b}{2}\right) f'(x) - (b-a)f(x) + \int_a^b f(t)dt,$$

from which, we get the integral identity

$$f(x)=\frac{1}{b-a}\int_a^b f(t)dt+\left(x-\frac{a+b}{2}\right)f'(x)-\frac{1}{b-a}\int_a^b k(x,t)f''(t)dt, \tag{1.2.29}$$

for $x\in[a,b]$. Similarly

$$g(x)=\frac{1}{b-a}\int_a^b g(t)dt+\left(x-\frac{a+b}{2}\right)g'(x)-\frac{1}{b-a}\int_a^b k(x,t)g''(t)dt, \tag{1.2.30}$$

for $x\in[a,b]$. Multiplying both sides of (1.2.29) and (1.2.30) by $g(x)$ and $f(x)$ respectively, adding and then integrating the resulting identity with respect to x from a to b and rewriting, we have

$$\frac{1}{b-a}\int_a^b f(x)g(x)dx=\frac{1}{2}\left[\left(\frac{1}{b-a}\int_a^b f(t)dt\right)\left(\frac{1}{b-a}\int_a^b g(x)dx\right)\right.$$

$$\left.+\left(\frac{1}{b-a}\int_a^b g(t)dt\right)\left(\frac{1}{b-a}\int_a^b f(x)dx\right)\right]+\frac{1}{2(b-a)}\int_a^b\left(x-\frac{a+b}{2}\right)(f(x)g(x))'dx$$

$$-\frac{1}{2(b-a)^2}\int_a^b\left[g(x)\int_a^b k(x,t)f''(t)dt+f(x)\int_a^b k(x,t)g''(t)dt\right]dx,$$

i.e.,

$$T(f,g)-\frac{1}{2(b-a)}\int_a^b\left(x-\frac{a+b}{2}\right)(fg)'(x)dx$$

$$=-\frac{1}{2(b-a)^2}\int_a^b\left[g(x)\int_a^b k(x,t)f''(t)dt+f(x)\int_a^b k(x,t)g''(t)dt\right]dx. \tag{1.2.31}$$

From (1.2.31) and using the properties of modulus, we have

$$\left|T(f,g)-\frac{1}{2(b-a)}\int_a^b\left(x-\frac{a+b}{2}\right)(fg)'(x)dx\right|$$

$$\leqslant\frac{1}{2(b-a)^2}\int_a^b\left[|g(x)|\int_a^b|k(x,t)||f''(t)|dt+|f(x)|\int_a^b|k(x,t)||g''(t)|dt\right]dx$$

$$\leqslant\frac{1}{2(b-a)^2}\int_a^b\left[|g(x)|\|f''\|_\infty+|f(x)|\|g''\|_\infty\right]B(x)dx,$$

and the inequality (1.2.27) is proved.

1.3 Čebyšev-type inequalities

In this section we offer some Čebyšev-type inequalities established in [42,72,112,113]. We shall make use of the notation set to define Čebyšev functional $T(f,g)$ in (2).

In [72], Matić, Pečarić and Ujević proved the following Čebyšev-type inequality.

Theorem 1.3.1. Let $f,\ g:[a,b]\to\mathbb{R}$ be two absolutely continuous functions on $[a,b]$. Let $f':[a,b]\to\mathbb{R}$ belong to $L_\infty[a,b]$. Then

$$|T(f,g)|\leqslant\frac{(b-a)}{2\sqrt{3}}\|f'\|_\infty\sqrt{T(g,g)}. \tag{1.3.1}$$

Proof. For the functions $f,\ g$ the Korkine's identity (1.2.2) holds. Following the proof of Theorem 1.2.1, we get (1.2.4). For any $s,\ t\in[a,b]$, we have

$$f(t)-f(s)=\int_s^t f'(\xi)d\xi.$$

Using this fact in (1.2.2), we observe that

$$\begin{aligned}|T(f,f)|&=\frac{1}{2(b-a)^2}\left|\int_a^b\int_a^b(f(t)-f(s))^2dtds\right|\\&=\frac{1}{2(b-a)^2}\left|\int_a^b\int_a^b\left(\int_s^t f'(\xi)d\xi\right)^2dtds\right|\\&\leqslant\frac{1}{2(b-a)^2}\int_a^b\int_a^b\left(\int_s^t|f'(\xi)|d\xi\right)^2dtds\\&\leqslant\frac{1}{2(b-a)^2}\|f'\|_\infty^2\int_a^b\int_a^b(t-s)^2dtds\\&=\frac{(b-a)^2}{12}\|f'\|_\infty^2.\end{aligned} \tag{1.3.2}$$

Using (1.3.2) in (1.2.4), we deduce the desired inequality in (1.3.1).

Remark 1.3.1. From the identity (1.2.2), it is easy to observe that

$$T(g,g)=\frac{1}{b-a}\int_a^b g^2(t)dt-\left(\frac{1}{b-a}\int_a^b g(t)dt\right)^2,$$

and using this, (1.3.1) can be written as

$$|T(f,g)|\leqslant\frac{b-a}{2\sqrt{3}}\|f'\|_\infty\left[\frac{1}{b-a}\int_a^b g^2(t)dt-\left(\frac{1}{b-a}\int_a^b g(t)dt\right)^2\right]^{\frac{1}{2}}. \tag{1.3.3}$$

As noted in [72], the inequality (1.3.1) (or (1.3.3)) is called the *pre-Čebyšev inequality*. A natural generalization of the Čebyšev inequality (1) established by Dragomir in [42] is given in the following theorem.

Theorem 1.3.2. Let $f,\ g:[a,b]\to\mathbb{R}$ be two Lipschitzian functions with constants $L_1>0$ and $L_2>0$, i.e.

$$|f(x)-f(y)|\leqslant L_1|x-y|,\quad |g(x)-g(y)|\leqslant L_2|x-y|, \tag{1.3.4}$$

for all $x,\ y\in[a,b]$. Then

$$|T(f,g)|\leqslant\frac{L_1L_2}{12}(b-a)^2. \tag{1.3.5}$$

The constant $\frac{1}{12}$ is the best possible.

Proof. From (1.3.4), we have

$$|(f(x)-f(y))(g(x)-g(y))| \leqslant L_1L_2(x-y)^2, \tag{1.3.6}$$

for all $x,\ y \in [a,b]$. For the functions $f,\ g$ the following Korkine's identity holds:

$$T(f,g) = \frac{1}{2(b-a)^2}\int_a^b\int_a^b (f(x)-f(y))(g(x)-g(y))dxdy. \tag{1.3.7}$$

From (1.3.7) and (1.3.6), we have

$$\begin{aligned}|T(f,g)| &\leqslant \frac{1}{2(b-a)^2}\int_a^b\int_a^b |(f(x)-f(y))(g(x)-g(y))|dxdy \\ &\leqslant \frac{1}{2(b-a)^2}\int_a^b\int_a^b L_1L_2(x-y)^2dxdy \\ &= \frac{L_1L_2}{12}(b-a)^2,\end{aligned}$$

and the inequality (1.3.5) is proved. Now, if we choose $f(x) = L_1x$, $g(x) = L_2x$, then f is L_1-Lipschitzian, g is L_2-Lipschitzian and the equality in (1.3.5) holds. The proof is complete.

Remark 1.3.2. We note that, if $f,\ g : [a,b] \to \mathbb{R}$ are two differentiable functions with derivatives of which are bounded on (a,b), then we get the Čebyšev inequality (1). For an interesting discussion to show that sometimes the estimation on $T(f,g)$ given by the Grüss inequality (3) is better than the estimation on $T(f,g)$ given by the Čebyšev inequality obtained in (1) and sometimes the other way around, see [42].

In [113], Pachpatte has established the following inequality similar to that of Čebyšev in (1).

Theorem 1.3.3. Let $f,\ g : [a,b] \to \mathbb{R}$ be absolutely continuous functions with derivatives $f',\ g' \in L_q[a,b]$, $q > 1$, then

$$|T(f,g)| \leqslant \frac{1}{(b-a)^3}\|f'\|_q\|g'\|_q\int_a^b (B(x))^{\frac{2}{r}}dx, \tag{1.3.8}$$

where

$$B(x) = \frac{1}{r+1}\left[(x-a)^{r+1}+(b-x)^{r+1}\right], \tag{1.3.9}$$

for $x \in [a,b]$ and $\frac{1}{q}+\frac{1}{r} = 1$.

Proof. As in the proof of Theorem 1.2.3, we have the following identities:

$$f(x) - \frac{1}{b-a}\int_a^b f(t)dt = \frac{1}{b-a}\int_a^b p(x,t)f'(t)dt, \tag{1.3.10}$$

$$g(x) - \frac{1}{b-a}\int_a^b g(t)dt = \frac{1}{b-a}\int_a^b p(x,t)g'(t)dt, \tag{1.3.11}$$

for $x \in [a,b]$, where $p(x,t)$ is given by (1.2.11). Multiplying the left hand sides and right hand sides of (1.3.10) and (1.3.11), we have

$$\begin{aligned} f(x)g(x) &- f(x)\left(\frac{1}{b-a}\int_a^b g(t)dt\right) - g(x)\left(\frac{1}{b-a}\int_a^b f(t)dt\right) \\ &+ \left(\frac{1}{b-a}\int_a^b f(t)dt\right)\left(\frac{1}{b-a}\int_a^b g(t)dt\right) \\ &= \frac{1}{(b-a)^2}\left(\int_a^b p(x,t)f'(t)dt\right)\left(\int_a^b p(x,t)g'(t)dt\right). \end{aligned} \tag{1.3.12}$$

Integrating both sides of (1.3.12) with respect to x from a to b and dividing both sides of the resulting identity by $(b-a)$, we get

$$T(f,g) = \frac{1}{(b-a)^3}\int_a^b \left(\int_a^b p(x,t)f'(t)dt\right)\left(\int_a^b p(x,t)g'(t)dt\right)dx. \tag{1.3.13}$$

From (1.3.13) and using the properties of modulus and Hölder's integral inequality, we have

$$\begin{aligned} |T(f,g)| &\leqslant \frac{1}{(b-a)^3}\int_a^b \left(\int_a^b |p(x,t)||f'(t)|dt\right)\left(\int_a^b |p(x,t)||g'(t)|dt\right)dx \\ &\leqslant \frac{1}{(b-a)^3}\int_a^b \left(\left\{\int_a^b |p(x,t)|^r dt\right\}^{\frac{1}{r}}\left\{\int_a^b |f'(t)|^q dt\right\}^{\frac{1}{q}}\right) \\ &\quad \times \left(\left\{\int_a^b |p(x,t)|^r dt\right\}^{\frac{1}{r}}\left\{\int_a^b |g'(t)|^q dt\right\}^{\frac{1}{q}}\right)dx \\ &= \frac{1}{(b-a)^3}\|f'\|_q\|g'\|_q\int_a^b \left(\left\{\int_a^b |p(x,t)|^r dt\right\}^{\frac{1}{r}}\right)^2 dx. \end{aligned} \tag{1.3.14}$$

A simple calculation shows that

$$\begin{aligned} \int_a^b |p(x,t)|^r dt &= \int_a^x |t-a|^r dt + \int_x^b |t-b|^r dt \\ &= \int_a^x (t-a)^r dt + \int_x^b (b-t)^r dt \\ &= \frac{(x-a)^{r+1} + (b-x)^{r+1}}{r+1} = B(x). \end{aligned} \tag{1.3.15}$$

Using (1.3.15) in (1.3.14), we get (1.3.8). The proof is complete.

In the proofs of the following Theorems we need the trapezoidal like representation formulas proved in [36] and [9].

Lemma 1.3.1 (see [36]). Let $f:[a,b]\to\mathbb{R}$ be an absolutely continuous function on $[a,b]$, then we have the identity

$$\frac{f(a)+f(b)}{2}-\frac{1}{b-a}\int_a^b f(x)dx=\frac{1}{2(b-a)^2}\int_a^b\int_a^b [f'(x)-f'(y)](x-y)dxdy.$$

Proof. We have successively

$$\int_a^b\int_a^b [f'(x)-f'(y)](x-y)dxdy=\int_a^b\int_a^b [xf'(x)+yf'(y)-xf'(y)-yf'(x)]dxdy$$

$$=2\int_a^b\int_a^b [xf'(x)-xf'(y)]dxdy=2\int_a^b\int_a^b xf'(x)dxdy-2\int_a^b\int_a^b xf'(y)dxdy$$

$$=2(b-a)\left[bf(b)-af(a)-\int_a^b f(x)dx\right]-\left(b^2-a^2\right)[f(b)-f(a)]$$

$$=(b-a)^2[f(a)+f(b)]-2(b-a)\int_a^b f(x)dx.$$

Dividing both sides by $2(b-a)^2$ yields the required result.

Lemma 1.3.2 (see [9]). Let $f:[a,b]\to\mathbb{R}$ be a differentiable function so that f' is absolutely continuous on $[a,b]$, then we have the identity

$$\int_a^b f(x)dx-\frac{b-a}{2}[f(a)+f(b)]+\frac{(b-a)^3}{12}[f';a,b]$$

$$=\frac{1}{2}\int_a^b (x-a)(b-x)\left\{[f';a,b]-f''(x)\right\}dx, \tag{1.3.16}$$

where

$$[f';a,b]=\frac{f'(b)-f'(a)}{b-a},$$

is the divided difference.

Proof. By applying the integration by parts formula twice, we have (see [9])

$$\int_a^b f(x)dx-\frac{b-a}{2}[f(a)+f(b)]=-\frac{1}{2}\int_a^b (x-a)(b-x)f''(x)dx. \tag{1.3.17}$$

On the other hand, by the simple identity:

$$\frac{1}{b-a}\int_a^b h(x)g(x)dx-\frac{1}{b-a}\int_a^b h(x)dx\frac{1}{b-a}\int_a^b g(x)dx$$

$$=\frac{1}{b-a}\int_a^b h(x)\left[g(x)-\frac{1}{b-a}\int_a^b g(y)dy\right]dx, \tag{1.3.18}$$

we may state that

$$\int_a^b (x-a)(b-x)f''(x)dx-\int_a^b (x-a)(b-x)dx\frac{1}{b-a}\int_a^b f''(x)dx$$

$$= \int_a^b (x-a)(b-x)\left[f''(x) - [f';a,b]\right] dx,$$

which is clearly equivalent to

$$\int_a^b (x-a)(b-x)f''(x)dx = \frac{(b-a)^2}{6}\left[f'(b) - f'(a)\right]$$

$$+ \int_a^b (x-a)(b-x)\left[f''(x) - [f';a,b]\right] dx. \tag{1.3.19}$$

Combining (1.3.17) with (1.3.19), we deduce (1.3.16).

We use the following notation to simplify the details of presentation. For suitable functions $f,\ g : [a,b] \to \mathbb{R}$, we set

$$L(f;a,b) = \frac{1}{2(b-a)^2}\int_a^b\int_a^b (f'(t) - f'(s))(t-s)dtds,$$

$$N(f',f'';a,b) = \frac{1}{2(b-a)}\int_a^b (t-a)(b-t)\left\{[f';a,b] - f''(t)\right\}dt,$$

$$P(f,g) = FG - \frac{1}{b-a}\left\{F\int_a^b g(t)dt + G\int_a^b f(t)dt\right\} + \left(\frac{1}{b-a}\int_a^b f(t)dt\right)\left(\frac{1}{b-a}\int_a^b g(t)dt\right),$$

$$S(f,g) = \overline{FG} - \frac{1}{b-a}\left\{\overline{F}\int_a^b g(t)dt + \overline{G}\int_a^b f(t)dt\right\} + \left(\frac{1}{b-a}\int_a^b f(t)dt\right)\left(\frac{1}{b-a}\int_a^b g(t)dt\right),$$

in which

$$F = \frac{f(a)+f(b)}{2}, \quad G = \frac{g(a)+g(b)}{2},$$

$$\overline{F} = \frac{f(a)+f(b)}{2} - \frac{(b-a)^2}{12}[f';a,b],$$

$$\overline{G} = \frac{g(a)+g(b)}{2} - \frac{(b-a)^2}{12}\left[g';a,b\right],$$

and define

$$\|f\|_2 = \left(\int_a^b |f(t)|^2\right)^{\frac{1}{2}} < \infty,$$

$$\|f\|_\infty = \sup_{t\in[a,b]} |f(t)| < \infty.$$

The following two Theorems established by Pachpatte in [112] deal with Čebyšev-type integral inequalities involving functions and their derivatives.

Theorem 1.3.4. Let $f,\ g : [a,b] \to \mathbb{R}$ be absolutely continuous functions on $[a,b]$ with $f',\ g' \in L_2[a,b]$, then

$$|P(f,g)| \leqslant \frac{(b-a)^2}{12}\left[\frac{1}{b-a}\|f'\|_2^2 - ([f;a,b])^2\right]^{\frac{1}{2}} \times \left[\frac{1}{b-a}\|g'\|_2^2 - ([g;a,b])^2\right]^{\frac{1}{2}}. \quad (1.3.20)$$

Proof. From the hypotheses, by Lemma 1.3.1 we have the following identities:

$$F - \frac{1}{b-a}\int_a^b f(t)dt = L(f;a,b), \quad (1.3.21)$$

$$G - \frac{1}{b-a}\int_a^b g(t)dt = L(g;a,b). \quad (1.3.22)$$

Multiplying the left hand sides and right hand sides of (1.3.21) and (1.3.22), we get

$$P(f,g) = L(f;a,b)L(g;a,b). \quad (1.3.23)$$

From (1.3.23), we have

$$|P(f,g)| = |L(f;a,b)|\,|L(g;a,b)|. \quad (1.3.24)$$

Using Schwarz inequality for double integrals, we have

$$|L(f;a,b)| \leqslant \frac{1}{2(b-a)^2}\int_a^b\int_a^b \left|(f'(t)-f'(s))(t-s)\right| dtds$$

$$\leqslant \left[\frac{1}{2(b-a)^2}\int_a^b\int_a^b (f'(t)-f'(s))^2 dtds\right]^{\frac{1}{2}} \times \left[\frac{1}{2(b-a)^2}\int_a^b\int_a^b (t-s)^2 dtds\right]^{\frac{1}{2}}. \quad (1.3.25)$$

By simple computation, we have

$$\frac{1}{2(b-a)^2}\int_a^b\int_a^b (f'(t)-f'(s))^2 dtds = \frac{1}{b-a}\int_a^b (f'(t))^2 dt - \left(\frac{1}{b-a}\int_a^b f'(t)dt\right)^2, \quad (1.3.26)$$

and

$$\frac{1}{2(b-a)^2}\int_a^b\int_a^b (t-s)^2 dtds = \frac{(b-a)^2}{12}. \quad (1.3.27)$$

Using (1.3.26), (1.3.27) in (1.3.25), we have

$$|L(f;a,b)| \leqslant \frac{b-a}{2\sqrt{3}}\left[\frac{1}{b-a}\|f'\|_2^2 - ([f;a,b])^2\right]^{\frac{1}{2}}. \quad (1.3.28)$$

Similarly, we obtain

$$|L(g;a,b)| \leqslant \frac{b-a}{2\sqrt{3}}\left[\frac{1}{b-a}\|g'\|_2^2 - ([g;a,b])^2\right]^{\frac{1}{2}}. \quad (1.3.29)$$

Using (1.3.28) and (1.3.29) in (1.3.24), we get the desired inequality in (1.3.20).

Theorem 1.3.5. Let $f,\ g:[a,b]\to\mathbb{R}$ be differentiable functions so that $f',\ g'$ are absolutely continuous on $[a,b]$, then

$$|S(f,g)|\leqslant\frac{(b-a)^4}{144}\left\|f''-[f';a,b]\right\|_\infty\left\|g''-[g';a,b]\right\|_\infty. \tag{1.3.30}$$

Proof. From the hypotheses, by Lemma 1.3.2, we have the following identities:

$$\frac{1}{b-a}\int_a^b f(t)dt-\overline{F}=N(f',f'';a,b), \tag{1.3.31}$$

$$\frac{1}{b-a}\int_a^b g(t)dt-\overline{G}=N(g',g'';a,b). \tag{1.3.32}$$

Multiplying the left sides and right sides of (1.3.31) and (1.3.32), we get

$$S(f,g)=N(f',f'';a,b)N(g',g'';a,b). \tag{1.3.33}$$

From (1.3.33) we have

$$|S(f,g)|=|N(f',f'';a,b)||N(g',g'';a,b)|. \tag{1.3.34}$$

By simple calculation, we have

$$\begin{aligned}|N(f',f'';a,b)|&\leqslant\frac{1}{2(b-a)}\int_a^b(t-a)(b-t)\left|[f';a,b]-f''(t)\right|dt\\&\leqslant\frac{1}{2(b-a)}\left\|f''-[f';a,b]\right\|_\infty\int_a^b(t-a)(b-t)dt\\&=\frac{(b-a)^2}{12}\left\|f''-[f';a,b]\right\|_\infty.\end{aligned} \tag{1.3.35}$$

Similarly, we obtain

$$|N(g',g'';a,b)|\leqslant\frac{(b-a)^2}{12}\left\|g''-[g';a,b]\right\|_\infty. \tag{1.3.36}$$

Using (1.3.35) and (1.3.36) in (1.3.34), we get the required inequality in (1.3.30).

1.4 Inequalities of the Grüss- and Čebyšev-type

In this section we present some recent inequalities of the Grüss-and Čebyšev-type established by Pachpatte [106,111,117,127].

For suitable functions $z,\ f,\ g:[a,b]\to\mathbb{R}$ and $w:[a,b]\to[0,\infty)$ an integrable function such that $\int_a^b w(x)dx>0$, we use the following notation to simplify the details of presentation:

$$D[z(x)]=\left[z(x)(1-\lambda)+\frac{z(a)+z(b)}{2}\lambda\right](b-a),\quad \lambda\in[0,1],$$

$$A(f,g)=\int_a^b[g(x)D[f(x)]+f(x)D[g(x)]]dx-2\left(\int_a^b f(x)dx\right)\left(\int_a^b g(x)dx\right),$$

$$B(f,g) = \int_a^b D[f(x)]D[g(x)]dx$$

$$-\left[\left(\int_a^b f(x)dx\right)\left(\int_a^b D[g(x)]dx\right) + \left(\int_a^b g(x)dx\right)\left(\int_a^b D[f(x)]dx\right)\right]$$

$$+(b-a)\left(\int_a^b f(x)dx\right)\left(\int_a^b g(x)dx\right),$$

$$G(f,g) = \int_a^b f(x)g(x)dx - \frac{1}{b^2-a^2}\left[\left(\int_a^b f(x)dx\right)\left(\int_a^b xg(x)dx\right)\right.$$

$$\left.+\left(\int_a^b g(x)dx\right)\left(\int_a^b xf(x)dx\right)\right],$$

$$H(f,g) = \int_a^b f(x)g(x)dx - \frac{3}{b^3-a^3}\left(\int_a^b xf(x)dx\right)\left(\int_a^b xg(x)dx\right),$$

$$T(w,f,g) = \int_a^b w(x)f(x)g(x)dx - \left(\int_a^b w(x)f(x)dx\right)\left(\int_a^b w(x)g(x)dx\right),$$

$$S(w,f,g) = \int_a^b w(x)f(x)g(x)dx - \frac{1}{\int_a^b w(x)dx}\left(\int_a^b w(x)f(x)dx\right)\left(\int_a^b w(x)g(x)dx\right),$$

and define $\|z\|_\infty = \sup_{t\in[a,b]} |z(t)| < \infty$.

The following Theorem deals with the inequalities established in [117].

Theorem 1.4.1. Let $f,\ g : [a,b] \to \mathbb{R}$ be continuous functions on $[a,b]$, differentiable on (a,b) and with derivatives $f',\ g' : (a,b) \to \mathbb{R}$ bounded on (a,b), then

$$|A(f,g)| \leqslant \int_a^b \left[|g(x)|\|f'\|_\infty + |f(x)|\|g'\|_\infty\right]\alpha(x)dx, \tag{1.4.1}$$

and

$$|B(f,g)| \leqslant \|f'\|_\infty\|g'\|_\infty \int_a^b \alpha^2(x)dx, \tag{1.4.2}$$

where

$$\alpha(x) = \frac{(b-a)^2}{4}\left[\lambda^2 + (\lambda-1)^2\right] + \left(x - \frac{a+b}{2}\right)^2, \tag{1.4.3}$$

for $a + \lambda\frac{b-a}{2} \leqslant x \leqslant b - \lambda\frac{b-a}{2}$, $\lambda \in [0,1]$.

Proof. Define the mapping

$$\overline{p}(x,t) = \begin{cases} t - \left[a + \lambda \dfrac{b-a}{2}\right], & t \in [a,x] \\ t - \left[b - \lambda \dfrac{b-a}{2}\right], & t \in (x,b] \end{cases} \tag{1.4.4}$$

for $\lambda \in [0,1]$. Integrating by parts, we have

$$\int_a^b \overline{p}(x,t) f'(t)dt = \int_a^x \left(t - \left[a + \lambda \frac{b-a}{2}\right]\right) f'(t)dt + \int_x^b \left(t - \left[b - \lambda \frac{b-a}{2}\right]\right) f'(t)dt$$

$$= \left[f(x)(1-\lambda) + \frac{f(a)+f(b)}{2}\lambda\right](b-a) - \int_a^b f(t)dt$$

i.e.,

$$D[f(x)] - \int_a^b f(t)dt = \int_a^b \overline{p}(x,t) f'(t)dt. \tag{1.4.5}$$

Similarly, we have

$$D[g(x)] - \int_a^b g(t)dt = \int_a^b \overline{p}(x,t) g'(t)dt. \tag{1.4.6}$$

Multiplying both sides of (1.4.5) and (1.4.6) by $g(x)$ and $f(x)$ respectively and adding the resulting identities, we have

$$g(x)D[f(x)] + f(x)D[g(x)] - g(x)\int_a^b f(t)dt - f(x)\int_a^b g(t)dt$$

$$= g(x)\int_a^b \overline{p}(x,t) f'(t)dt + f(x)\int_a^b \overline{p}(x,t) g'(t)dt. \tag{1.4.7}$$

Integrating both sides of (1.4.7) with respect to x from a to b, we have

$$A(f,g) = \int_a^b \left[g(x)\int_a^b \overline{p}(x,t) f'(t)dt + f(x)\int_a^b \overline{p}(x,t) g'(t)dt\right]dx. \tag{1.4.8}$$

Using the properties of modulus, from (1.4.8), we have

$$|A(f,g)| \leqslant \int_a^b \left[|g(x)|\int_a^b |\overline{p}(x,t)||f'(t)|dt + |f(x)|\int_a^b |\overline{p}(x,t)||g'(t)|dt\right]dx$$

$$\leqslant \int_a^b \left\{\left[|g(x)|\|f'\|_\infty + |f(x)|\|g'\|_\infty\right]\int_a^b |\overline{p}(x,t)dt|\right\}dx. \tag{1.4.9}$$

On the other hand,

$$\int_a^b |\overline{p}(x,t)|\,dt = \int_a^x \left|t - \left(a + \lambda \frac{b-a}{2}\right)\right| dt + \int_x^b \left|t - \left(b - \lambda \frac{b-a}{2}\right)\right| dt. \tag{1.4.10}$$

Now, we observe that

$$\int_p^r |t-q|\,dt = \int_p^q (q-t)\,dt + \int_q^r (t-q)\,dt$$

$$= \frac{1}{2}\left[(q-p)^2+(r-q)^2\right] = \frac{1}{4}(p-r)^2+\left(q-\frac{r+p}{2}\right)^2, \tag{1.4.11}$$

for all r, p, q such that $p \leqslant q \leqslant r$. Using (1.4.11), we have that

$$\int_a^x \left|t-\left(a+\lambda\frac{b-a}{2}\right)\right| dt = \frac{1}{4}(x-a)^2+\left[\left(a+\lambda\frac{b-a}{2}\right)-\frac{a+x}{2}\right]^2, \tag{1.4.12}$$

$$\int_x^b \left|t-\left(b-\lambda\frac{b-a}{2}\right)\right| dt = \frac{1}{4}(b-x)^2+\left[\left(b-\lambda\frac{b-a}{2}\right)-\frac{x+b}{2}\right]^2. \tag{1.4.13}$$

Using (1.4.12), (1.4.13) in (1.4.10) we get

$$\int_a^b |\overline{p}(x,t)|dt = \frac{1}{2}\frac{(x-a)^2+(b-x)^2}{2}+\left(\lambda\frac{b-a}{2}-\frac{x-a}{2}\right)^2+\left(\frac{b-x}{2}-\lambda\frac{b-a}{2}\right)^2$$

$$= \frac{(b-a)^2}{4}\left[\lambda^2+(\lambda-1)^2\right]+\left(x-\frac{a+b}{2}\right)^2 = \alpha(x). \tag{1.4.14}$$

Using (1.4.14) in (1.4.9), we get the required inequality in (1.4.1).

Multiplying the left hand sides and right hand sides of (1.4.5) and (1.4.6), we get

$$D[f(x)]D[g(x)] - D[g(x)]\int_a^b f(t)dt - D[f(x)]\int_a^b g(t)dt + \left(\int_a^b f(t)dt\right)\left(\int_a^b g(t)dt\right)$$

$$= \left(\int_a^b \overline{p}(x,t)f'(t)dt\right)\left(\int_a^b \overline{p}(x,t)g'(t)dt\right). \tag{1.4.15}$$

Integrating both sides of (1.4.15) from a to b, we have

$$B(f,g) = \int_a^b \left(\int_a^b \overline{p}(x,t)f'(t)dt\right)\left(\int_a^b \overline{p}(x,t)g'(t)dt\right)dx. \tag{1.4.16}$$

Using the properties of modulus, from (1.4.16), we get

$$|B(f,g)| \leqslant \|f'\|_\infty \|g'\|_\infty \int_a^b \left(\int_a^b |\overline{p}(x,t)|\,dt\right)^2 dx. \tag{1.4.17}$$

Using (1.4.14) in (1.4.17), we get the required inequality in (1.4.2). The proof is complete.

Remark 1.4.1. If we take $\lambda = 0$ in Theorem 1.4.1, then by simple calculations,

$$|T(f,g)| \leqslant \int_a^b \left[|g(x)|\|f'\|_\infty + |f(x)|\|g'\|_\infty\right]\alpha_1(x)dx, \tag{1.4.18}$$

and

$$|T(f,g)| \leqslant \|f'\|_\infty \|g'\|_\infty \frac{1}{(b-a)^3}\int_a^b \alpha_2^2(x)dx, \tag{1.4.19}$$

where

$$\alpha_1(x) = \frac{1}{2}\left[\frac{1}{4}+\frac{\left(x-\frac{a+b}{2}\right)^2}{(b-a)^2}\right],$$

$$\alpha_2(t) = \frac{1}{4}(b-a)^2+\left(x-\frac{a+b}{2}\right)^2,$$

for $x \in [a,b]$.

In proving the inequalities in the next theorem, established in [106], we make use of the following variant of the well-known Lagrange's mean value theorem given by Pompeiu in [145].

Lemma 1.4.1 (see [145]). For every real valued function f differentiable on an interval $[a,b]$ not containing 0 and for all pairs $x_1 \neq x_2$ in $[a,b]$ there exists a point c in (x_1,x_2) such that

$$\frac{x_1 f(x_2) - x_2 f(x_1)}{x_1 - x_2} = f(c) - cf'(c).$$

For the proof of Pompeiu's mean value theorem, we refer the interested readers to [57,147].

Theorem 1.4.2. Let $f,\ g : [a,b] \to \mathbb{R}$ be continuous functions on $[a,b]$ and differentiable on (a,b) with $[a,b]$ not containing 0, then

$$|G(f,g)| \leqslant \|f - lf'\|_\infty \int_a^b |g(x)| \left|\frac{1}{2} - \frac{x}{a+b}\right| dx + \|g - lg'\|_\infty \int_a^b |f(x)| \left|\frac{1}{2} - \frac{x}{a+b}\right| dx, \tag{1.4.20}$$

and

$$|H(f,g)| \leqslant \|f - lf'\|_\infty \|g - lg'\|_\infty |M|, \tag{1.4.21}$$

where $l(t) = t,\ t \in [a,b]$ and

$$M = (b-a)\left[1 - \frac{3}{4}\frac{(a+b)^2}{a^2+ab+b^2}\right]. \tag{1.4.22}$$

Proof. From the hypotheses, for $x,\ t \in [a,b]$, $t \neq x$ there exist points c and d between x and t such that

$$tf(x) - xf(t) = [f(c) - cf'(c)](t-x), \tag{1.4.23}$$

and

$$tg(x) - xg(t) = [g(d) - dg'(d)](t-x). \tag{1.4.24}$$

Multiplying both sides of (1.4.23) and (1.4.24) by $g(x)$ and $f(x)$ respectively and adding the resulting identities, we have

$$2tf(x)g(x) - xg(x)f(t) - xf(x)g(t)$$

$$= [f(c) - cf'(c)](t-x)g(x) + [g(d) - dg'(d)](t-x)f(x). \tag{1.4.25}$$

Integrating both sides of (1.4.25) with respect to t over $[a,b]$, we have

$$(b^2 - a^2)f(x)g(x) - xg(x)\int_a^b f(t)dt - xf(x)\int_a^b g(t)dt$$

$$= [f(c) - cf'(c)]\left\{\frac{b^2 - a^2}{2}g(x) - xg(x)(b-a)\right\}$$

$$+[g(d)-dg'(d)]\left\{\frac{b^2-a^2}{2}f(x)-xf(x)(b-a)\right\}. \quad (1.4.26)$$

Now, integrating both sides of (1.4.26) with respect to x over $[a,b]$, we have

$$(b^2-a^2)\int_a^b f(x)g(x)dx-\left(\int_a^b f(t)dt\right)\left(\int_a^b xg(x)dx\right)-\left(\int_a^b g(t)dt\right)\left(\int_a^b xf(x)dx\right)$$

$$=[f(c)-cf'(c)]\left\{\frac{b^2-a^2}{2}\int_a^b g(x)dx-(b-a)\int_a^b xg(x)dx\right\}$$

$$+[g(d)-dg'(d)]\left\{\frac{b^2-a^2}{2}\int_a^b f(x)dx-(b-a)\int_a^b xf(x)dx\right\}. \quad (1.4.27)$$

Rewriting (1.4.27), we have

$$G(f,g)=[f(c)-cf'(c)]\int_a^b g(x)\left\{\frac{1}{2}-\frac{x}{a+b}\right\}dx$$

$$+[g(d)-dg'(d)]\int_a^b f(x)\left\{\frac{1}{2}-\frac{x}{a+b}\right\}dx. \quad (1.4.28)$$

Using the properties of modulus, from (1.4.28), we have

$$|G(f,g)|\leqslant\|f-lf'\|_\infty\int_a^b |g(x)|\left|\frac{1}{2}-\frac{x}{a+b}\right|dx+\|g-lg'\|_\infty\int_a^b |f(x)|\left|\frac{1}{2}-\frac{x}{a+b}\right|dx,$$

and the inequality (1.4.20) is proved.

Multiplying the left hand sides and right hand sides of (1.4.23) and (1.4.24), we get

$$t^2f(x)g(x)-(xf(x))(tg(t))-(xg(x))(tf(t))+x^2f(t)g(t)$$

$$=[f(c)-cf'(c)][g(d)-dg'(d)](t-x)^2. \quad (1.4.29)$$

Integrating both sides of (1.4.29) with respect to t over $[a,b]$, we have

$$\frac{b^3-a^3}{3}f(x)g(x)-xf(x)\int_a^b tg(t)dt-xg(x)\int_a^b tf(t)dt+x^2\int_a^b f(t)g(t)dt$$

$$=[f(c)-cf'(c)][g(d)-dg'(d)]\times\left\{\frac{b^3-a^3}{3}-x(b^2-a^2)+x^2(b-a)\right\}. \quad (1.4.30)$$

Now, integrating both sides of (1.4.30) with respect to x over $[a,b]$, we have

$$\frac{b^3-a^3}{3}\int_a^b f(x)g(x)dx-\left(\int_a^b xf(x)dx\right)\left(\int_a^b tg(t)dt\right)$$

$$-\left(\int_a^b xg(x)dx\right)\left(\int_a^b tf(t)dt\right)+\frac{b^3-a^3}{3}\int_a^b f(t)g(t)dt$$

$$=[f(c)-cf'(c)][g(d)-dg'(d)]$$

$$\times\left\{\frac{b^3-a^3}{3}(b-a)-(b^2-a^2)\frac{b^2-a^2}{2}+(b-a)\frac{b^3-a^3}{3}\right\}. \tag{1.4.31}$$

Rewriting (1.4.31), we have

$$H(f,g)=[f(c)-cf'(c)][g(d)-dg'(d)]M. \tag{1.4.32}$$

Using the properties of modulus, from (1.4.32), we have

$$|H(f,g)|\leqslant\|f-lf'\|_\infty\|g-lg'\|_\infty|M|,$$

which is the required inequality in (1.4.21). The proof is complete.

Let $h:[a,b]\to\mathbb{R}$ be a differentiable function on $[a,b]$ and $h':[a,b]\to\mathbb{R}$ be integrable on $[a,b]$. Let $w:[a,b]\to[0,\infty)$ be some probability density function, that is, an integrable function satisfying $\int_a^b w(t)dt=1$ with $W(t)=\int_a^t w(x)dx$ for $t\in[a,b]$, $W(t)=0$ for $t<a$ and $W(t)=1$ for $t>b$. In the proof of the following Theorem given in [111] we use the Pečarić's extension (see [142]) of Montgomery's identity:

$$h(x)=\int_a^b w(t)h(t)dt+\int_a^b P_w(x,t)h'(t)dt, \tag{1.4.33}$$

where $P_w(x,t)$ is the weighted Peano kernel defined by

$$P_w(x,t)=\begin{cases}W(t), & t\in[a,x]\\ W(t)-1, & t\in(x,b]\end{cases} \tag{1.4.34}$$

Identity (1.4.33) can be proved easily by considering $\int_a^b P_w(x,t)h'(t)dt$ and integrating by parts.

Theorem 1.4.3. Let $f,\ g:[a,b]\to\mathbb{R}$ be differentiable functions on $[a,b]$ with $f',\ g':[a,b]\to\mathbb{R}$ integrable on $[a,b]$. Let $w:[a,b]\to[0,\infty)$ be an integrable function satisfying $\int_a^b w(t)dt=1$ and $W(t)$ be as defined above, then

$$|T(w,f,g)|\leqslant\frac{1}{2}\int_a^b w(x)\left[|g(x)|\|f'\|_\infty+|f(x)|\|g'\|_\infty\right]E(x)dx, \tag{1.4.35}$$

$$|T(w,f,g)|\leqslant\|f'\|_\infty\|g'\|_\infty\int_a^b w(x)E^2(x)dx, \tag{1.4.36}$$

where

$$E(x)=\int_a^b |P_w(x,t)|dt, \tag{1.4.37}$$

for $x\in[a,b]$ and $P_w(x,t)$ is given by (1.4.34).

Proof. From the hypotheses, the following identities hold:

$$f(x)=\int_a^b w(t)f(t)dt+\int_a^b P_w(x,t)f'(t)dt, \tag{1.4.38}$$

and

$$g(x)=\int_a^b w(t)g(t)dt+\int_a^b P_w(x,t)g'(t)dt. \tag{1.4.39}$$

Multiplying both sides of (1.4.38) and (1.4.39) by $w(x)g(x)$ and $w(x)f(x)$, adding the resulting identities and rewriting, we have

$$\begin{aligned} w(x)f(x)g(x)&=\frac{1}{2}\left[w(x)g(x)\int_a^b w(t)f(t)dt+w(x)f(x)\int_a^b w(t)g(t)dt\right]\\ &+\frac{1}{2}\left[w(x)g(x)\int_a^b P_w(x,t)f'(t)dt+w(x)f(x)\int_a^b P_w(x,t)g'(t)dt\right]. \end{aligned} \tag{1.4.40}$$

Integrating both sides of (1.4.40) with respect to x over $[a,b]$ and rewriting, we have

$$T(w,f,g)=\frac{1}{2}\int_a^b\left[w(x)g(x)\int_a^b P_w(x,t)f'(t)dt+w(x)f(x)\int_a^b P_w(x,t)g'(t)dt\right]dx. \tag{1.4.41}$$

From (1.4.41) and using the properties of modulus, we have

$$\begin{aligned} |T(w,f,g)|&\leqslant\frac{1}{2}\int_a^b\left[w(x)|g(x)|\int_a^b |P_w(x,t)||f'(t)|dt+w(x)|f(x)|\int_a^b |P_w(x,t)||g'(t)|dt\right]dx\\ &\leqslant\frac{1}{2}\int_a^b w(x)\left[|g(x)|\|f'\|_\infty+|f(x)|\|g'\|_\infty\right]E(x)dx, \end{aligned}$$

and the inequality (1.4.35) is proved.

From (1.4.38) and (1.4.39), we observe that

$$\begin{aligned} &\left(f(x)-\int_a^b w(t)f(t)dt\right)\left(g(x)-\int_a^b w(t)g(t)dt\right)\\ &=\left(\int_a^b P_w(x,t)f'(t)dt\right)\left(\int_a^b P_w(x,t)g'(t)dt\right), \end{aligned}$$

i.e.,

$$\begin{aligned} f(x)g(x)-f(x)\int_a^b w(t)g(t)dt-g(x)\int_a^b w(t)f(t)dt+\left(\int_a^b w(t)f(t)dt\right)\left(\int_a^b w(t)g(t)dt\right)\\ =\left(\int_a^b P_w(x,t)f'(t)dt\right)\left(\int_a^b P_w(x,t)g'(t)dt\right). \end{aligned} \tag{1.4.42}$$

Multiplying both sides of (1.4.42) by $w(x)$, integrating the resulting identity with respect to x over $[a,b]$ and using $\int_a^b w(x)dx=1$,, we have

$$T(w,f,g)=\int_a^b w(x)\left(\int_a^b P_w(x,t)f'(t)dt\right)\times\left(\int_a^b P_w(x,t)g'(t)dt\right)dx. \tag{1.4.43}$$

From (1.4.43) and using the properties of modulus, we have

$$\begin{aligned} |T(w,f,g)|&\leqslant\int_a^b w(x)\left(\int_a^b |P_w(x,t)||f'(t)|dt\right)\times\left(\int_a^b |P_w(x,t)||g'(t)|dt\right)dx\\ &\leqslant\|f'\|_\infty\|g'\|_\infty\int_a^b w(x)E^2(x)dx, \end{aligned}$$

which is the required inequality in (1.4.36). The proof is complete.

We end this section with the following Theorem which deals with the inequalities proved in [127].

Theorem 1.4.4. Let $f,\ g,\ h : [a,b] \to \mathbb{R}$ be continuous functions on $[a,b]$ and differentiable on (a,b) and $w : [a,b] \to [0,\infty)$ be an integrable function such that $\int_a^b w(x)dx > 0$. If $h'(t) \neq 0$ for each $t \in (a,b)$, then

$$|S(w,f,g)| \leqslant \frac{1}{2}\left[|S(w,g,h)|\left\|\frac{f'}{h'}\right\|_\infty + |S(w,f,h)|\left\|\frac{g'}{h'}\right\|_\infty\right], \tag{1.4.44}$$

$$|S(w,f,g)| \leqslant \left\|\frac{f'}{h'}\right\|_\infty \left\|\frac{g'}{h'}\right\|_\infty \times \left|\int_a^b w(x)h^2(x)dx - \frac{\left(\int_a^b w(x)h(x)dx\right)^2}{\int_a^b w(x)dx}\right|. \tag{1.4.45}$$

Proof. Let $x,\ y \in [a,b]$ with $y \neq x$. Applying Cauchy's mean value theorem, there exist points c and d between y and x such that (see [146])

$$f(x) - f(y) = \frac{f'(c)}{h'(c)}(h(x) - h(y)), \tag{1.4.46}$$

$$g(x) - g(y) = \frac{g'(d)}{h'(d)}(h(x) - h(y)). \tag{1.4.47}$$

Multiplying both sides of (1.4.46) and (1.4.47) by $g(x)$ and $f(x)$ respectively and adding the resulting identities, we get

$$2f(x)g(x) - g(x)f(y) - f(x)g(y) = \frac{f'(c)}{h'(c)}(g(x)h(x) - g(x)h(y))$$

$$+\frac{g'(d)}{h'(d)}(f(x)h(x) - f(x)h(y)). \tag{1.4.48}$$

Multiplying both sides of (1.4.48) by $w(y)$ and integrating the resulting identity with respect to y over $[a,b]$, we have

$$2\left(\int_a^b w(y)dy\right)f(x)g(x) - g(x)\int_a^b w(y)f(y)dy - f(x)\int_a^b w(y)g(y)dy$$

$$= \frac{f'(c)}{h'(c)}\left(\left(\int_a^b w(y)dy\right)g(x)h(x) - g(x)\int_a^b w(y)h(y)dy\right)$$

$$+\frac{g'(d)}{h'(d)}\left(\left(\int_a^b w(y)dy\right)f(x)h(x) - f(x)\int_a^b w(y)h(y)dy\right). \tag{1.4.49}$$

Next, multiplying both sides of (1.4.49) by $w(x)$ and integrating the resulting identity with respect to x over $[a,b]$, we have

$$2\left(\int_a^b w(y)dy\right)\int_a^b w(x)f(x)g(x)dx - \left(\int_a^b w(x)g(x)dx\right)\left(\int_a^b w(y)f(y)dy\right)$$

$$-\left(\int_a^b w(x)f(x)dx\right)\left(\int_a^b w(y)g(y)dy\right)$$

$$= \frac{f'(c)}{h'(c)} \left(\left(\int_a^b w(y)dy \right) \int_a^b w(x)g(x)h(x)dx - \left(\int_a^b w(x)g(x)dx \right) \left(\int_a^b w(y)h(y)dy \right) \right)$$

$$+ \frac{g'(d)}{h'(d)} \left(\left(\int_a^b w(y)dy \right) \int_a^b w(x)f(x)h(x)dx - \left(\int_a^b w(x)f(x)dx \right) \left(\int_a^b w(y)h(y)dy \right) \right). \tag{1.4.50}$$

From (1.4.50), it is easy to observe that

$$S(w,f,g) = \frac{1}{2} \left[\frac{f'(c)}{h'(c)} S(w,g,h) + \frac{g'(d)}{h'(d)} S(w,f,h) \right]. \tag{1.4.51}$$

Using the properties of modulus, from (1.4.51),we have

$$|S(w,f,g)| \leqslant \frac{1}{2} \left[|S(w,g,h)| \left\| \frac{f'}{h'} \right\|_\infty + |S(w,f,h)| \left\| \frac{g'}{h'} \right\|_\infty \right],$$

and the inequality (1.4.44) is proved.

Multiplying both sides of (1.4.46) and (1.4.47) by $w(y)$ and integrating the resulting identities with respect to y over $[a,b]$, we get

$$\left(\int_a^b w(y)dy \right) f(x) - \int_a^b w(y)f(y)dy$$

$$= \frac{f'(c)}{h'(c)} \left[\left(\int_a^b w(y)dy \right) h(x) - \int_a^b w(y)h(y)dy \right], \tag{1.4.52}$$

and

$$\left(\int_a^b w(y)dy \right) g(x) - \int_a^b w(y)g(y)dy$$

$$= \frac{g'(d)}{h'(d)} \left[\left(\int_a^b w(y)dy \right) h(x) - \int_a^b w(y)h(y)dy \right]. \tag{1.4.53}$$

Multiplying the left hand sides and right hand sides of (1.4.52) and (1.4.53), we get

$$\left(\int_a^b w(y)dy \right)^2 f(x)g(x) - \left(\int_a^b w(y)dy \right) f(x) \left(\int_a^b w(y)g(y)dy \right)$$

$$- \left(\int_a^b w(y)dy \right) g(x) \left(\int_a^b w(y)f(y)dy \right) + \left(\int_a^b w(y)f(y)dy \right) \left(\int_a^b w(y)g(y)dy \right)$$

$$= \left(\frac{f'(c)}{h'(c)} \right) \left(\frac{g'(d)}{h'(d)} \right) \left[\left(\int_a^b w(y)dy \right)^2 h^2(x) + \left(\int_a^b w(y)h(y)dy \right)^2 \right.$$

$$\left. -2 \left(\int_a^b w(y)dy \right) h(x) \left(\int_a^b w(y)h(y)dy \right) \right]. \tag{1.4.54}$$

Multiplying both sides of (1.4.54) by $w(x)$ and integrating the resulting identity with respect to x over $[a,b]$, we get

$$\left(\int_a^b w(y)dy\right)^2 \int_a^b w(x)f(x)g(x)dx$$

$$-\left(\int_a^b w(y)dy\right)\left(\int_a^b w(x)f(x)dx\right)\left(\int_a^b w(y)g(y)dy\right)$$

$$-\left(\int_a^b w(y)dy\right)\left(\int_a^b w(x)g(x)dx\right)\left(\int_a^b w(y)f(y)dy\right)$$

$$+\left(\int_a^b w(x)dx\right)\left(\int_a^b w(y)f(y)dy\right)\left(\int_a^b w(y)g(y)dy\right)$$

$$=\left(\frac{f'(c)}{h'(c)}\right)\left(\frac{g'(d)}{h'(d)}\right)\left[\left(\int_a^b w(y)dy\right)^2\left(\int_a^b w(x)h^2(x)dx\right)\right.$$

$$+\left(\int_a^b w(x)dx\right)\left(\int_a^b w(y)h(y)dy\right)^2$$

$$\left.-2\left(\int_a^b w(y)dy\right)\left(\int_a^b w(x)h(x)dx\right)\left(\int_a^b w(y)h(y)dy\right)\right]. \tag{1.4.55}$$

From (1.4.55), it is easy to observe that

$$S(w,f,g)=\left(\frac{f'(c)}{h'(c)}\right)\left(\frac{g'(d)}{h'(d)}\right)\times\left[\int_a^b w(x)h^2(x)dx-\frac{\left(\int_a^b w(x)h(x)dx\right)^2}{\int_a^b w(x)dx}\right]. \tag{1.4.56}$$

Using the properties of modulus, from (1.4.56), we get the desired inequality in (1.4.45). The proof is complete.

Remark 1.4.2. We note that in [42], Dragomir has given a number of inequalities similar to (1) and (3) by using different hypotheses on the functions and their derivatives. For earlier discussion on such inequalities, see [79,144].

1.5 More inequalities of the Grüss-and Čebyšev-type

This section deals with some more inequalities of the Grüss- and Čebyšev-types involving functions and their higher order derivatives, established by Pachpatte in [118,120,121].

First, we introduce some notation to simplify the details of presentation. For some suitable functions f, g, h and their derivatives $f^{(n)}$, $g^{(n)}$, $h^{(n)}$ ($n \geqslant 1$ is an integer) defined on $[a,b]$ and a harmonic sequence of polynomials $\{P_n(t)\}, t \in [a,b]$, we set

$$M[h(x)] = h(x) + \frac{h(a)+h(b)}{2} - \left(x - \frac{a+b}{2}\right)h'(x), \tag{1.5.1}$$

$$N[h(x)] = h(x) - \frac{h(b)-h(a)}{b-a}\left(x - \frac{a+b}{2}\right), \tag{1.5.2}$$

$$A[h(x)] = h(x) + \frac{1}{b-a}\sum_{k=1}^{n-1}\left[\frac{(b-x)^{k+1}+(-1)^k(x-a)^{k+1}}{(k+1)!}\right]h^{(k)}(x), \tag{1.5.3}$$

$$B[h(x)] = \frac{1}{n}\left[h(x) + \sum_{k=1}^{n-1}(-1)^k P_k(x)h^{(k)}(x) + \sum_{k=1}^{n-1}\overline{H}_k\right], \tag{1.5.4}$$

$$\overline{H}_k = \frac{(-1)^k(n-k)}{b-a}\left[P_k(a)h^{(k-1)}(a) - P_k(b)h^{(k-1)}(b)\right], \tag{1.5.5}$$

and

$$F_k(x) = \frac{n-k}{k!}\frac{f^{(k-1)}(a)(x-a)^k - f^{(k-1)}(b)(x-b)^k}{b-a}, \tag{1.5.6}$$

$$G_k(x) = \frac{n-k}{k!}\frac{g^{(k-1)}(a)(x-a)^k - g^{(k-1)}(b)(x-b)^k}{b-a}, \tag{1.5.7}$$

$$H_k(x) = \frac{n-k}{k!}\frac{h^{(k-1)}(a)(x-a)^k - h^{(k-1)}(b)(x-b)^k}{b-a}, \tag{1.5.8}$$

$$I_k = \frac{1}{k!}\int_a^b f^{(k)}(y)(x-y)^k dy,\ I_0 = \int_a^b f(y)dy, \tag{1.5.9}$$

$$J_k = \frac{1}{k!}\int_a^b g^{(k)}(y)(x-y)^k dy,\ J_0 = \int_a^b g(y)dy, \tag{1.5.10}$$

$$L_k = \frac{1}{k!}\int_a^b h^{(k)}(y)(x-y)^k dy, \quad L_0 = \int_a^b h(y)dy, \tag{1.5.11}$$

for $1 \leqslant k \leqslant n-1$. We use the usual convention that an empty sum is taken to be zero and define $\|h\|_\infty = \sup_{t\in[a,b]}|h(t)| < \infty$.

We begin with proving some auxiliary results.

Lemma 1.5.1 (see [35]). Let $h : [a,b] \to \mathbb{R}$ be a function with first derivative absolutely continuous on $[a,b]$ and assume that the second derivative $h'' \in L_\infty[a,b]$, then

$$M[h(x)] - \frac{2}{b-a}\int_a^b h(t)dt = -\frac{1}{b-a}\int_a^b p(x,t)\left(t - \frac{a+b}{2}\right)h''(t)dt, \tag{1.5.12}$$

$x \in [a,b]$, where $M[h(x)]$ and $p(x,t)$ are given by (1.5.1) and (1.2.11) respectively.

Proof. The following identity holds (see, the proof of Theorem 1.2.3):

$$f(x)=\frac{1}{b-a}\left[\int_a^b f(t)dt+\int_a^b p(x,t)f'(t)dt\right], \tag{1.5.13}$$

for $x\in[a,b]$, provided f is absolutely continuous on $[a,b]$. Choose in (1.5.13) $f(x)=\left(x-\frac{a+b}{2}\right)h'(x)$, to get

$$\left(x-\frac{a+b}{2}\right)h'(x)=\frac{1}{b-a}\left[\int_a^b\left(t-\frac{a+b}{2}\right)h'(t)dt\right.$$

$$\left.+\int_a^b p(x,t)\left[h'(t)+\left(t-\frac{a+b}{2}\right)h''(t)\right]dt\right]. \tag{1.5.14}$$

Integrating by parts, we have

$$\int_a^b\left(t-\frac{a+b}{2}\right)h'(t)dt=\frac{(h(a)+h(b))(b-a)}{2}-\int_a^b h(t)dt. \tag{1.5.15}$$

Also, upon using (1.5.13), we have

$$\int_a^b p(x,t)\left[h'(t)+\left(t-\frac{a+b}{2}\right)h''(t)\right]dt$$

$$=\int_a^b p(x,t)h'(t)dt+\int_a^b p(x,t)\left(t-\frac{a+b}{2}\right)h''(t)dt$$

$$=(b-a)h(x)-\int_a^b h(t)dt+\int_a^b p(x,t)\left(t-\frac{a+b}{2}\right)h''(t)dt. \tag{1.5.16}$$

Using (1.5.15) and (1.5.16) in (1.5.14), we deduce that

$$(b-a)\left(x-\frac{a+b}{2}\right)h'(x)=\frac{(h(a)+h(b))(b-a)}{2}-\int_a^b h(t)dt$$

$$+(b-a)h(x)-\int_a^b h(t)dt+\int_a^b p(x,t)\left(t-\frac{a+b}{2}\right)h''(t)dt. \tag{1.5.17}$$

Rewriting (1.5.17) we get (1.5.12).

Lemma 1.5.2 (see [28]). Let $h:[a,b]\to\mathbb{R}$ be a continuous function on $[a,b]$ and twice differentiable on (a,b) with second derivative $h'':(a,b)\to\mathbb{R}$ being bounded on (a,b), then

$$N[h(x)]-\frac{1}{b-a}\int_a^b h(t)dt=\frac{1}{(b-a)^2}\int_a^b\int_a^b p(x,t)p(t,s)h''(s)ds, \tag{1.5.18}$$

for $x\in[a,b]$, where $N[h(x)]$ and $p(x,t)$ are given by (1.5.2) and (1.2.11) respectively.

Proof. From the hypotheses, the identity

$$h(x) = \frac{1}{b-a}\int_a^b h(t)dt + \frac{1}{b-a}\int_a^b p(x,t)h'(t)dt, \tag{1.5.19}$$

holds for $x \in [a,b]$. Applying the identity (1.5.19) for $h'(t)$, we can state

$$h'(t) = \frac{1}{b-a}\int_a^b h'(s)ds + \frac{1}{b-a}\int_a^b p(t,s)h''(s)ds$$

$$= \frac{h(b)-h(a)}{b-a} + \frac{1}{b-a}\int_a^b p(t,s)h''(s)ds.$$

Using this in the right hand side of (1.5.19), we get

$$h(x) = \frac{1}{b-a}\int_a^b h(t)dt + \frac{1}{b-a}\int_a^b p(x,t)\left[\frac{h(b)-h(a)}{b-a} + \frac{1}{b-a}\int_a^b p(t,s)h''(s)ds\right]dt$$

$$= \frac{1}{b-a}\int_a^b h(t)dt + \frac{h(b)-h(a)}{(b-a)^2}\int_a^b p(x,t)dt$$

$$+ \frac{1}{(b-a)^2}\int_a^b\int_a^b p(x,t)p(t,s)h''(s)dsdt. \tag{1.5.20}$$

It is easy to observe that

$$\int_a^b p(x,t)dt = \int_a^x (t-a)dt + \int_x^b (t-b)dt = (b-a)\left(x - \frac{a+b}{2}\right). \tag{1.5.21}$$

Using (1.5.21) in (1.5.20) and rewriting, we get (1.5.18).

Lemma 1.5.3 (see [16]). Let $h : [a,b] \to \mathbb{R}$ be a function such that $h^{(n-1)}$ is absolutely continuous on $[a,b]$, then

$$\int_a^b h(t)dt = \sum_{k=0}^{n-1}\left[\frac{(b-x)^{k+1} + (-1)^k(x-a)^{k+1}}{(k+1)!}\right]h^{(k)}(x) + (-1)^n\int_a^b E_n(x,t)h^{(n)}(t)dt, \tag{1.5.22}$$

for $x \in [a,b]$, where

$$E_n(x,t) = \begin{cases} \dfrac{(t-a)^n}{n!} & \text{if } t \in [a,x] \\ \dfrac{(t-b)^n}{n!} & \text{if } t \in (x,b] \end{cases} \tag{1.5.23}$$

for $x \in [a,b]$ and $n \geqslant 1$ is a natural number.

Proof. The proof is by mathematical induction. For $n = 1$, we have to prove the identity

$$\int_a^b h(t)dt = (b-a)h(x) - \int_a^b E_1(x,t)h^{(1)}(t)dt. \tag{1.5.24}$$

The equality (1.5.24) can be proved by following the proof of identity (1.2.12) given in Theorem 1.2.3. Assume that (1.5.22) holds for n and let us prove it for $n+1$. That is, we have to prove the equality

$$\int_a^b h(t)dt = \sum_{k=0}^{n} \left[\frac{(b-x)^{k+1} + (-1)^k (x-a)^{k+1}}{(k+1)!} \right] h^{(k)}(x)$$

$$+(-1)^{n+1} \int_a^b E_{n+1}(x,t)h^{(n+1)}(t)dt. \tag{1.5.25}$$

It is easy to observe that

$$\int_a^b E_{n+1}(x,t)h^{(n+1)}(t)dt = \int_a^x \frac{(t-a)^{n+1}}{(n+1)!} h^{(n+1)}(t)dt + \int_x^b \frac{(t-b)^{n+1}}{(n+1)!} h^{(n+1)}(t)dt$$

$$= \frac{(t-a)^{n+1}}{(n+1)!} h^{(n)}(t) \bigg|_a^x - \frac{1}{n!} \int_a^x (t-a)^n h^{(n)}(t)dt$$

$$+ \frac{(t-b)^{n+1}}{(n+1)!} h^{(n)}(t) \bigg|_x^b - \frac{1}{n!} \int_x^b (t-b)^n h^{(n)}(t)dt$$

$$= \frac{(x-a)^{n+1} + (-1)^{n+2}(b-x)^{n+1}}{(n+1)!} h^{(n)}(x) - \int_a^b E_n(x,t)h^{(n)}(t)dt.$$

That is

$$\int_a^b E_n(x,t)h^{(n)}(t)dt$$

$$= \frac{(x-a)^{n+1} + (-1)^{n+2}(b-x)^{n+1}}{(n+1)!} h^{(n)}(x) - \int_a^b E_{n+1}(x,t)h^{(n+1)}(t)dt. \tag{1.5.26}$$

Now by using induction hypotheses and (1.5.26), we get

$$\int_a^b h(t)dt = \sum_{k=0}^{n-1} \left[\frac{(b-x)^{k+1} + (-1)^k (x-a)^{k+1}}{(k+1)!} \right] h^{(k)}(x)$$

$$+ \frac{(b-x)^{n+1} + (-1)^n (x-a)^{n+1}}{(n+1)!} h^{(n)}(x) - (-1)^n \int_a^b E_{n+1}(x,t)h^{(n+1)}(t)dt$$

$$= \sum_{k=0}^{n} \left[\frac{(b-x)^{k+1} + (-1)^k (x-a)^{k+1}}{(} k+1)! \right] h^{(k)}(x) + (-1)^{n+1} \int_a^b E_{n+1}(x,t)h^{(n+1)}(t)dt,$$

which is the identity (1.5.25) and the proof is complete.

Lemma 1.5.4 (see [22]). Let $\{P_n(t)\}$, $t \in [a,b]$ be a harmonic sequence of polynomials, that is $P_n'(t) = P_{n-1}(t)$, $n \in \mathbb{N}$, $P_0(t) = 1$. Further, let $h : [a,b] \to \mathbb{R}$ be such that $h^{(n-1)}$ is absolutely continuous for some $n \in \mathbb{N}$, then

$$B[h(x)] - \frac{1}{b-a}\int_a^b h(t)dt = \frac{(-1)^{n-1}}{n(b-a)}\int_a^b P_{n-1}(t)p(x,t)h^{(n)}(t)dt, \tag{1.5.27}$$

for $x \in [a,b]$, where $B[h(x)]$ and $p(x,t)$ are given by (1.5.4) and (1.2.11) respectively.

Proof. Integrating by parts, we have (see also [21])

$$(-1)^{n-1}\int_y^x P_{n-1}(t)h^{(n)}(t)dt = (-1)^{n-1}P_{n-1}(t)h^{(n-1)}(t)\Big|_y^x + (-1)^{n-2}\int_y^x P_{n-2}(t)h^{(n-1)}(t)dt$$

$$= (-1)^{n-1}\left[P_{n-1}(x)h^{(n-1)}(x) - P_{n-1}(y)h^{(n-1)}(y)\right] + (-1)^{n-2}\int_y^x P_{n-2}(t)h^{(n-1)}(t)dt,$$

for $x,\ y \in [a,b]$. By applying the same procedure to the last integral, we successively get the relation

$$(-1)^{n-1}\int_y^x P_{n-1}(t)h^{(n)}(t)dt = \sum_{k=1}^{n-1}(-1)^k\left[P_k(x)h^{(k)}(x) - P_k(y)h^{(k)}(y)\right] + h(x) - h(y),$$

i.e.,

$$h(y) = h(x) + \sum_{k=1}^{n-1}(-1)^k\left[P_k(x)h^{(k)}(x) - P_k(y)h^{(k)}(y)\right] + (-1)^n\int_y^x P_{n-1}(t)h^{(n)}(t)dt, \tag{1.5.28}$$

for $x,\ y \in [a,b]$. If we set $x = a$ and $y = b$, $n = m+1$ and replace $h(t)$ by $\int_a^t h(u)du$ in (1.5.28), we get

$$\int_a^b h(t)dt = \sum_{k=1}^{m}(-1)^k\left[P_k(a)h^{(k-1)}(a) - P_k(b)h^{(k-1)}(b)\right]$$
$$+(-1)^m\int_a^b P_m(t)h^{(m)}(t)dt. \tag{1.5.29}$$

Integrating both sides of (1.5.28) with respect to y over $[a,b]$, we have

$$\int_a^b h(y)dy = (b-a)\left[h(x) + \sum_{k=1}^{n-1}(-1)^k P_k(x)h^{(k)}(x)\right]$$
$$-\sum_{k=1}^{n-1}(-1)^k\int_a^b P_k(y)h^{(k)}(y)dy + (-1)^n\int_a^b\int_y^x P_{n-1}(t)h^{(n)}(t)dtdy. \tag{1.5.30}$$

Using (1.5.29), we have

$$\int_a^b h(y)dy = (b-a)\left[h(x) + \sum_{k=1}^{n-1}(-1)^k P_k(x)h^{(k)}(x)\right]$$

$$-\sum_{k=1}^{n-1}\left[\sum_{j=1}^{k}(-1)^{j}\left[P_j(b)h^{(j-1)}(b)-P_j(a)h^{(j-1)}(a)\right]+\int_a^b h(t)dt\right]$$

$$+(-1)^n\int_a^b\int_y^x P_{n-1}(t)h^{(n)}(t)dtdy,$$

i.e.,

$$n\int_a^b h(y)dy=(b-a)\left[h(x)+\sum_{k=1}^{n-1}(-1)^k P_k(x)h^{(k)}(x)\right]$$

$$-\sum_{k=1}^{n-1}(-1)^k(n-k)\left[P_k(b)h^{(k-1)}(b)-P_k(a)h^{(k-1)}(a)\right]$$

$$+(-1)^n\int_a^b\int_y^x P_{n-1}(t)h^{(n)}(t)dtdy. \tag{1.5.31}$$

By making use of (1.5.4) and (1.5.5) and (1.2.11) in (1.5.31) and rewriting, we get (1.5.27).

Remark 1.5.1. We note that, for the harmonic sequence of polynomials

$$P_k(t)=\frac{(t-x)^k}{k!},\quad k\geqslant 0,$$

the relation (1.5.27) reduces to the identity given by Fink in [59].

We are now ready to state and prove the following Theorems which deal with the inequalities proved in [118].

Theorem 1.5.1. Let $f,\ g:[a,b]\to\mathbb{R}$ be functions with first derivatives absolutely continuous on $[a,b]$ and assume that the second derivatives $f'',\ g'':(a,b)\to\mathbb{R}$ are bounded on (a,b). Then the inequalities

$$\left|\frac{1}{b-a}\int_a^b[g(x)M[f(x)]+f(x)M[g(x)]]\,dx-\frac{4}{(b-a)^2}\left(\int_a^b f(x)dx\right)\left(\int_a^b g(x)dx\right)\right|$$

$$\leqslant\frac{1}{(b-a)^2}\int_a^b\left[|g(x)|\|f''\|_\infty+|f(x)|\|g''\|_\infty\right]I(x)dx, \tag{1.5.32}$$

and

$$\left|\frac{1}{b-a}\int_a^b M[f(x)]M[g(x)]\,dx-\frac{2}{(b-a)^2}\left[\left(\int_a^b M[g(x)]dx\right)\left(\int_a^b f(x)dx\right)\right.\right.$$

$$\left.\left.+\left(\int_a^b M[f(x)]dx\right)\left(\int_a^b g(x)dx\right)\right]+\frac{4}{(b-a)^2}\left(\int_a^b f(x)dx\right)\left(\int_a^b g(x)dx\right)\right|$$

$$\leqslant\frac{1}{(b-a)^3}\|f''\|_\infty\|g''\|_\infty\int_a^b I^2(x)dx, \tag{1.5.33}$$

hold, where

$$I(x)=\int_a^b|p(x,t)|\left|t-\frac{a+b}{2}\right|dt, \tag{1.5.34}$$

in which $p(x,t)$ is given by (1.2.11).

Theorem 1.5.2. Let $f,\ g:[a,b]\to\mathbb{R}$ be twice differentiable functions on (a,b) and $f'',\ g'':(a,b)\to\mathbb{R}$ be bounded, then the inequalities

$$\left|\frac{1}{b-a}\int_a^b [g(x)N[f(x)]+f(x)N[g(x)]]\,dx-\frac{2}{(b-a)^2}\left(\int_a^b f(x)dx\right)\left(\int_a^b g(x)dx\right)\right|$$
$$\leqslant\frac{1}{(b-a)^3}\int_a^b [|g(x)|\|f''\|_\infty+|f(x)|\|g''\|_\infty]H(x)dx, \tag{1.5.35}$$

and

$$\left|\frac{1}{b-a}\int_a^b N[f(x)]N[g(x)]\,dx-\frac{1}{(b-a)^2}\left[\left(\int_a^b N[g(x)]dx\right)\left(\int_a^b f(x)dx\right)\right.\right.$$
$$\left.\left.+\left(\int_a^b N[f(x)]dx\right)\left(\int_a^b g(x)dx\right)\right]+\frac{1}{(b-a)^2}\left(\int_a^b f(x)dx\right)\left(\int_a^b g(x)dx\right)\right|$$
$$\leqslant\frac{1}{(b-a)^5}\|f''\|_\infty\|g''\|_\infty\int_a^b H^2(x)dx, \tag{1.5.36}$$

hold, where

$$H(x)=\int_a^b\int_a^b |p(x,t)|\,|p(t,s)|\,dsdt, \tag{1.5.37}$$

in which $p(x,t)$ is given by (1.2.11).

Proofs of Theorems 1.5.1 and 1.5.2. From the hypotheses of Theorem 1.5.1, we have the following identities (see, Lemma 1.5.1):

$$M[f(x)]-\frac{2}{b-a}\int_a^b f(t)dt=-\frac{1}{b-a}\int_a^b p(x,t)\left(t-\frac{a+b}{2}\right)f''(t)dt, \tag{1.5.38}$$

$$M[g(x)]-\frac{2}{b-a}\int_a^b g(t)dt=-\frac{1}{b-a}\int_a^b p(x,t)\left(t-\frac{a+b}{2}\right)g''(t)dt, \tag{1.5.39}$$

for $x\in[a,b]$. Multiplying both sides of (1.5.38) and (1.5.39) by $g(x)$ and $f(x)$ respectively and adding the resulting identities, we have

$$g(x)M[f(x)]+f(x)M[g(x)]-\frac{2}{b-a}\left[g(x)\int_a^b f(t)dt+f(x)\int_a^b g(t)dt\right]$$
$$=-\frac{1}{b-a}\left[g(x)\int_a^b p(x,t)\left(t-\frac{a+b}{2}\right)f''(t)dt\right.$$
$$\left.+f(x)\int_a^b p(x,t)\left(t-\frac{a+b}{2}\right)g''(t)dt\right]. \tag{1.5.40}$$

Dividing both sides of (1.5.40) by $(b-a)$ and integrating with respect to x over $[a,b]$, we get

$$\frac{1}{b-a}\int_a^b [g(x)M[f(x)]+f(x)M[g(x)]]\,dx-\frac{4}{(b-a)^2}\left(\int_a^b f(x)dx\right)\left(\int_a^b g(x)dx\right)$$

$$= -\frac{1}{(b-a)^2}\int_a^b \left[g(x)\int_a^b p(x,t)\left(t-\frac{a+b}{2}\right)f''(t)dt\right.$$

$$\left. +f(x)\int_a^b p(x,t)\left(t-\frac{a+b}{2}\right)g''(t)dt\right]dx. \tag{1.5.41}$$

From (1.5.41) and using the properties of modulus, we have

$$\left|\frac{1}{b-a}\int_a^b [g(x)M[f(x)]+f(x)M[g(x)]]\,dx - \frac{4}{(b-a)^2}\left(\int_a^b f(x)dx\right)\left(\int_a^b g(x)dx\right)\right|$$

$$\leqslant \frac{1}{(b-a)^2}\int_a^b \left[|g(x)|\int_a^b |p(x,t)|\left|t-\frac{a+b}{2}\right||f''(t)|dt\right.$$

$$\left. +|f(x)|\int_a^b |p(x,t)|\left|t-\frac{a+b}{2}\right||g''(t)|\,dt\right]dx$$

$$\leqslant \frac{1}{(b-a)^2}\int_a^b [|g(x)|\|f''\|_\infty + |f(x)|\|g''\|_\infty]\left(\int_a^b |p(x,t)|\left|t-\frac{a+b}{2}\right|dt\right)dx$$

$$= \frac{1}{(b-a)^2}\int_a^b [|g(x)|\|f''\|_\infty + |f(x)|\|g''\|_\infty]\, I(x)dx,$$

and the inequality (1.5.32) is proved.

Multiplying the left hand sides and right hand sides of (1.5.38) and (1.5.39), we have

$$M[f(x)]M[g(x)] - \frac{2}{b-a}\left[M[g(x)]\int_a^b f(t)dt + M[f(x)]\int_a^b g(t)dt\right]$$

$$+\frac{4}{(b-a)^2}\left(\int_a^b f(t)dt\right)\left(\int_a^b g(t)dt\right)$$

$$= \frac{1}{(b-a)^2}\left(\int_a^b p(x,t)\left(t-\frac{a+b}{2}\right)f''(t)dt\right)$$

$$\times\left(\int_a^b p(x,t)\left(t-\frac{a+b}{2}\right)g''(t)dt\right). \tag{1.5.42}$$

Dividing both sides of (1.5.42) by $(b-a)$ and then integrating with respect to x over $[a,b]$, we get

$$\frac{1}{b-a}\int_a^b M[f(x)]M[g(x)]dx - \frac{2}{(b-a)^2}\left[\left(\int_a^b M[g(x)]dx\right)\left(\int_a^b f(t)dt\right)\right.$$

$$\left.+\left(\int_a^b M[f(x)]dx\right)\left(\int_a^b g(t)dt\right)\right] + \frac{4}{(b-a)^2}\left(\int_a^b f(t)dt\right)\left(\int_a^b g(t)dt\right)$$

$$= \frac{1}{(b-a)^3}\int_a^b \left(\int_a^b p(x,t)\left(t-\frac{a+b}{2}\right)f''(t)dt\right)$$

$$\times\left(\int_a^b p(x,t)\left(t-\frac{a+b}{2}\right)g''(t)dt\right)dx. \tag{1.5.43}$$

From (1.5.43) and using the properties of modulus, we have

$$\begin{aligned}&\left|\frac{1}{b-a}\int_a^b M[f(x)]M[g(x)]dx-\frac{2}{(b-a)^2}\left[\left(\int_a^b M[g(x)]dx\right)\left(\int_a^b f(t)dt\right)\right.\right.\\&\left.\left.+\left(\int_a^b M[f(x)]dx\right)\left(\int_a^b g(t)dt\right)\right]+\frac{4}{(b-a)^2}\left(\int_a^b f(t)dt\right)\left(\int_a^b g(t)dt\right)\right|\\&\leqslant\frac{1}{(b-a)^3}\int_a^b\left(\int_a^b|p(x,t)|\left|t-\frac{a+b}{2}\right||f''(t)|dt\right)\left(\int_a^b|p(x,t)|\left|t-\frac{a+b}{2}\right||g''(t)|dt\right)\\&\leqslant\frac{1}{(b-a)^3}\|f''\|_\infty\|g''\|_\infty\int_a^b\left(\int_a^b|p(x,t)|\left|t-\frac{a+b}{2}\right|dt\right)^2dx\\&=\frac{1}{(b-a)^3}\|f''\|_\infty\|g''\|_\infty\int_a^b I^2(x)dx,\end{aligned}$$

which is the required inequality in (1.5.33).

From the hypotheses of Theorem 1.5.2 we have the following identities (see, Lemma 1.5.2):

$$N[f(x)]-\frac{1}{b-a}\int_a^b f(t)dt=\frac{1}{(b-a)^2}\int_a^b\int_a^b p(x,t)p(t,s)f''(s)dsdt, \tag{1.5.44}$$

and

$$N[g(x)]-\frac{1}{b-a}\int_a^b g(t)dt=\frac{1}{(b-a)^2}\int_a^b\int_a^b p(x,t)p(t,s)g''(s)dsdt, \tag{1.5.45}$$

for $x\in[a,b]$. Multiplying both sides of (1.5.44) and (1.5.45) by $g(x)$ and $f(x)$ respectively and adding the resulting identities, we have

$$\begin{aligned}&g(x)N[f(x)]+f(x)N[g(x)]-\frac{1}{b-a}\left[g(x)\int_a^b f(t)dt+f(x)\int_a^b g(t)dt\right]\\&=\frac{1}{(b-a)^2}\left[g(x)\int_a^b\int_a^b p(x,t)p(t,s)f''(s)dsdt\right.\\&\left.+f(x)\int_a^b\int_a^b p(x,t)p(t,s)g''(s)dsdt\right].\end{aligned} \tag{1.5.46}$$

Multiplying the left hand sides and right hand sides of (1.5.44) and (1.5.45), we have

$$\begin{aligned}&N[f(x)]N[g(x)]-\frac{1}{b-a}\left[N[g(x)]\int_a^b f(t)dt+N[f(x)]\int_a^b g(t)dt\right]\\&+\frac{1}{(b-a)^2}\left(\int_a^b f(t)dt\right)\left(\int_a^b g(t)dt\right)\\&=\frac{1}{(b-a)^4}\left(\int_a^b\int_a^b p(x,t)p(t,s)f''(s)dsdt\right)\\&\times\left(\int_a^b\int_a^b p(x,t)p(t,s)g''(s)dsdt\right).\end{aligned} \tag{1.5.47}$$

From (1.5.46) and (1.5.47) and following similar arguments as in the proof of Theorem 1.5.1, below (1.5.40) and (1.5.42) with suitable changes we get the desired inequalities in (1.5.35) and (1.5.36).

Next, we give the inequalities established in [120].

Theorem 1.5.3. Let $f,\ g:[a,b]\to\mathbb{R}$ be functions such that $f^{(n-1)},\ g^{(n-1)}$ are absolutely continuous on $[a,b]$ and $f^{(n)},\ g^{(n)}\in L_\infty[a,b]$ for some $n\in\mathbb{N}$, then

$$\begin{aligned}&\left|\frac{1}{b-a}\int_a^b [g(x)A[f(x)]+f(x)A[g(x)]]\,dx\right.\\&\left.-\frac{2}{(b-a)^2}\left(\int_a^b f(x)dx\right)\left(\int_a^b g(x)dx\right)\right|\\&\leqslant \frac{1}{(b-a)^2}\int_a^b \left[|g(x)|\|f^{(n)}\|_\infty+|f(x)|\|g^{(n)}\|_\infty\right]H_n(x)dx,\end{aligned}\tag{1.5.48}$$

and

$$\begin{aligned}&\left|\frac{1}{b-a}\int_a^b A[f(x)]A[g(x)]\,dx\right.\\&-\frac{1}{(b-a)^2}\left[\left(\int_a^b A[g(x)]dx\right)\left(\int_a^b f(x)dx\right)+\left(\int_a^b A[f(x)]dx\right)\left(\int_a^b g(x)dx\right)\right]\\&\left.+\frac{1}{(b-a)^2}\left(\int_a^b f(x)dx\right)\left(\int_a^b g(x)dx\right)\right|\\&\leqslant \frac{1}{(b-a)^3}\|f^{(n)}\|_\infty\|g^{(n)}\|_\infty\int_a^b H_n^2(x)dx,\end{aligned}\tag{1.5.49}$$

where

$$H_n(x)=\int_a^b |E_n(x,t)|dt,\tag{1.5.50}$$

in which $E_n(x,t)$ is given by (1.5.23).

Theorem 1.5.4. Let $\{P_n(t)\}$, $t\in[a,b]$ be a harmonic sequence of polynomials and $f,\ g:[a,b]\to\mathbb{R}$ be functions such that $f^{(n-1)},\ g^{(n-1)}$ are absolutely continuous on $[a,b]$ and $f^{(n)},\ g^{(n)}\in L_\infty[a,b]$ for some $n\in\mathbb{N}$, then

$$\begin{aligned}&\left|\frac{1}{b-a}\int_a^b [g(x)B[f(x)]+f(x)B[g(x)]]\,dx\right.\\&\left.-\frac{2}{(b-a)^2}\left(\int_a^b f(x)dx\right)\left(\int_a^b g(x)dx\right)\right|\\&\leqslant \frac{1}{(b-a)^2}\int_a^b \left[|g(x)|\|f^{(n)}\|_\infty+|f(x)|\|g^{(n)}\|_\infty\right]D_n(x)dx,\end{aligned}\tag{1.5.51}$$

and

$$\begin{aligned}&\left|\frac{1}{b-a}\int_a^b B[f(x)]B[g(x)]\,dx\right.\\&-\frac{1}{(b-a)^2}\left[\left(\int_a^b B[g(x)]dx\right)\left(\int_a^b f(x)dx\right)+\left(\int_a^b B[f(x)]dx\right)\left(\int_a^b g(x)dx\right)\right]\\&\left.+\frac{1}{(b-a)^2}\left(\int_a^b f(x)dx\right)\left(\int_a^b g(x)dx\right)\right|\\&\leqslant \frac{1}{(b-a)^3}\|f^{(n)}\|_\infty\|g^{(n)}\|_\infty\int_a^b D_n^2(x)dx,\end{aligned}\tag{1.5.52}$$

where

$$D_n(x)=\frac{1}{n}\int_a^b |P_{n-1}(t)p(x,t)|dt,\tag{1.5.53}$$

in which $p(x,t)$ is given by (1.2.11).

Proofs of Theorems 1.5.3 and 1.5.4. From the hypotheses of Theorem 1.5.3, we have the following identities (see, Lemma 1.5.3):

$$A[f(x)] - \frac{1}{b-a}\int_a^b f(t)dt = \frac{(-1)^{n+1}}{b-a}\int_a^b E_n(x,t)f^{(n)}(t)dt, \tag{1.5.54}$$

$$A[g(x)] - \frac{1}{b-a}\int_a^b g(t)dt = \frac{(-1)^{n+1}}{b-a}\int_a^b E_n(x,t)g^{(n)}(t)dt, \tag{1.5.55}$$

for $x \in [a,b]$. Multiplying both sides of (1.5.54) and (1.5.55) by $g(x)$ and $f(x)$ respectively and adding the resulting identities, we have

$$g(x)A[f(x)] + f(x)A[g(x)] - \frac{1}{b-a}\left[g(x)\int_a^b f(t)dt + f(x)\int_a^b g(t)dt\right]$$

$$= \frac{(-1)^{n+1}}{b-a}\left[g(x)\int_a^b E_n(x,t)f^{(n)}(t)dt + f(x)\int_a^b E_n(x,t)g^{(n)}(t)dt\right]. \tag{1.5.56}$$

Multiplying the left hand sides and right hand sides of (1.5.54) and (1.5.55), we get

$$A[f(x)]A[g(x)] - \frac{1}{b-a}\left[A[g(x)]\int_a^b f(t)dt + A[f(x)]\int_a^b g(t)dt\right]$$

$$+\frac{1}{(b-a)^2}\left(\int_a^b f(t)dt\right)\left(\int_a^b g(t)dt\right)$$

$$= \frac{(-1)^{2n+2}}{(b-a)^2}\left(\int_a^b E_n(x,t)f^{(n)}(t)dt\right)\left(\int_a^b E_n(x,t)g^{(n)}(t)dt\right). \tag{1.5.57}$$

Dividing both sides of (1.5.56) and (1.5.57) by $(b-a)$ and then integrating both sides with respect to x over $[a,b]$ and following closely the proof of Theorem 1.5.1 with suitable changes, we get (1.5.48) and (1.5.49).

From the hypotheses of Theorem 1.5.4, the following identities hold (see, Lemma 1.5.4):

$$B[f(x)] - \frac{1}{b-a}\int_a^b f(t)dt = \frac{(-1)^{n-1}}{n(b-a)}\int_a^b P_{n-1}(t)p(x,t)f^{(n)}(t)dt, \tag{1.5.58}$$

and

$$B[g(x)] - \frac{1}{b-a}\int_a^b g(t)dt = \frac{(-1)^{n-1}}{n(b-a)}\int_a^b P_{n-1}(t)p(x,t)g^{(n)}(t)dt, \tag{1.5.59}$$

for $x \in [a,b]$. The proofs of the inequalities (1.5.51) and (1.5.52) can be completed by following the proofs of Theorems 1.5.1–1.5.3, we leave the details to the reader.

Remark 1.5.2. We note that, one can very easily obtain bounds on the right hand sides in (1.5.32), (1.5.33), (1.5.35), (1.5.36) and (1.5.48), (1.5.49), (1.5.51), (1.5.52) when f'', g'' and $f^{(n)}$, $g^{(n)}$ belong to $L_q[a,b]$ for $q > 1$, $\frac{1}{q}+\frac{1}{r} = 1$ or $L_1[a,b]$. The precise formulation of such results are very close to those given in Theorems 1.5.1–1.5.4 with suitable changes. We omit the details.

To complete this section we present the Grüss-type inequality established in [121].

Theorem 1.5.5. Let $f,\ g,\ h : [a,b] \to \mathbb{R}$ be continuous functions on $[a,b]$ and n-times differentiable on (a,b) and with derivatives $f^{(n)},\ g^{(n)},\ h^{(n)} : (a,b) \to \mathbb{R}$ being bounded on (a,b) for some $n \in \mathbb{N}$, then

$$\left| \frac{1}{b-a}\int_a^b f(x)g(x)h(x)dx - \frac{1}{3(b-a)^2}\int_a^b \left[g(x)h(x)\left\{ I_0 + \sum_{k=1}^{n-1} I_k \right\} \right.\right.$$
$$\left.\left. + h(x)f(x)\left\{ J_0 + \sum_{k=1}^{n-1} J_k \right\} + f(x)g(x)\left\{ L_0 + \sum_{k=1}^{n-1} L_k \right\}\right] dx \right|$$
$$\leqslant \frac{1}{3(b-a)^2}\int_a^b \left[|g(x)||h(x)|\|f^{(n)}\|_\infty + |h(x)||f(x)|\|g^{(n)}\|_\infty \right.$$
$$\left. + |f(x)||g(x)| \left\| h^{(n)} \right\|_\infty \right] M_n(x)dx, \tag{1.5.60}$$

where

$$M_n(x) = \frac{(x-a)^{n+1} + (b-x)^{n+1}}{(n+1)!}, \tag{1.5.61}$$

for $x \in [a,b]$.

Proof. Let $x \in [a,b]$, $y \in (a,b)$. From the hypotheses on $f,\ g,\ h$ and Taylor's formula with the Lagrange form of reminder (see [77]), we have

$$f(x) = f(y) + \sum_{k=1}^{n-1} \frac{f^{(k)}(y)}{k!}(x-y)^k + \frac{1}{n!} f^{(n)}(\xi)(x-y)^n, \tag{1.5.62}$$

$$g(x) = g(y) + \sum_{k=1}^{n-1} \frac{g^{(k)}(y)}{k!}(x-y)^k + \frac{1}{n!} g^{(n)}(\eta)(x-y)^n, \tag{1.5.63}$$

$$h(x) = h(y) + \sum_{k=1}^{n-1} \frac{h^{(k)}(y)}{k!}(x-y)^k + \frac{1}{n!} h^{(n)}(\sigma)(x-y)^n, \tag{1.5.64}$$

where $\xi = y + \alpha(x-y)$ $(0 < \alpha < 1)$, $\eta = y + \beta(x-y)$ $(0 < \beta < 1)$, $\sigma = y + \gamma(x-y)$ $(0 < \gamma < 1)$. From the definitions of $I_k, J_k, L_k, I_0, J_0, L_0$ and integration by parts (see, [77]), we have the relations

$$I_0 + \sum_{k=1}^{n-1} I_k = nI_0 - (b-a)\sum_{k=1}^{n-1} F_k(x), \tag{1.5.65}$$

$$J_0 + \sum_{k=1}^{n-1} J_k = nJ_0 - (b-a)\sum_{k=1}^{n-1} G_k(x), \tag{1.5.66}$$

$$L_0 + \sum_{k=1}^{n-1} L_k = nL_0 - (b-a)\sum_{k=1}^{n-1} H_k(x). \tag{1.5.67}$$

Multiplying both sides of (1.5.62), (1.5.63) and (1.5.64) by $g(x)h(x)$, $h(x)f(x)$ and $f(x)g(x)$ respectively and adding the resulting identities, we get

$$3f(x)g(x)h(x) = g(x)h(x)\left\{f(y)+\sum_{k=1}^{n-1}\frac{f^{(k)}(y)}{k!}(x-y)^k+\frac{1}{n!}f^{(n)}(\xi)(x-y)^n\right\}$$

$$+h(x)f(x)\left\{g(y)+\sum_{k=1}^{n-1}\frac{g^{(k)}(y)}{k!}(x-y)^k+\frac{1}{n!}g^{(n)}(\eta)(x-y)^n\right\}$$

$$+f(x)g(x)\left\{h(y)+\sum_{k=1}^{n-1}\frac{h^{(k)}(y)}{k!}(x-y)^k+\frac{1}{n!}h^{(n)}(\sigma)(x-y)^n\right\}. \tag{1.5.68}$$

Integrating both sides of (1.5.68) with respect to y over (a,b) and rewriting, we obtain

$$f(x)g(x)h(x) = \frac{1}{3(b-a)}\left[g(x)h(x)\left\{I_0+\sum_{k=1}^{n-1}I_k\right\}\right.$$

$$\left.+h(x)f(x)\left\{J_0+\sum_{k=1}^{n-1}J_k\right\}+f(x)g(x)\left\{L_0+\sum_{k=1}^{n-1}L_k\right\}\right]$$

$$+\frac{1}{3(b-a)(n!)}\left[g(x)h(x)f^{(n)}(\xi)+h(x)f(x)g^{(n)}(\eta)\right.$$

$$\left.+f(x)g(x)h^{(n)}(\sigma)\right]\int_a^b(x-y)^n dy. \tag{1.5.69}$$

Integrating both sides of (1.5.69) with respect to x over $[a,b]$ and rewriting, we get

$$\frac{1}{b-a}\int_a^b f(x)g(x)h(x)dx-\frac{1}{3(b-a)^2}\int_a^b\left[g(x)h(x)\left\{I_0+\sum_{k=1}^{n-1}I_k\right\}\right.$$

$$\left.+h(x)f(x)\left\{J_0+\sum_{k=1}^{n-1}J_k\right\}+f(x)g(x)\left\{L_0+\sum_{k=1}^{n-1}L_k\right\}\right]dx$$

$$=\frac{1}{3(b-a)^2(n!)}\int_a^b\left[g(x)h(x)f^{(n)}(\xi)+h(x)f(x)g^{(n)}(\eta)\right.$$

$$\left.+f(x)g(x)h^{(n)}(\sigma)\right]\left(\int_a^b(x-y)^n dy\right)dx. \tag{1.5.70}$$

From (1.5.70) and using the properties of modulus, we get the desired inequality in (1.5.60). The proof is complete.

Remark 1.5.3. Taking $h(x) = 1$ and hence $h^{(k)}(x) = 0$, $k = 1,\ldots,n$, in Theorem 1.5.5 and by simple computations, it is easy to see that the inequality (1.5.60), reduces to

$$\left|\frac{1}{b-a}\int_a^b f(x)g(x)dx - \frac{1}{2(b-a)^2}\int_a^b \left[g(x)\left\{I_0 + \sum_{k=1}^{n-1} I_k\right\} + f(x)\left\{J_0 + \sum_{k=1}^{n-1} J_k\right\}\right] dx\right|$$

$$\leqslant \frac{1}{2(b-a)^2}\int_a^b \left[|g(x)|\|f^{(n)}\|_\infty + |f(x)|\ \|g^{(n)}\|_\infty\right] M_n(x)dx. \tag{1.5.71}$$

We note that, here we have used Taylor's formula with Lagrange form of remainder to establish (1.5.60). Instead of this, one can use Taylor's formula with integral remainder, to obtain a result in the framework of Theorem 1.5.5.

1.6 Discrete Inequalities of the Grüss-and Čebyšev-type

A number of Grüss-and Čebyšev-type discrete inequalities
have been investigated by different researchers, see [79,144], where further references are also given. In this section we deal with the recent results established by Pachpatte in [128,133,138].
We begin with a discrete version of the premature Grüss-type inequality proved in [128].

Theorem 1.6.1. Let $f = (f_1,\ldots,f_n)$, $g = (g_1,\ldots,g_n)$ be two n-tuples of real numbers and $\gamma \leqslant g_i \leqslant \Gamma$ for $i = 1,\ldots,n$, where $\gamma, \Gamma \in \mathbb{R}$ are constants, then

$$|C_n(f,g)| \leqslant \frac{1}{2}(\Gamma - \gamma)\sqrt{C_n(f,f)}, \tag{1.6.1}$$

where

$$C_n(f,g) = \frac{1}{n}\sum_{i=1}^n f_i g_i - \left(\frac{1}{n}\sum_{i=1}^n f_i\right)\left(\frac{1}{n}\sum_{i=1}^n g_i\right). \tag{1.6.2}$$

Proof. By direct computation it is easy to observe that the following discrete Korkine's type identity holds:

$$C_n(f,g) = \frac{1}{2n^2}\sum_{i=1}^n\sum_{j=1}^n (f_i - f_j)(g_i - g_j). \tag{1.6.3}$$

It is easy to observe that

$$C_n(f,f) = \frac{1}{n}\sum_{i=1}^n f_i^2 - \left(\frac{1}{n}\sum_{i=1}^n f_i\right)^2. \tag{1.6.4}$$

Furthermore, by using the Cauchy-Schwarz inequality for sums, we observe that $C_n(f,f) \geqslant 0$. Similarly, $C_n(g,g) \geqslant 0$. From (1.6.3) and using the Cauchy-Schwarz inequality for double sums, we have

$$|C_n(f,g)|^2 = \left\{ \frac{1}{2n^2} \sum_{i=1}^{n} \sum_{j=1}^{n} (f_i - f_j)(g_i - g_j) \right\}^2$$

$$\leqslant \left\{ \frac{1}{2n^2} \sum_{i=1}^{n} \sum_{j=1}^{n} (f_i - f_j)^2 \right\} \left\{ \frac{1}{2n^2} \sum_{i=1}^{n} \sum_{j=1}^{n} (g_i - g_j)^2 \right\}$$

$$= C_n(f,f) C_n(g,g). \tag{1.6.5}$$

The following identity also holds:

$$C_n(g,g) = \left(\Gamma - \frac{1}{n} \sum_{i=1}^{n} g_i \right) \left(\frac{1}{n} \sum_{i=1}^{n} g_i - \gamma \right) - \frac{1}{n} \sum_{i=1}^{n} (\Gamma - g_i)(g_i - \gamma). \tag{1.6.6}$$

Using the fact that $(\Gamma - g_i)(g_i - \gamma) \geqslant 0$ in (1.6.6) and then the elementary inequality $cd \leqslant \left(\frac{c+d}{2}\right)^2$, c, $d \in \mathbb{R}$, we have

$$C_n(g,g) \leqslant \left(\Gamma - \frac{1}{n} \sum_{i=1}^{n} g_i \right) \left(\frac{1}{n} \sum_{i=1}^{n} g_i - \gamma \right) \leqslant \left(\frac{\Gamma - \gamma}{2} \right)^2. \tag{1.6.7}$$

Using (1.6.7) in (1.6.5), we get (1.6.1). The proof is complete.

Remark 1.6.1. In Theorem 1.6.1, if we assume that $\phi \leqslant f_i \leqslant \Phi$ for $i = 1, \ldots, n$, where ϕ, $\Phi \in \mathbb{R}$ are constants, then by following the same arguments used to obtain (1.6.7), we get

$$C_n(f,f) \leqslant \left(\frac{\Phi - \phi}{2} \right)^2. \tag{1.6.8}$$

Using (1.6.8) in (1.6.1), we get

$$|C_n(f,g)| \leqslant \frac{1}{4} (\Phi - \phi)(\Gamma - \gamma), \tag{1.6.9}$$

which in turn can be considered a discrete version of the well-known Grüss inequality.

The discrete Grüss-type inequality given in [138] is embodied in the following theorem.

Theorem 1.6.2. Let $f = (f_1, \ldots, f_n)$, $g = (g_1, \ldots, g_n)$ be two n-tuples of real numbers and $p = (p_1, \ldots, p_n)$ be an n-tuple of nonnegative real numbers such that $P_n = \sum_{i=1}^{n} p_i > 0$, then

$$|C_n(p,f,g)| \leqslant \frac{1}{P_n} \sum_{i=1}^{n} p_i \left| \left(f_i - \frac{1}{P_n} \sum_{j=1}^{n} p_j f_j \right) \left(g_i - \frac{1}{P_n} \sum_{j=1}^{n} p_j g_j \right) \right|, \tag{1.6.10}$$

where

$$C_n(p,f,g) = \frac{1}{P_n} \sum_{i=1}^{n} p_i f_i g_i - \left(\frac{1}{P_n} \sum_{i=1}^{n} p_i f_i \right) \left(\frac{1}{P_n} \sum_{i=1}^{n} p_i g_i \right). \tag{1.6.11}$$

Proof. First we observe that

$$\frac{1}{P_n}\sum_{i=1}^{n} p_i\left(f_i-\frac{1}{P_n}\sum_{j=1}^{n} p_j f_j\right)\left(g_i-\frac{1}{P_n}\sum_{j=1}^{n} p_j g_j\right)$$

$$=\frac{1}{P_n}\sum_{i=1}^{n} p_i\left\{f_i g_i-f_i\frac{1}{P_n}\sum_{j=1}^{n} p_j g_j-g_i\frac{1}{P_n}\sum_{j=1}^{n} p_j f_j+\frac{1}{P_n^2}\sum_{j=1}^{n} p_j f_j\sum_{j=1}^{n} p_j g_j\right\}$$

$$=\frac{1}{P_n}\sum_{i=1}^{n} p_i f_i g_i-\left(\frac{1}{P_n}\sum_{i=1}^{n} p_i f_i\right)\left(\frac{1}{P_n}\sum_{j=1}^{n} p_j g_j\right)-\left(\frac{1}{P_n}\sum_{i=1}^{n} p_i g_i\right)\left(\frac{1}{P_n}\sum_{j=1}^{n} p_j f_j\right)$$

$$+\frac{1}{P_n}\sum_{i=1}^{n} p_i\left(\frac{1}{P_n}\sum_{j=1}^{n} p_j f_j\right)\left(\frac{1}{P_n}\sum_{j=1}^{n} p_j g_j\right)$$

$$=\frac{1}{P_n}\sum_{i=1}^{n} p_i f_i g_i-\left(\frac{1}{P_n}\sum_{i=1}^{n} p_i f_i\right)\left(\frac{1}{P_n}\sum_{i=1}^{n} p_i g_i\right)$$

$$=C_n(p,f,g). \tag{1.6.12}$$

From (1.6.12) and using the properties of modulus, we have

$$|C_n(p,f,g)|\leqslant\frac{1}{P_n}\sum_{i=1}^{n} p_i\left|\left(f_i-\frac{1}{P_n}\sum_{j=1}^{n} p_j f_j\right)\left(g_i-\frac{1}{P_n}\sum_{j=1}^{n} p_j g_j\right)\right|,$$

which is the required inequality in (1.6.10) and the proof is complete.

Remark 1.6.2. By taking $p_i=1$ for $i=1,\dots,n$ and hence $P_n=n$, $C_n(p,f,g)=C_n(f,g)$ in (1.6.10), we get

$$|C_n(f,g)|\leqslant\frac{1}{n}\sum_{i=1}^{n}\left|\left(f_i-\frac{1}{n}\sum_{j=1}^{n} f_j\right)\left(g_i-\frac{1}{n}\sum_{j=1}^{n} g_j\right)\right|. \tag{1.6.13}$$

We note that the inequality (1.6.13) can be considered as the discrete version of the Grüss-type integral inequality given by Dragomir and McAndrew in [34].

The discrete Čebyšev-type inequality established in [128] is given in the following theorem.

Theorem 1.6.3. Let $f=(f_1,\dots,f_n)$, $g=(g_1,\dots,g_n)$ be two n-tuples of real numbers, then

$$|C_n(f,g)|\leqslant\frac{\sqrt{n^2-1}}{2\sqrt{3}}\max_{1\leqslant k\leqslant n-1}|\Delta f_k|\sqrt{C_n(g,g)}, \tag{1.6.14}$$

where $C_n(f,g)$ is given by (1.6.2) and $\Delta f_k=f_{k+1}-f_k$.

Proof. First we recall that the discrete Korkine-type identity (1.6.3) holds. Following the proof of Theorem 1.6.1 we get (1.6.5). It is easy to observe that the following identity holds:

$$f_i - f_j = \sum_{k=j}^{i-1} (f_{k+1} - f_k) = \sum_{k=j}^{i-1} \Delta f_k. \tag{1.6.15}$$

We also observe that

$$|C_n(f,f)| = \frac{1}{2n^2} \left| \sum_{i=1}^{n} \sum_{j=1}^{n} (f_i - f_j)^2 \right| \leqslant \frac{1}{2n^2} \sum_{i=1}^{n} \sum_{j=1}^{n} \left(\sum_{k=j}^{i-1} |\Delta f_k| \right)^2$$

$$\leqslant \frac{1}{2n^2} \left(\max_{1 \leqslant k \leqslant n-1} |\Delta f_k| \right)^2 \sum_{i=1}^{n} \sum_{j=1}^{n} (i-j)^2. \tag{1.6.16}$$

By simple computation, we get

$$\sum_{i=1}^{n} \sum_{j=1}^{n} (i-j)^2 = \frac{n^2 \left(n^2 - 1\right)}{6}. \tag{1.6.17}$$

Using (1.6.17) in (1.6.16) and the fact that $C_n(f,f) \geqslant 0$, we get

$$C_n(f,f) \leqslant \frac{n^2-1}{12} \left(\max_{1 \leqslant k \leqslant n-1} |\Delta f_k| \right)^2. \tag{1.6.18}$$

Using (1.6.18) in (1.6.5), we get (1.6.14) and the proof is complete.

Remark 1.6.3. We note that the inequality (1.6.14) can be considered as a discrete version of the inequality given in Theorem 1.3.1. By following a similar arguments as in the proof of (1.6.18), we obtain

$$C_n(g,g) \leqslant \frac{n^2-1}{12} \left(\max_{1 \leqslant k \leqslant n-1} |\Delta g_k| \right)^2. \tag{1.6.19}$$

Using (1.6.19) in (1.6.14), we get

$$C_n(f,g) \leqslant \frac{n^2-1}{12} \max_{1 \leqslant k \leqslant n-1} |\Delta f_k| \max_{1 \leqslant k \leqslant n-1} |\Delta g_k|. \tag{1.6.20}$$

The inequality (1.6.20) can be considered a discrete version of the Čebyšev inequality given in (1).

The following Theorem deals with the discrete weighted Čebyšev-type inequality proved in [138].

Theorem 1.6.4. Let f, g, p, P_n be as in Theorem 1.6.2, then

$$|C_n(p,f,g)| \leqslant \max_{1 \leqslant k \leqslant n-1} |\Delta f_k| \max_{1 \leqslant k \leqslant n-1} |\Delta g_k| \times \left[\frac{1}{P_n} \sum_{i-1}^{n} p_i i^2 - \left(\frac{\sum_{i=1}^{n} i p_i}{P_n} \right)^2 \right], \tag{1.6.21}$$

where $C_n(p,f,g)$ is given by (1.6.11).

Proof. By direct computation, it is easy to observe that the following discrete Korkine identity holds:

$$C_n(p,f,g) = \frac{1}{2P_n^2} \sum_{i=1}^{n} \sum_{j=1}^{n} p_i p_j (f_i - f_j)(g_i - g_j). \tag{1.6.22}$$

We observe that the following identities also hold:

$$f_i - f_j = \sum_{k=j}^{i-1} (f_{k+1} - f_k) = \sum_{k=j}^{i-1} \Delta f_k, \tag{1.6.23}$$

$$g_i - g_j = \sum_{k=j}^{i-1} (g_{k+1} - g_k) = \sum_{k=j}^{i-1} \Delta g_k. \tag{1.6.24}$$

Using (1.6.23) and (1.6.24) in (1.6.22), we get

$$C_n(p,f,g) = \frac{1}{2P_n^2} \sum_{i=1}^{n} \sum_{j=1}^{n} p_i p_j \left(\sum_{k=j}^{i-1} \Delta f_k \right) \left(\sum_{k=j}^{i-1} \Delta g_k \right). \tag{1.6.25}$$

From (1.6.25) and using the properties of modulus, we have

$$\begin{aligned}
|C_n(p,f,g)| &\leqslant \frac{1}{2P_n^2} \sum_{i=1}^{n} \sum_{j=1}^{n} p_i p_j \left(\sum_{k=j}^{i-1} |\Delta f_k| \right) \left(\sum_{k=j}^{i-1} |\Delta g_k| \right) \\
&\leqslant \frac{1}{2P_n^2} \max_{1 \leqslant k \leqslant n-1} |\Delta f_k| \max_{1 \leqslant k \leqslant n-1} |\Delta g_k| \sum_{i=1}^{n} \sum_{j=1}^{n} p_i p_j (i-j)^2 \\
&= \frac{1}{2P_n^2} \max_{1 \leqslant k \leqslant n-1} |\Delta f_k| \max_{1 \leqslant k \leqslant n-1} |\Delta g_k| 2 \left[P_n \sum_{i=1}^{n} p_i i^2 - \left(\sum_{i=1}^{n} i p_i \right)^2 \right] \\
&= \max_{1 \leqslant k \leqslant n-1} |\Delta f_k| \max_{1 \leqslant k \leqslant n-1} |\Delta g_k| \times \left[\frac{1}{P_n} \sum_{i=1}^{n} p_i i^2 - \left(\frac{\sum_{i=1}^{n} i p_i}{P_n} \right)^2 \right],
\end{aligned}$$

and the inequality (1.6.21) is proved.

Remark 1.6.4. We note that A. Lupas [79, Chapter X] proved some results similar to that of the inequality (1.6.21) when $f = (f_1, \ldots, f_n)$, $g = (g_1, \ldots, g_n)$ are two monotonic n-tuples in the same sense and $p = (p_1, \ldots, p_n)$ is a positive n-tuple. In the special case, when $p_i = 1$ for $i = 1, \ldots, n$ and hence $P_n = n$, $C_n(p,f,g) = C_n(f,g)$, the inequality (1.6.21) reduces to

$$|C_n(f,g)| \leqslant \max_{1 \leqslant k \leqslant n-1} |\Delta f_k| \max_{1 \leqslant k \leqslant n-1} |\Delta g_k| \times \left[\frac{1}{n} \sum_{i=1}^{n} i^2 - \left(\frac{\sum_{i=1}^{n} i}{n} \right)^2 \right]. \tag{1.6.26}$$

In the proof of the next theorem, we need the following representation formula given in [1].

Lemma 1.6.1. Let $\{x_1,\ldots,x_n\}$ be a finite sequence of real numbers and $\{w_1,\ldots,w_n\}$ be a finite sequence of positive real numbers, then

$$x_k = \frac{1}{W_n}\sum_{i=1}^{n} w_i x_i + \sum_{i=1}^{n-1} D_w(k,i)\Delta x_i, \tag{1.6.27}$$

where $\Delta x_i = x_{i+1} - x_i$ and

$$D_w(k,i) = \frac{1}{W_n}\begin{cases} W_i, & 1 \leqslant i \leqslant k-1, \\ (-\overline{W}_i), & k \leqslant i \leqslant n, \end{cases} \tag{1.6.28}$$

is the discrete weighted Peano kernel, in which

$$W_k = \sum_{i=1}^{k} w_i, \quad \overline{W}_k = \sum_{i=k+1}^{n} w_i = W_n - W_k.$$

Proof. By direct computation it is easy to observe that the following discrete identity holds (see [1]):

$$\sum_{i=1}^{n} w_i x_i = x_k W_n + \sum_{i=1}^{k-1} W_i(x_i - x_{i+1}) + \sum_{i=k}^{n-1} \overline{W}_i(x_{i+1} - x_i), \tag{1.6.29}$$

for $1 \leqslant k \leqslant n$. Rewriting (1.6.29) by using (1.6.28) we get the desired identity in (1.6.27).

Remark 1.6.5. If we take $w_i = 1$ for $i = 1,\ldots,n$, then $W_i = i$ and $\overline{W}_i = n - i$ and (1.6.27) reduces to the discrete Montgomery identity,

$$x_k = \frac{1}{n}\sum_{i=1}^{n} x_i + \sum_{i=1}^{n-1} D_n(k,i)\Delta x_i, \tag{1.6.30}$$

where

$$D_n(k,i) = \begin{cases} \dfrac{i}{n}, & 1 \leqslant i \leqslant k-1, \\ \dfrac{i}{n} - 1, & k \leqslant i \leqslant n. \end{cases} \tag{1.6.31}$$

Finally, we present the Grüss-type discrete inequalities given in [133].

Theorem 1.6.5. Let $\{u_k\}$, $\{v_k\}$ for $k = 1,\ldots,n$ be two finite sequences of real numbers such that $\max_{1\leqslant k\leqslant n-1}\{|\Delta u_k|\} = A$, $\max_{1\leqslant k\leqslant n-1}\{|\Delta v_k|\} = B$, where A, B are nonnegative constants, then the following inequalities hold:

$$|J_n(u_k,v_k)| \leqslant \frac{1}{2n}\sum_{k=1}^{n}[|v_k|A + |u_k|B]H_n(k), \tag{1.6.32}$$

and

$$|J_n(u_k,v_k)| \leqslant \frac{AB}{n}\sum_{k=1}^{n}(H_n(k))^2, \tag{1.6.33}$$

where

$$J_n(u_k,v_k) = \frac{1}{n}\sum_{k=1}^{n} u_k v_k - \left(\frac{1}{n}\sum_{k=1}^{n} u_k\right)\left(\frac{1}{n}\sum_{k=1}^{n} v_k\right), \tag{1.6.34}$$

$$H_n(k) = \sum_{i=1}^{n-1} |D_n(k,i)|, \tag{1.6.35}$$

in which $D_n(k,i)$ is defined by (1.6.31).

Proof. From the hypotheses, we have the following identities (see, Remark 1.6.5):

$$u_k - \frac{1}{n}\sum_{i=1}^{n} u_i = \sum_{i=1}^{n-1} D_n(k,i)\Delta u_i, \tag{1.6.36}$$

$$v_k - \frac{1}{n}\sum_{i=1}^{n} v_i = \sum_{i=1}^{n-1} D_n(k,i)\Delta v_i, \tag{1.6.37}$$

for $k = 1,\dots,n$. Multiplying both sides of (1.6.36) and (1.6.37) by v_k and u_k respectively, adding the resulting identities and rewriting, we get

$$u_k v_k - \frac{1}{2n}\left[v_k\sum_{i=1}^{n} u_i + u_k\sum_{i=1}^{n} v_i\right] = \frac{1}{2}\left[v_k\sum_{i=1}^{n-1} D_n(k,i)\Delta u_i + u_k\sum_{i=1}^{n-1} D_n(k,i)\Delta v_i\right]. \tag{1.6.38}$$

Summing both sides of (1.6.38) over k from 1 to n and rewriting, we get

$$J_n(u_k,v_k) = \frac{1}{2n}\sum_{k=1}^{n}\left[v_k\sum_{i=1}^{n-1} D_n(k,i)\Delta u_i + u_k\sum_{i=1}^{n-1} D_n(k,i)\Delta v_i\right]. \tag{1.6.39}$$

From (1.6.39) and using the properties of modulus, we get the required inequality in (1.6.32).

Multiplying the left hand sides and right hand sides of (1.6.36) and (1.6.37), we get

$$u_k v_k - \frac{1}{n}\left[v_k\sum_{i=1}^{n} u_i + u_k\sum_{i=1}^{n} v_i\right] + \frac{1}{n^2}\left(\sum_{i=1}^{n} u_i\right)\left(\sum_{i=1}^{n} v_i\right)$$

$$= \left[\sum_{i=1}^{n-1} D_n(k,i)\Delta u_i\right]\left[\sum_{i=1}^{n-1} D_n(k,i)\Delta v_i\right]. \tag{1.6.40}$$

Summing both sides of (1.6.40) over k from 1 to n and rewriting, we have

$$J_n(u_k,v_k) = \frac{1}{n}\sum_{k=1}^{n}\left[\sum_{i=1}^{n-1} D_n(k,i)\Delta u_i\right]\left[\sum_{i=1}^{n-1} D_n(k,i)\Delta v_i\right]. \tag{1.6.41}$$

From (1.6.41) and using the properties of modulus, we get the desired inequality in (1.6.33). The proof is complete.

1.7 Applications

In this section we present applications of a few of the inequalities given in earlier sections which have been investigated during the past few years.

1.7.1 *Estimation of the remainder in the Trapezoid formula*

As an application of Theorem 1.2.2, in this section, we present a version of the Trapezoid-type inequality and it's application given in [34].

We start with the following Trapezoid-type inequality.

Theorem 1.7.1. Let $f : [a,b] \to \mathbb{R}$ be a differentiable mapping on (a,b) having first derivative $f' : (a,b) \to \mathbb{R}$ bounded on (a,b), then

$$\left|\frac{f(a)+f(b)}{2} - \frac{1}{b-a}\int_a^b f(x)dx\right| \leqslant \frac{b-a}{4} \max_{x\in(a,b)} \left|f'(x) - \frac{f(b)-f(a)}{b-a}\right|. \tag{1.7.1}$$

Proof. A simple integration by parts, gives

$$\frac{f(a)+f(b)}{2}(b-a) - \int_a^b f(x)dx = \int_a^b \left(x - \frac{a+b}{2}\right) f'(x)dx. \tag{1.7.2}$$

Applying the inequality (1.2.7) given in Theorem 1.2.2, we get

$$\left|\frac{1}{b-a}\int_a^b \left(x-\frac{a+b}{2}\right) f'(x)dx - \left(\frac{1}{b-a}\int_a^b \left(x-\frac{a+b}{2}\right)dx\right)\left(\frac{1}{b-a}\int_a^b f'(x)dx\right)\right|$$

$$\leqslant \frac{1}{b-a}\int_a^b \left|\left(x-\frac{a+b}{2} - \frac{1}{b-a}\int_a^b \left(y-\frac{a+b}{2}\right)dy\right) \times \left(f'(x) - \frac{1}{b-a}\int_a^b f'(y)dy\right)\right| dx.$$

As

$$\int_a^b \left(x-\frac{a+b}{2}\right)dx = 0,$$

we get

$$\begin{aligned}\left|\int_a^b \left(x-\frac{a+b}{2}\right) f'(x)dx\right| &\leqslant \int_a^b \left|\left(x-\frac{a+b}{2}\right)\left(f'(x) - \frac{f(b)-f(a)}{b-a}\right)\right| dx \\ &\leqslant \max_{x\in(a,b)} \left|f'(x) - \frac{f(b)-f(a)}{b-a}\right| \int_a^b \left|x-\frac{a+b}{2}\right| dx \\ &= \frac{(b-a)^2}{4} \max_{x\in(a,b)} \left|f'(x) - \frac{f(b)-f(a)}{b-a}\right|. \end{aligned} \tag{1.7.3}$$

Now using (1.7.2) in (1.7.3), we obtain (1.7.1).

In the following Theorem we assume that $f : I \subseteq \mathbb{R} \to \mathbb{R}$ is a differentiable mapping with derivative satisfying the following condition:

$$|f(b) - f(a) - (b-a)f'(x)| \leqslant \Omega(b-a)^2, \quad \Omega > 0, \tag{1.7.4}$$

for all $a,\ b \in I$ and $x \in (a,b)$.

If f' is M-lipschitzian, i.e.,

$$|f'(u) - f'(v)| \leqslant M|u-v|, \quad M > 0,$$

then

$$|f(b) - f(a) - (b-a)f'(x)| = |f'(c) - f'(x)||b-a| \leqslant M|c-x||b-a| \leqslant M(b-a)^2,$$

where $c \in (a,b)$. Consequently, the mappings having the first derivative lipschitzian satisfy the condition (1.7.4).

The following trapezoid formula holds.

Theorem 1.7.2. Let $f : [a,b] \to \mathbb{R}$ be a differentiable mapping on (a,b) with derivative $f' : (a,b) \to \mathbb{R}$ satisfying the above condition (1.7.4) on (a,b). If $I_h : a = x_0 < x_1 < \cdots < x_{n-1} < x_n = b$ is a division of $[a,b]$ and $h_i = x_{i+1} - x_i$, $i = 0,1,\ldots,n-1$, then we have:

$$\int_a^b f(t)dt = A_{T,I_h}(f) + R_{T,I_h}(f), \tag{1.7.5}$$

where

$$A_{T,I_h}(f) = \sum_{i=0}^{n-1} \frac{f(x_i) + f(x_{i+1})}{2} h_i, \tag{1.7.6}$$

and the remainder $R_{T,I_h}(f)$ satisfies the estimation:

$$\left| R_{T,I_h}(f) \right| \leqslant \frac{\Omega}{4} \sum_{i=0}^{n-1} h_i^3. \tag{1.7.7}$$

Proof. Applying Theorem 1.7.1 on the interval $[x_i, x_{i+1}]$, we can write

$$\begin{aligned}\left| (x_{i+1} - x_i) \frac{f(x_i) + f(x_{i+1})}{2} - \int_{x_i}^{x_{i+1}} f(t)dt \right| &\leqslant \frac{x_{i+1} - x_i}{4} \max_{x \in (x_i, x_{i+1})} \left| f'(x) - \frac{f(x_{i+1}) - f(x_i)}{x_{i+1} - x_i} \right| \\ &\leqslant \frac{\Omega (x_{i+1} - x_i)^3}{4},\end{aligned}$$

i.e.,

$$\left| \frac{f(x_i) + f(x_{i+1})}{2} h_i - \int_{x_i}^{x_{i+1}} f(t)dt \right| \leqslant \frac{\Omega h_i^3}{4}, \tag{1.7.8}$$

for all $i = 0,1,\ldots,n-1$.

Summing both sides of (1.7.8) over i from $i = 0$ to $n-1$ and using the generalized triangle inequality, we get the approximation (1.7.5) and the remainder satisfies the estimation (1.7.7).

Remark 1.7.1. We note that the trapezoid formula given above works for a class larger than the class $C^2[a,b]$ for which the usual trapezoid formula works with the remainder term satisfying the estimation

$$\left| R_{T,I_h}(f) \right| \leqslant \frac{\|f''\|_\infty}{12} \sum_{i=0}^{n-1} h_i^3,$$

where $\|f''\|_\infty = \sup_{t \in (a,b)} |f''(t)| < \infty$.

1.7.2 *Bounds for a perturbed generalized Taylor's formula*

In [31], Dragomir was the first to introduce the perturbed Taylor formula, with the idea to estimate the remainder using Grüss and Čebyšev-type inequalities. We state the following result given in [31].

Theorem 1.7.3. Let $f : I \to \mathbb{R}$ ($I \subset \mathbb{R}$ is a closed interval, $a \in I$) be such that $f^{(n)}$ is absolutely continuous. Then we have the Taylor's perturbed formula:

$$f(x) = T_n(f;a,x) + \frac{(x-a)^{n+1}}{(n+1)!}\left[f^{(n)};a,x\right] + G_n(f;a,x), \tag{1.7.9}$$

where

$$T_n(f;a,x) := \sum_{k=0}^{n} \frac{(x-a)^k}{k!} f^{(k)}(a), \tag{1.7.10}$$

and

$$\left[f^{(n)};a,x\right] := \frac{f^{(n)}(x) - f^{(n)}(a)}{x-a}. \tag{1.7.11}$$

The remainder $G_n(f;a,x)$ satisfies the estimation:

$$|G_n(f;a,x)| \leqslant \frac{(x-a)^{n+1}}{4(n!)}\left[\Gamma(x) - \gamma(x)\right], \tag{1.7.12}$$

where

$$\Gamma(x) := \sup_{t\in[a,x]} f^{(n+1)}(t), \quad \gamma(x) := \inf_{t\in[a,x]} f^{(n+1)}(t), \tag{1.7.13}$$

for all $x \geqslant a$, $x \in I$.

In this section we present some generalizations and improvements of Theorem 1.7.3, as well as some other results from [72].

In [72], Matić, Pečarić and Ujević proved the following generalized Taylor formula.

Theorem 1.7.4. Let $\{P_n(x)\}$ be a harmonic sequence of polynomials, that is

$$P_n'(x) = P_{n-1}(x), \quad n \in \mathbb{N}; \quad P_0(x) = 1.$$

Further, let $I \subset \mathbb{R}$ be a closed interval and $a \in I$. If $f : I \to \mathbb{R}$ is any function such that, for some $n \in \mathbb{N}$, $f^{(n)}$ is absolutely continuous, then for any $x \in I$,

$$f(x) = f(a) + \sum_{k=1}^{n} (-1)^{k+1}\left[P_k(x)f^{(k)}(x) - P_k(a)f^{(k)}(a)\right] + R_n(f;a,x), \tag{1.7.14}$$

where

$$R_n(f;a,x) = (-1)^n \int_a^x P_n(t) f^{(n+1)}(t)dt. \tag{1.7.15}$$

Proof. Integrating by parts, we have

$$\begin{aligned}(-1)^n \int_a^x P_n(t)f^{(n+1)}(t)dt &= (-1)^n P_n(t)f^{(n)}(t)\Big|_a^x + (-1)^{n-1}\int_a^x P_{n-1}(t)f^{(n)}(t)dt \\ &= (-1)^n\left[P_n(x)f^{(n)}(x) - P_n(a)f^{(n)}(a)\right] + (-1)^{n-1}\int_a^x P_{n-1}(t)f^{(n)}(t)dt.\end{aligned}$$

Clearly, we can apply the same procedure to the term $(-1)^{n-1}\int_a^x P_{n-1}(t)f^{(n)}(t)dt$. So, by successive integration by parts, we obtain

$$(-1)^n \int_a^x P_n(t)f^{(n+1)}(t)dt = \sum_{k=1}^{n}(-1)^k\left[P_k(x)f^{(k)}(x) - P_k(a)f^{(k)}(a)\right] + f(x) - f(a),$$

and this is equivalent to (1.7.14).

Remark 1.7.2. The formula (1.7.14) can be called a generalized Taylor's formula. Since, if we set $P_n(t) = \frac{(t-x)^n}{n!}$ in (1.7.14), then we get the classical Taylor's formula:

$$f(x) = f(a) + \sum_{k=1}^{n} \frac{(x-a)^k}{k!} f^{(k)}(a) + R_n^T(f;a,x), \tag{1.7.16}$$

where

$$R_n^T(f;a,x) = \frac{1}{n!} \int_a^x (x-t)^n f^{(n+1)}(t)dt. \tag{1.7.17}$$

The following theorem, proved in [72], deals with the generalization of the result stated in Theorem 1.7.3 which also improves the estimation (1.7.12).

Theorem 1.7.5. Let $\{P_n(x)\}$ be a harmonic sequence of polynomials. Let $I \subset \mathbb{R}$ be a closed interval and $a \in I$. Suppose $f : I \to \mathbb{R}$, is such that, for some $n \in \mathbb{N}$, $f^{(n)}$ is absolutely continuous, then for any $x \in I$ we have the generalized Taylor's perturbed formula:

$$f(x) = \overline{T}_n(f;a,x) + (-1)^n [P_{n+1}(x) - P_{n+1}(a)] \left[f^{(n)};a,x\right] + \overline{G}_n(f;a,x), \tag{1.7.18}$$

where

$$\overline{T}_n(f;a,x) = f(a) + \sum_{k=1}^{n} (-1)^{k+1} \left[P_k(x) f^{(k)}(x) - P_k(a) f^{(k)}(a)\right], \tag{1.7.19}$$

and $\left[f^{(n)};a,x\right]$ is defined by (1.7.11). For $x \geqslant a$ the remainder $\overline{G}_n(f;a,x)$ satisfies the estimation

$$\left|\overline{G}_n(f;a,x)\right| \leqslant \frac{x-a}{2} \sqrt{T(P_n,P_n)}\,[\Gamma(x) - \gamma(x)], \tag{1.7.20}$$

where $\Gamma(x)$ and $\gamma(x)$ are defined by (1.7.13).

Proof. Taylor's generalized formula (1.7.14) can be rewritten as

$$f(x) = \overline{T}_n(f;a,x) + (-1)^n [P_{n+1}(x) - P_{n+1}(a)] \left[f^{(n)};a,x\right] + \overline{G}(f;a,x),$$

where

$$\overline{G}_n(f;a,x) := R_n(f;a,x) - (-1)^n [P_{n+1}(x) - P_{n+1}(a)] \left[f^{(n)};a,x\right],$$

and this is just the representation (1.7.18). By (1.7.15), we have

$$\overline{G}_n(f;a,x) = (-1)^n \left\{ \int_a^x P_n(t) f^{(n+1)}(t)dt - [P_{n+1}(x) - P_{n+1}(a)] \left[f^{(n)};a,x\right] \right\}. \tag{1.7.21}$$

On the other hand, setting $f = P_n$ and $g = f^{(n+1)}$ in Theorem 1.2.1, we get

$$\left| \frac{1}{x-a} \int_a^x P_n(t) f^{(n+1)}(t)dt - \frac{1}{(x-a)^2} \int_a^x P_n(t)dt \int_a^x f^{(n+1)}(t)dt \right|$$

$$\leqslant \frac{1}{2}[\Gamma(x)-\gamma(x)]\sqrt{T(P_n,P_n)}. \tag{1.7.22}$$

Note that

$$\int_a^x P_n(t)dt = \int_a^x P'_{n+1}(t)dt = P_{n+1}(x) - P_{n+1}(a),$$

and

$$\int_a^x f^{(n+1)}(t)dt = f^{(n)}(x) - f^{(n)}(a) = (x-a)\left[f^{(n)};a,x\right],$$

so that, after multiplying (1.7.22) by $(x-a)$, we have

$$\left|\int_a^x P_n(t)f^{(n+1)}(t)dt - [P_{n+1}(x)-P_{n+1}(a)]\left[f^{(n)};a,x\right]\right| \leqslant \frac{x-a}{2}[\Gamma(x)-\gamma(x)]\sqrt{T(P_n,P_n)}.$$

Combining this with (1.7.21), we get the estimation (1.7.20).

The above result gives the following improvement of the estimation (1.7.12).

Corollary 1.7.1. Let the assumptions of Theorem 1.7.3 be satisfied, then the remainder $G_n(f;a,x)$ defined by (1.7.9) satisfies the estimation

$$|G_n(f;a,x)| \leqslant \frac{n(x-a)^{n+1}}{2[(n+1)!]\sqrt{2n+1}}[\Gamma(x)-\gamma(x)]. \tag{1.7.23}$$

Proof. If $P_n(t) = \frac{(t-x)^n}{n!}$, then it is easy to see that $\overline{T}_n(f;a,x) = T_n(f;a,x)$ and $(-1)^n[P_{n+1}(x)-P_{n+1}(a)] = \frac{(x-a)^{n+1}}{(n+1)!}$, so that (1.7.18) becomes (1.7.9), that is $\overline{G}_n(f;a,x) = G_n(f;a,x)$. Also, we have

$$\begin{aligned} T(P_n,P_n) &= \frac{1}{x-a}\int_a^x \frac{(t-x)^{2n}}{(n!)^2}dt - \frac{1}{(x-a)^2}\left(\int_a^x \frac{(t-x)^n}{n!}dt\right)^2 \\ &= \frac{1}{(n!)^2}\left[\frac{1}{x-a}\left(\frac{(t-x)^{2n+1}}{2n+1}\bigg|_a^x\right) - \frac{1}{(x-a)^2}\left(\frac{(t-x)^{n+1}}{n+1}\bigg|_a^x\right)^2\right] \\ &= \frac{n^2(x-a)^{2n}}{[(n+1)!]^2(2n+1)}, \end{aligned}$$

that is,

$$\sqrt{T(P_n,P_n)} = \frac{n(x-a)^n}{(n+1)!\sqrt{2n+1}}. \tag{1.7.24}$$

Now, we apply the inequality (1.7.20) to obtain the desired result.

Remark 1.7.3. Denote by Δ_n and $\overline{\Delta}_n$ the right hand sides of (1.7.12) and (1.7.23) respectively, then, we have

$$\overline{\Delta}_n = \frac{2n}{(n+1)\sqrt{2n+1}}\Delta_n < \Delta_n,$$

since obviously $\frac{2n}{(n+1)\sqrt{2n+1}} < 1$ for all $n \in \mathbb{N}$. Moreover, $\frac{2n}{(n+1)\sqrt{2n+1}}$ tends to zero when, n tends to ∞. So the estimation in (1.7.23) is much better than the estimation in (1.7.12).

We next give another estimation obtained in [72] for the remainder term in the representation formula (1.7.18).

Theorem 1.7.6. Suppose that the assumptions of Theorem 1.7.5 are satisfied. Additionally, suppose $f^{(n+1)}$ is differentiable and such that

$$M^{(n+2)}(x) := \sup_{t\in[a,x]} \left| f^{(n+2)}(t) \right| < \infty,$$

for $x \geqslant a$. The remainder $\overline{G}_n(f;a,x)$ satisfies the estimation

$$\left|\overline{G}_n(f;a,x)\right| \leqslant \frac{(x-a)^2 M^{(n+2)}(x)}{\sqrt{12}} \sqrt{T\left(P_n,P_n\right)}, \tag{1.7.25}$$

for all $x \geqslant a$, $x \in I$.

Proof. From the proof of Theorem 1.7.5, it is easy to see that

$$\overline{G}_n(f;a,x) = (-1)^n (x-a) T\left(f^{(n+1)}, P_n\right). \tag{1.7.26}$$

From (1.7.26) and making use of the inequality given in Theorem 1.3.1 when $f = f^{(n+1)}$ and $g = P_n$, we observe that

$$\left|\overline{G}_n(f;a,x)\right| = (x-a)\left|T\left(f^{(n+1)},P_n\right)\right| \leqslant \frac{(x-a)^2 M^{(n+2)}(x)}{\sqrt{12}} \sqrt{T(P_n,P_n)},$$

and the proof is complete.

Corollary 1.7.2. Let the assumptions of Theorem 1.7.6 be satisfied, then we have the representation (1.7.9) and the remainder $G_n(f;a,x)$ satisfies the estimation:

$$|G_n(f;a,x)| \leqslant \overline{\Delta}(n) := \frac{n(x-a)^{n+2} M^{(n+2)}(x)}{\sqrt{12}(n+1)!\sqrt{2n+1}}. \tag{1.7.27}$$

Proof. Set $P_n(t) = \frac{(t-x)^n}{n!}$ and apply Theorem 1.7.6, then use (1.7.24) to obtain the desired result.

Remark 1.7.4. In [31, Theorem 2.4] the following estimation can be obtained

$$|G_n(f;a,x)| \leqslant \Delta_n := \frac{(x-a)^{n+2} M^{(n+2)}(x)}{12[(n+1)!]}.$$

We have

$$\overline{\Delta}_n = \frac{\sqrt{12}}{(n+1)\sqrt{2n+1}} \Delta_n,$$

and $\frac{\sqrt{12}}{(n+1)\sqrt{2n+1}} < 1$, for $n > 1$. So, the estimation established in Corollary 1.7.2 is better than the one given in [31, p.187]. For some other estimations by using Taylor's perturbed formula, see [72].

1.7.3 *Some inequalities for expectation and the cumulative distribution functions*

In the present section we apply Theorem 1.3.1 to obtain some inequalities recently given in [5] for the expectation and cumulative distribution function of a random variable having probability density function defined on a finite interval.

Let $f : [a,b] \to \mathbb{R}$ be the probability density function (*p.d.f.*) of the random variable X, that is, an integrable function satisfying $\int_a^b f(t)dt = 1$,

$$E(x) := \int_a^b tf(t)dt,$$

its expectation and the cumulative distribution function $F : [a,b] \to [0,1]$, i.e.,

$$F(x) = \int_a^x f(t)dt, \quad x \in [a,b].$$

We start with the following result for expectation.

Theorem 1.7.7. Let X be a random variable having the probability density function $f : [a,b] \to \mathbb{R}$. Assume that f is absolutely continuous on $[a,b]$ and $f' \in L_\infty[a,b]$, then

$$\left| E(x) - \frac{a+b}{2} \right| \leqslant \frac{1}{12}(b-a)^3 \|f'\|_\infty, \tag{1.7.28}$$

where $E(X)$ is the expectation of the random variable X.

Proof. If we put $g(t) = t$ in the inequality (1.3.3), then we have

$$\left| \frac{1}{b-a} \int_a^b tf(t)dt - \left(\frac{1}{b-a} \int_a^b f(t)dt \right) \left(\frac{1}{b-a} \int_a^b tdt \right) \right|$$

$$\leqslant \frac{1}{2\sqrt{3}}(b-a)\|f'\|_\infty \left[\frac{1}{b-a} \int_a^b t^2 dt - \left(\frac{1}{b-a} \int_a^b tdt \right)^2 \right]^{\frac{1}{2}}.$$

However,

$$\frac{1}{b-a} \int_a^b t^2 dt - \left(\frac{1}{b-a} \int_a^b tdt \right)^2 = \frac{(b-a)^2}{12},$$

and so (1.7.28) is true.

The following Theorems provide inequalities that connect the expectation $E(X)$ and the cumulative distribution function $F(x)$ of a random variable X having *p.d.f.*, f,

Theorem 1.7.8. Let X be a random variable with *p.d.f.*, $f : [a,b] \to \mathbb{R}$ absolutely continuous on $[a,b]$ and $f' \in L_\infty[a,b]$, then

$$\left| E(x) + (b-a)F(x) - x - \frac{b-a}{2} \right| \leqslant \frac{1}{12}(b-a)^3 \|f'\|_\infty, \tag{1.7.29}$$

for all $x \in [a,b]$.

Proof. We use the following identity established by Barnett and Dragomir in [4]:

$$(b-a)F(x)+E(x)-b=\int_a^b p(x,t)dF(t)=\int_a^b p(x,t)f(t)dt, \tag{1.7.30}$$

for all $x\in[a,b]$, where

$$p(x,t)=\begin{cases} t-a, & a\leqslant t\leqslant x\leqslant b,\\ t-b, & a\leqslant x<t\leqslant b.\end{cases}$$

Indeed, the Riemann-Stieltjes integral $\int_a^b p(x,t)dF(t)$ exists for any $x\in[a,b]$ and the formula of integration by parts for Riemann-Stieltjes integral gives

$$\int_a^b p(x,t)dF(t)=\int_a^x (t-a)dF(t)+\int_x^b (t-b)dF(t)$$

$$=(t-a)F(t)\Big|_a^x-\int_a^x F(t)dt+(t-b)F(t)\Big|_x^b-\int_x^b F(t)dt$$

$$=(b-a)F(x)-\int_a^b f(t)dt. \tag{1.7.31}$$

On the other hand, the integration by parts formula for Riemann-Stieltjes integral also gives

$$E(x):=\int_a^b t dF(t)=tF(t)\Big|_a^b-\int_a^b f(t)dt$$

$$=bF(b)-aF(a)-\int_a^b f(t)dt=b-\int_a^b f(t)dt. \tag{1.7.32}$$

Now, using (1.7.31) and (1.7.32), we get (1.7.30).

Now, if we apply the inequality (1.3.3) given in Theorem 1.3.1 for $g(t)=p(x,t)$, we obtain

$$\left|\frac{1}{b-a}\int_a^b p(x,t)f(t)dt-\left(\frac{1}{b-a}\int_a^b p(x,t)dt\right)\left(\frac{1}{b-a}\int_a^b f(t)dt\right)\right|$$

$$\leqslant\frac{1}{2\sqrt{3}}(b-a)\|f'\|_\infty\left[\frac{1}{b-a}\int_a^b p^2(x,t)dt-\left(\frac{1}{b-a}\int_a^b p(x,t)dt\right)^2\right]^{\frac{1}{2}}. \tag{1.7.33}$$

Observe that

$$\frac{1}{b-a}\int_a^b p(x,t)dt=x-\frac{a+b}{2}, \tag{1.7.34}$$

and

$$\frac{1}{b-a}\int_a^b p^2(x,t)dt-\left(\frac{1}{b-a}\int_a^b p(x,t)dt\right)^2=\frac{1}{b-a}\left[\frac{(b-x)^3+(x-a)^3}{3}\right]-\left(x-\frac{a+b}{2}\right)^2$$

$$=\frac{1}{12}(b-a)^2. \tag{1.7.35}$$

Using (1.7.30), (1.7.34), (1.7.35) in (1.7.33), we get the required inequality in (1.7.29).

Theorem 1.7.9. Let X, F and f be as in Theorem 1.7.8, then

$$\left|E(x)+\frac{b-a}{2}F(x)-\frac{x+b}{2}\right| \leqslant \frac{1}{4}(b-a)\|f'\|_\infty\left[\left(x-\frac{a+b}{2}\right)^2+\frac{1}{12}(b-a)^2\right], \quad (1.7.36)$$

for all $x \in [a,b]$.

Proof. Applying the inequality given in Theorem 1.3.1 in the form (1.3.3), we get

$$\left|\frac{1}{x-a}\int_a^x (t-a)f(t)dt-\left(\frac{1}{x-a}\int_a^x (t-a)dt\right)\left(\frac{1}{x-a}\int_a^x f(t)dt\right)\right|$$

$$\leqslant \frac{1}{2\sqrt{3}}(x-a)\|f'\|_\infty\left[\frac{1}{x-a}\int_a^x (t-a)^2dt-\left(\frac{1}{x-a}\int_a^x (t-a)dt\right)^2\right]^{\frac{1}{2}}$$

$$=\frac{1}{12}(x-a)^2\|f'\|_\infty, \quad (1.7.37)$$

and similarly,

$$\left|\frac{1}{b-x}\int_x^b (t-b)f(t)dt-\left(\frac{1}{b-x}\int_x^b (t-b)dt\right)\left(\frac{1}{b-x}\int_x^b f(t)dt\right)\right|$$

$$\leqslant \frac{1}{12}(b-x)^2\|f'\|_\infty, \quad (1.7.38)$$

for all $x \in [a,b]$. From (1.7.37) and (1.7.38), we can write

$$\left|\int_a^x (t-a)f(t)dt-\frac{x-a}{2}F(x)\right| \leqslant \frac{1}{12}(x-a)^3\|f'\|_\infty, \quad (1.7.39)$$

and

$$\left|\int_x^b (t-b)f(t)dt+\frac{b-x}{2}(1-F(x))\right| \leqslant \frac{1}{12}(b-x)^3\|f'\|_\infty, \quad (1.7.40)$$

for all $x \in [a,b]$. Summing (1.7.39) and (1.7.40) and using the triangle inequality, we deduce

$$\left|\int_a^x (t-a)f(t)dt+\int_x^b (t-b)f(t)dt-\frac{b-a}{2}F(x)+\frac{b-x}{2}\right|$$

$$\leqslant \frac{1}{12}\|f'\|_\infty\left[(x-a)^3+(b-x)^3\right]$$

$$=\frac{1}{12}(b-a)\|f'\|_\infty\left[3\left(x-\frac{a+b}{2}\right)^2+\frac{1}{4}(b-a)^2\right]$$

$$=\frac{1}{4}(b-a)\|f'\|_\infty\left[\left(x-\frac{a+b}{2}\right)^2+\frac{1}{12}(b-a)^2\right]. \quad (1.7.41)$$

Using the identity (1.7.30) in (1.7.41), we deduce (1.7.36).

Remark 1.7.5. If we take $x = a$ or $x = b$ in (1.7.29) and (1.7.36), then we recapture the inequality given in (1.7.28) and if we take $x = \frac{a+b}{2}$ in (1.7.29) and (1.7.36), then we get inequalities that are of independent interest

1.8 Miscellaneous inequalities

1.8.1 *Dragomir [42]*

Let f and g be two functions defined and integrable on $[a,b]$. If $\phi \leqslant f(x) \leqslant \Phi$, $\gamma \leqslant g(x) \leqslant \Gamma$ for all $x \in [a,b]$, where ϕ, Φ, γ, Γ are given real constants, and $h : [a,b] \to [0,\infty)$ is integrable and $\int_a^b h(x)dx > 0$, then

$$\left| \int_a^b h(x)dx \int_a^b f(x)g(x)h(x)dx - \int_a^b h(x)f(x)dx \int_a^b h(x)g(x)dx \right| \leqslant \frac{1}{4}(\Phi - \phi)(\Gamma - \gamma)\left(\int_a^b h(x)dx \right)^2,$$

and the constant $\frac{1}{4}$ is the best possible.

1.8.2 *Dragomir [42]*

Let f, $g : [a,b] \to \mathbb{R}$ be two differentiable mappings on (a,b) and $p : [a,b] \to [0,\infty)$ is integrable on $[a,b]$. If f', $g' \in L_\infty[a,b]$, then

$$\begin{aligned} &\left| \int_a^b p(x)dx \int_a^b p(x)f(x)g(x)dx - \int_a^b p(x)f(x)dx \int_a^b p(x)g(x)dx \right| \\ &\leqslant \frac{1}{2} \int_a^b \int_a^b p(x)p(y) \left| \int_x^y |f'(t)|dt \right| \left| \int_x^y |g'(t)|dt \right| dxdy \\ &\leqslant \|f'\|_\infty \|g'\|_\infty \left[\int_a^b p(x)dx \int_a^b p(x)x^2dx - \left(\int_a^b p(x)xdx \right)^2 \right], \end{aligned}$$

and the inequality is sharp.

1.8.3 *Pachpatte [113]*

Let f, $g : [a,b] \to \mathbb{R}$ be absolutely continuous functions whose derivatives f', $g' \in L_p[a,b]$, $p > 1$, then

$$|T(f,g)| \leqslant \frac{1}{2(b-a)^2} \int_a^b \left[|g(x)| \|f'\|_p + |f(x)| \|g'\|_p \right] (B(x))^{\frac{1}{q}} dx,$$

where

$$B(x) = \frac{1}{q+1}\left[(x-a)^{q+1} + (b-x)^{q+1} \right],$$

for $x \in [a,b]$, $\frac{1}{p} + \frac{1}{q} = 1$ and $T(f,g)$ is the notation set in (2).

1.8.4 *Pachpatte [137]*

Let $f,\ g,\ h : [a,b] \to \mathbb{R}$ be continuous functions on $[a,b]$ and differentiable on (a,b). If $h'(t) \neq 0$ for each $t \in (a,b)$ and $\frac{f'}{h'}, \frac{g'}{h'} : (a,b) \to \mathbb{R}$ are bounded on (a,b), then

$$|T(f,g)| \leqslant \frac{1}{(b-a)^2}\left\|\frac{f'}{h'}\right\|_\infty \left\|\frac{g'}{h'}\right\|_\infty |M|,$$

where

$$M = (b-a)\int_a^b h^2(x)dx - \left(\int_a^b h(x)dx\right)^2,$$

and $T(f,g)$ is the notation set in (2).

1.8.5 *Cerone and Dragomir [18]*

Let $f,\ g : I \subseteq \mathbb{R} \to \mathbb{R}$ be measurable functions on I and the intervals $[a,b],\ [c,d] \subset I$. Define the functional

$$T(f,g;a,b,c,d) = \frac{1}{b-a}\int_a^b f(x)g(x)dx + \frac{1}{d-c}\int_c^d f(y)g(y)dy$$
$$-\frac{1}{b-a}\int_a^b f(x)dx \frac{1}{d-c}\int_c^d g(y)dy - \frac{1}{b-a}\int_a^b g(x)dx \frac{1}{d-c}\int_c^d f(y)dy, \tag{1.8.1}$$

and assume that the integrals involved in (1.8.1) exist, then

$$|T(f,g;a,b,c,d)| \leqslant \left[T(f;a,b) + T(f;c,d) + (M(f;a,b) - M(f;c,d))^2\right]^{\frac{1}{2}}$$
$$\times \left[T(g;a,b) + T(g;c,d) + (M(g;a,b) - M(g;c,d))^2\right]^{\frac{1}{2}}, \tag{1.8.2}$$

where for a measurable function $h : I \to \mathbb{R}$ on $[a,b] \subset I$ we set the notations

$$T(h;a,b) = \frac{1}{b-a}\int_a^b h^2(x)dx - \left(\frac{1}{b-a}\int_a^b h(x)dx\right)^2$$
$$M(h;a,b) = \frac{1}{b-a}\int_a^b h(x)dx,$$

and the integrals involved in the right membership of (1.8.2) exist.

1.8.6 *Pachpatte [137]*

Let $f,\ g : [a,b] \to \mathbb{R}$ be continuous functions on $[a,b]$ and differentiable on (a,b), with derivatives $f',\ g' : (a,b) \to \mathbb{R}$ being bounded on (a,b), then

$$|P(f,g)| \leqslant \frac{1}{16}(b-a)^2 \|f'\|_\infty \|g'\|_\infty,$$

where

$$P(f,g) = FG - \frac{1}{b-a}\left[F\int_a^b g(x)dx + G\int_a^b f(x)dx\right]$$
$$+\left(\frac{1}{b-a}\int_a^b f(x)dx\right)\left(\frac{1}{b-a}\int_a^b g(x)dx\right), \tag{1.8.3}$$

in which

$$F = \frac{f(a)+f(b)}{2}, \quad G = \frac{g(a)+g(b)}{2}.$$

1.8.7 *Pachpatte [113]*

Let $f,\ g : [a,b] \to \mathbb{R}$ be absolutely continuous functions with derivatives $f',\ g' \in L_p[a,b]$, $p > 1$, then we have the inequalities

$$|S(f,g)| \leqslant \frac{1}{(b-a)^2} M^{\frac{2}{q}} \|f'\|_p \|g'\|_p,$$

$$|H(f,g)| \leqslant \frac{1}{(b-a)^2} M^{\frac{1}{q}} \int_a^b \left[|g(x)| \|f'\|_p + |f(x)| \|g'\|_p \right] dx,$$

where $\frac{1}{p} + \frac{1}{q} = 1$,

$$M = \frac{\left(2^{q+1}+1\right)(b-a)^{q+1}}{3(q+1)6^q},$$

and

$$S(f,g) = \overline{F}\overline{G} - \frac{1}{b-a}\left[\overline{F}\int_a^b g(x)dx + \overline{G}\int_a^b f(x)dx\right]$$

$$+\left(\frac{1}{b-a}\int_a^b f(x)dx\right)\left(\frac{1}{b-a}\int_a^b g(x)dx\right),$$

$$H(f,g) = \frac{1}{b-a}\int_a^b \left[\overline{F}g(x) + \overline{G}f(x)\right] dx - 2\left(\frac{1}{b-a}\int_a^b f(x)dx\right)\left(\frac{1}{b-a}\int_a^b g(x)dx\right),$$

in which

$$\overline{F} = \frac{1}{3}\left[\frac{f(a)+f(b)}{2} + 2f\left(\frac{a+b}{2}\right)\right],$$

$$\overline{G} = \frac{1}{3}\left[\frac{g(a)+g(b)}{2} + 2g\left(\frac{a+b}{2}\right)\right].$$

1.8.8 *Pachpatte [137]*

Let $f,\ g : [a,b] \to \mathbb{R}$ be continuous functions on $[a,b]$ and twice differentiable on (a,b), with second derivatives $f'',\ g'' : (a,b) \to \mathbb{R}$ being bounded on (a,b), then

$$|P(f,g)| \leqslant \frac{1}{144}(b-a)^4 \|f''\|_\infty \|g''\|_\infty,$$

where $P(f,g)$ is the notation set in (1.8.3).

1.8.9 *Pachpatte [137]*

Let $f,\ g : [a,b] \to \mathbb{R}$ be continuous functions on $[a,b]$ and twice differentiable on (a,b), whose second derivatives $f'',\ g'' : (a,b) \to \mathbb{R}$ are bounded on (a,b). Then

$$|Q(f,g)| \leqslant \frac{1}{576}(b-a)^4 \|f''\|_\infty \|g''\|_\infty,$$

where

$$Q(f,g) = AB - \frac{1}{b-a}\left[A\int_a^b g(x)dx + B\int_a^b f(x)dx\right] + \left(\frac{1}{b-a}\int_a^b f(x)dx\right)\left(\frac{1}{b-a}\int_a^b g(x)dx\right), \qquad (1.8.4)$$

in which

$$A = f\left(\frac{a+b}{2}\right), \quad B = g\left(\frac{a+b}{2}\right).$$

1.8.10 *Pachpatte [112]*

Let $f,\ g : [a,b] \to \mathbb{R}$ be absolutely continuous functions on $[a,b]$ with $f', g' \in L_2[a,b]$, then

$$|Q(f,g)| \leqslant \frac{(b-a)^2}{12}\left[\frac{1}{b-a}\|f'\|_2^2 - ([f;a,b])^2\right]^{\frac{1}{2}} \left[\frac{1}{b-a}\|g'\|_2^2 - ([g;a,b])^2\right]^{\frac{1}{2}},$$

where $Q(f,g)$ is the notation set in (1.8.4) and

$$[f;a,b] = \frac{f(b)-f(a)}{b-a}, \quad [g;a,b] = \frac{g(b)-g(a)}{b-a}.$$

1.8.11 *Pachpatte [124]*

Let $f,\ g,\ h : [a,b] \to \mathbb{R}$ be twice differentiable functions on (a,b) and $f'',\ g'',\ h'' : (a,b) \to \mathbb{R}$ are bounded, then

$$\left|\frac{1}{b-a}\int_a^b f(x)g(x)h(x)dx - \frac{1}{3(b-a)^2}D[f,g,h] - \frac{1}{3(b-a)}\int_a^b \left(x - \frac{a+b}{2}\right)(f(x)g(x)h(x))'dx\right| \leqslant \frac{1}{3(b-a)^2}\int_a^b B[f,g,h](x)E(x)dx,$$

where

$$D[f,g,h] = \left(\int_a^b g(x)h(x)dx\right)\left(\int_a^b f(x)dx\right) + \left(\int_a^b h(x)f(x)dx\right)\left(\int_a^b g(x)dx\right) + \left(\int_a^b f(x)g(x)dx\right)\left(\int_a^b h(x)dx\right), \qquad (1.8.5)$$

$$B[f,g,h](x) = |g(x)||h(x)|\|f''\|_\infty + |h(x)||f(x)|\|g''\|_\infty + |f(x)||g(x)|\|h''\|_\infty, \qquad (1.8.6)$$

$$E(x) = \int_a^b |k(x,t)|dt,$$

in which $k(x,t)$ is given by (1.2.28).

1.8.12 *Pachpatte [124]*

Let $f,\ g,\ h : [a,b] \to \mathbb{R}$ be functions with first derivatives absolutely continuous on $[a,b]$ and assume that the second derivatives $f'',\ g'',\ h'' : (a,b) \to \mathbb{R}$ are bounded on (a,b), then

$$\left|\frac{1}{b-a}\int_a^b f(x)g(x)h(x)dx - \frac{1}{3(b-a)^2}D[f,g,h]\right.$$

$$\left. -\frac{2}{3(b-a)}\int_a^b \left(x-\frac{a+b}{2}\right)(f(x)g(x)h(x))'dx + \frac{1}{3(b-a)}\int_a^b L[f,g,h](x)dx\right|$$

$$\leqslant \frac{1}{3(b-a)^2}\int_a^b B[f,g,h](x)I(x)dx,$$

where $D[f,g,h]$, $B[f,g,h](x)$ are the notations set in (1.8.5), (1.8.6) and

$$L[f,g,h](x) = \frac{f(a)+f(b)}{2}g(x)h(x) + \frac{g(a)+g(b)}{2}h(x)f(x) + \frac{h(a)+h(b)}{2}f(x)g(x),$$

$$I(x) = \int_a^b |p(x,t)|\left|t-\frac{a+b}{2}\right|dt,$$

in which $p(x,t)$ is given by (1.2.11).

1.8.13 *Pachpatte [124]*

Let $f,\ g,\ h : [a,b] \to \mathbb{R}$ be continuous on $[a,b]$ and twice differentiable on (a,b) with second derivatives $f'',\ g'',\ h'' : (a,b) \to \mathbb{R}$ bounded on (a,b), then

$$\left|\frac{1}{b-a}\int_a^b f(x)g(x)h(x)dx - \frac{1}{3(b-a)^2}D[f,g,h]\right.$$

$$\left. -\frac{1}{3(b-a)}\int_a^b \left(x-\frac{a+b}{2}\right)M[f,g,h](x)dx\right|$$

$$\leqslant \frac{1}{3(b-a)^3}\int_a^b B[f,g,h](x)H(x)dx,$$

where $D[f,g,h], B[f,g,h](x)$ are the notations set in (1.8.5), (1.8.6) and

$$M[f,g,h](x) = \frac{f(b)-f(a)}{b-a}g(x)h(x) + \frac{g(b)-g(a)}{b-a}h(x)f(x) + \frac{h(b)-h(a)}{b-a}f(x)g(x),$$

$$H(x) = \int_a^b\int_a^b |p(x,t)||p(t,s)|dsdt,$$

in which $p(x,t)$ is given by (1.2.11).

1.8.14 *Pachpatte [132]*

Let $f,\ g : [a,b] \to \mathbb{R}$ be continuous functions on $[a,b]$ and twice differential on (a,b), with first and second derivatives $f', f'', g', g'' : (a,b) \to \mathbb{R}$ bounded on (a,b), then

$$|T(f,g)| \leqslant \frac{1}{b-a}\int_a^b A(x)dx,$$

where $T(f,g)$ is the notation set in (2) and

$$A(x) = \left(x - \frac{a+b}{2}\right)^2 \|f'\|_\infty \|g'\|_\infty$$

$$+\left|x - \frac{a+b}{2}\right| \{\|f'\|_\infty \|g''\|_\infty + \|g'\|_\infty \|f''\|_\infty\} L(x) + \|f''\|_\infty \|g''\|_\infty L^2(x),$$

for $x \in [a,b]$,

$$L(x) = \frac{1}{b-a}\int_a^b |k(x,t)|dt = \frac{(b-a)^2}{24} + \frac{1}{2}\left(x - \frac{a+b}{2}\right)^2,$$

in which $k(x,t) : [a,b]^2 \to \mathbb{R}$ is given by

$$k(x,t) = \begin{cases} \dfrac{(t-a)^2}{2}, & t \in [a,x], \\ \dfrac{(t-b)^2}{2}, & t \in (x,b]. \end{cases}$$

1.8.15 *Pachpatte [132]*

Let $f,\ g : [a,b] \to \mathbb{R}$ be continuous functions on $[a,b]$ and twice differentiable on (a,b), with second derivatives $f'',\ g'' : (a,b) \to \mathbb{R}$ bounded on (a,b), then

$$|T(f,g)| \leqslant \frac{1}{b-a}\int_a^b B(x)dx,$$

where $T(f,g)$ is the notation set in (2) and

$$B(x) = |FG|\left(x - \frac{a+b}{2}\right)^2$$

$$+\left|x - \frac{a+b}{2}\right| \{|F|\|g''\|_\infty + |G|\|f''\|_\infty\} M(x) + \|f''\|_\infty \|g''\|_\infty M^2(x),$$

for $x \in [a,b]$,

$$F = \frac{f(b)-f(a)}{b-a}, \quad G = \frac{g(b)-g(a)}{b-a},$$

$$M(x) = \frac{1}{(b-a)^2}\int_a^b \int_a^b |p(x,t)||p(t,s)|dsdt = \frac{1}{2}\left\{\left[\frac{\left(x-\frac{a+b}{2}\right)^2}{(b-a)^2} + \frac{1}{4}\right]^2 + \frac{1}{12}\right\}(b-a)^2,$$

in which $p(x,t) : [a,b]^2 \to \mathbb{R}$ is given by

$$p(x,t) = \begin{cases} t-a, & t \in [a,x], \\ t-b, & t \in (x,b]. \end{cases}$$

1.8.16 *Pachpatte [90]*

Let $f,\ g:[a,b]\to\mathbb{R}$ be functions such that $f^{(n-1)},\ g^{(n-1)}$ are absolutely continuous on $[a,b]$ and $f^{(n)},\ g^{(n)}\in L_\infty[a,b]$ for some $n\in\mathbb{N}$, then

$$\left|\frac{1}{b-a}\int_a^b f(x)g(x)dx-\left(\frac{1}{b-a}\int_a^b f(x)dx\right)\left(\frac{1}{b-a}\int_a^b g(x)dx\right)\right.$$
$$\left.-\frac{1}{2(b-a)^2}\int_a^b\left[\left(\sum_{k=1}^{n-1}\overline{F}_k(x)\right)g(x)+\left(\sum_{k=1}^{n-1}\overline{G}_k(x)\right)f(x)\right]dx\right|$$
$$\leqslant\frac{1}{2(b-a)^2}\int_a^b\left[|g(x)|\|f^{(n)}\|_\infty+|f(x)|\|g^{(n)}\|_\infty\right]A_n(x)dx,$$

where

$$\overline{F}_k(x)=\left[\frac{(b-x)^{k+1}+(-1)^k(x-a)^{k+1}}{(k+1)!}\right]f^{(k)}(x),$$
$$\overline{G}_k(x)=\left[\frac{(b-x)^{k+1}+(-1)^k(x-a)^{k+1}}{(k+1)!}\right]g^{(k)}(x),$$
$$A_n(x)=\int_a^b|K_n(x,t)|dt,$$

in which $K_n(x,t):[a,b]^2\to\mathbb{R}$ is given by

$$K_n(x,t)=\begin{cases}\dfrac{(t-a)^n}{n!}, & t\in[a,x],\\ \dfrac{(t-b)^n}{n!}, & t\in(x,b].\end{cases}$$

for $x\in[a,b]$.

1.8.17 *Pachpatte [119]*

Let the functions $f,\ g:[a,b]\to\mathbb{R}$ be such that $f^{(n-1)},g^{(n-1)}$ are absolutely continuous on $[a,b]$ and $f^{(n)},\ g^{(n)}\in L_\infty[a,b]$ for some $n\in\mathbb{N}$, then

$$|A[f,g;a,b;n]|\leqslant\frac{1}{[(n-1)!]^2(b-a)^3}\|f^{(n)}\|_\infty\|f^{(n)}\|_\infty\int_a^b[E_n(x)]^2dx,$$

where

$$A[f,g;a,b;n]=\frac{1}{b-a}\int_a^b\left[f(x)+\sum_{k=1}^{n-1}F_k(x)\right]\left[g(x)+\sum_{k=1}^{n-1}G_k(x)\right]dx$$
$$-n\left[\left(\frac{1}{b-a}\int_a^b\left[f(x)+\sum_{k=1}^{n-1}F_k(x)\right]dx\right)\left(\frac{1}{b-a}\int_a^b g(x)dx\right)\right.$$
$$\left.+\left(\frac{1}{b-a}\int_a^b\left[g(x)+\sum_{k=1}^{n-1}G_k(x)\right]dx\right)\left(\frac{1}{b-a}\int_a^b f(x)dx\right)\right]$$
$$+n^2\left(\frac{1}{b-a}\int_a^b f(x)dx\right)\left(\frac{1}{b-a}\int_a^b g(x)dx\right),$$

$$E_n(x)=\int_a^b\left|(x-t)^{n-1}p(x,t)\right|dt, \tag{1.8.7}$$

in which $F_k(x),\ G_k(x)$ are given by (1.5.6), (1.5.7) and $p(x,t)$ is given by (1.2.11).

1.8.18 *Pachpatte [119]*

Let the functions $f,\ g : [a,b] \to \mathbb{R}$ be such that $f^{(n-1)}, g^{(n-1)}$ are absolutely continuous on $[a,b]$ and $f^{(n)},\ g^{(n)} \in L_\infty[a,b]$ for some $n \in \mathbb{N}$, then

$$|B[f,g;a,b;n]| \leqslant \frac{1}{2(n-1)!(b-a)^2} \int_a^b \left[\left| g(x) + \sum_{k=1}^{n-1} G_k(x) \right| \|f^{(n)}\|_\infty \right.$$

$$\left. + \left| f(x) + \sum_{k=1}^{n-1} F_k(x) \right| \|g^{(n)}\|_\infty \right] E_n(x)dx,$$

where

$$B[f,g;a,b;n] = \frac{1}{b-a} \int_a^b \left[f(x) + \sum_{k=1}^{n-1} F_k(x) \right] \left[g(x) + \sum_{k=1}^{n-1} G_k(x) \right] dx$$

$$-\frac{n}{2} \left[\left(\frac{1}{b-a} \int_a^b \left[f(x) + \sum_{k=1}^{n-1} F_k(x) \right] dx \right) \left(\frac{1}{b-a} \int_a^b g(x)dx \right) \right.$$

$$\left. + \left(\frac{1}{b-a} \int_a^b \left[g(x) + \sum_{k=1}^{n-1} G_k(x) \right] dx \right) \left(\frac{1}{b-a} \int_a^b f(x)dx \right) \right],$$

and $E_n(x)$ is given by (1.8.7), and $F_k(x),\ G_k(x)$ are given by (1.5.6), (1.5.7).

1.8.19 *Dragomir and Khan [51]*

Let $\overline{a} = (a_1, \ldots, a_n)$ and $\overline{b} = (b_1, \ldots, b_n)$ be two sequences of positive real numbers with

$$0 < a \leqslant a_i \leqslant A < \infty, \quad 0 < b \leqslant b_i \leqslant B < \infty, \tag{1.8.8}$$

for each $i \in \{1, \ldots, n\}$ and let $C_n(\overline{a},\overline{b})$ is given by (1.6.2) replacing $f,\ g$ by $\overline{a},\ \overline{b}$, then

$$\left| C_n(\overline{a},\overline{b}) \right| \leqslant \frac{1}{4} \frac{(A-a)(B-b)}{\sqrt{aAbB}} \left(\frac{1}{n} \sum_{i=1}^n a_i \right) \left(\frac{1}{n} \sum_{i=1}^n b_i \right),$$

and the constant $\frac{1}{4}$ is best possible.

1.8.20 *Dragomir and Khan [51]*

Let $\overline{a} = (a_1, \ldots, a_n)$ and$\overline{b} = (b_1, \ldots, b_n)$ be two sequences of positive real numbers satisfying (1.8.8) and let $C_n(\overline{a},\overline{b})$ is given by (1.6.2) by replacing $f,\ g$ by $\overline{a},\ \overline{b}$, then

$$|C_n(\overline{a},\overline{b})| \leqslant \left(\sqrt{A} - \sqrt{a} \right) \left(\sqrt{B} - \sqrt{b} \right) \sqrt{\left(\frac{1}{n} \sum_{i=1}^n a_i \right) \left(\frac{1}{n} \sum_{i=1}^n b_i \right)},$$

and the inequality is sharp.

1.8.21 *Pachpatte [99]*

Let $f(n)$, $g(n)$ be real-valued functions defined on $\mathbb{N}_{a,b} = \{a, a+1, \ldots, a+m = b\}$, $a \in \mathbb{R}$, $m \in \mathbb{N}$, and equal to 0 if $n \notin \mathbb{N}_{a,b}$, for which $\Delta f(n)$,
$Deltag(n)$ exist and $|\Delta f(n)| \leqslant A$, $|\Delta g(n)| \leqslant B$, for $n \in \mathbb{N}_{a,b}$, where A, B are nonnegative real constants. Then

$$\left| \frac{1}{b-a} \sum_{n=a}^{b-1} f(n)g(n) - \frac{1}{(b-a)^2} \left[\left(\sum_{n=a}^{b-1} g(n) \right) \left(\sum_{n=a}^{b-1} f(n+1) \right) \right. \right.$$
$$\left. \left. + \left(\sum_{n=a}^{b-1} f(n) \right) \left(\sum_{n=a}^{b-1} g(n+1) \right) \right] + \frac{1}{(b-a)^2} \left(\sum_{n=a}^{b-1} f(n+1) \right) \left(\sum_{n=a}^{b-1} g(n+1) \right) \right|$$
$$\leqslant \frac{AB}{(b-a)^3} \sum_{n=a}^{b-1} (H(n))^2,$$

where

$$H(n) = \sum_{s=a}^{b-1} |r(n,s)|,$$

in which

$$r(n,s) = \begin{cases} s-a, & s \in [a, n-1], \\ s-b, & s \in [n, b]. \end{cases}$$

for n, $s \in \mathbb{N}_{a,b}$.

1.8.22 *Pachpatte [105]*

Let $f(n)$, $g(n)$, $h(n)$ be real-valued functions defined on $\mathbb{N}_{a,b} = \{a, a+1, \ldots, a+m = b\}$, $a \in \mathbb{R}$, $m \in \mathbb{N}$, and equal to 0 if $n \notin \mathbb{N}_{a,b}$, for which $\Delta f(n)$, $\Delta g(n)$, $\Delta h(n)$ exist and $|\Delta f(n)| \leqslant M_1$, $|\Delta g(n)| \leqslant M_2$, $|\Delta h(n)| \leqslant M_3$, for $n \in \mathbb{N}_{a,b}$, where M_1, M_2, M_3 are nonnegative constants. Then

$$\left| \frac{1}{b-a} \sum_{n=a}^{b-1} f(n)g(n)h(n) - \frac{1}{3} \left[\left(\frac{1}{b-a} \sum_{n=a}^{b-1} g(n)h(n) \right) \left(\frac{1}{b-a} \sum_{n=a}^{b-1} f(n) \right) \right. \right.$$
$$\left. \left. + \left(\frac{1}{b-a} \sum_{n=a}^{b-1} h(n)f(n) \right) \left(\frac{1}{b-a} \sum_{n=a}^{b-1} g(n) \right) + \left(\frac{1}{b-a} \sum_{n=a}^{b-1} f(n)g(n) \right) \left(\frac{1}{b-a} \sum_{n=a}^{b-1} h(n) \right) \right] \right|$$
$$\leqslant \frac{1}{3(b-a)} \sum_{n=a}^{b-1} \left[|g(n)||h(n)|M_1 + |h(n)||f(n)|M_2 + |f(n)||g(n)|M_3 \right] B(n),$$

where

$$B(n) = \left[\frac{1}{2} + \left| n - \frac{a+b}{2} \right| \right],$$

for $n \in \mathbb{N}_{a,b}$.

1.9 Notes

The inequality in Theorem 1.2.1 is taken from Matić,Pečarić and Ujević [72]. In [72] it is shown that, if a factor is known, say $g(t)$, $t \in [a,b]$, then instead of using the Grüss inequality (3) to estimate the difference given by $T(f,g)$, it is better to use the inequality (1.2.1). They demonstrated this by improving some results given by Dragomir in [31] related to Taylor's formula with integral remainder. Theorem 1.2.2 deals with a Grüss-type inequality proved by Dragomir and McAndrew [34]. The inequalities in Theorems 1.2.3–1.2.5 deal with Grüss-type integral inequalities involving functions and their derivatives and taken from Pachpatte [96,105].

The Čebyšev-type inequality in Theorem 1.3.1 is taken from Matić, Pečarić and Ujević [72]. Theorem 1.3.2 is a generalization of the Čebyšev inequality and taken from Dragomir [42]. Theorems 1.3.3–1.3.5 deal with Čebyšev-type inequalities established by Pachpatte in [112,113]. Section 1.4 contains inequalities of the Grüss-and Čebyšev-type investigated by Pachpatte in [106,111,117,127]. Section 1.5 deals with some more inequalities of the Grüss-and Čebyšev-type involving functions and their higher order derivatives and taken from Pachpatte [118,120,121]. The discrete Grüss-and Čebyšev-type inequalities in Theorems 1.6.1–1.6.5 are due to Pachpatte [128,133,138]. The results in Theorems 1.7.1 and 1.7.2 are due to Dragomir and McAndrew [34]. Theorem 1.7.3 is taken from Dragomir [31]. Theorems 1.7.4–1.7.6 are due to Matić, Pečarić and Ujević [72] and the results in Theorems 1.7.7–1.7.9 are taken from Barnett and Dragomir [5]. Section 1.9 contains some useful miscellaneous inequalities.

Chapter 2

Multidimensional Grüss-Čebyšev and-Trapezoid-type inequalities

2.1 Introduction

During the last two decades many researchers have given considerable attention to the famous inequalities (1), (3) and (4) associated to the names of Čebyšev, Grüss and Trapezoid. In view of the usefulness of these inequalities and their applications, many authors have investigated a large number of new multidimensional, Grüss, Čebyšev and Trapezoid type inequalities. Some of these results provide simple and elegant extensions of the inequalities (1), (3) and (4) and have a wider scope of applicability. These results did not just add new objects of study, but also brought new insights and techniques to handle such inequalities. This chapter deals with a number of new multidimensional inequalities discovered by various investigators, which claim their origin to the well-known inequalities in (1), (3) and (4). Some applications are given to illustrate the usefulness of certain inequalities.

2.2 Some Grüss-type inequalities in inner product spaces

In this section we offer some fundamental Grüss-type inequalities established by Dragomir [32,43,53] and Dragomir, Pečarić and Tepeš [56] in inner product spaces.
We start with the following Grüss-type inequality investigated in [32].

Theorem 2.2.1. Let $(H,(\cdot,\cdot))$ be an inner product space over $\mathbb{K}$ ($\mathbb{K}=\mathbb{R}$, $\mathbb{C}$) and $e \in H$, $\|e\|=1$. If ϕ, γ, Φ, Γ are real or complex numbers and x, y are vectors in H such that the condition

$$\operatorname{Re}(\Phi e - x, x-\phi e) \geqslant 0, \quad \operatorname{Re}(\Gamma e - y, y-\gamma e) \geqslant 0, \tag{2.2.1}$$

holds, then we have the inequality

$$|(x,y)-(x,e)(e,y)| \leqslant \frac{1}{4}|\Phi-\phi||\Gamma-\gamma|. \tag{2.2.2}$$

The constant $\frac{1}{4}$ is the best possible.

Proof. It is obvious that (see [156])

$$(x,y)-(x,e)(e,y)=(x-(x,e)e,y-(y,e)e). \tag{2.2.3}$$

Using Schwarz's inequality in inner product spaces, we have

$$\begin{aligned}|(x,y)-(x,e)(e,y)|^2 &= |(x-(x,e)e,y-(y,e)e)|^2\\ &\leqslant \|x-(x,e)e\|^2\|y-(y,e)e\|^2\\ &= \left(\|x\|^2-|(x,e)|^2\right)\left(\|y\|^2-|(y,e)|^2\right).\end{aligned} \tag{2.2.4}$$

On the other hand, a simple computation shows that

$$(\Phi-(x,e))\left(\overline{(x,e)}-\overline{\phi}\right)-(\Phi e-x,x-\phi e)=\|x\|^2-|(x,e)|^2, \tag{2.2.5}$$

and

$$(\Gamma-(y,e))\left(\overline{(y,e)}-\overline{\gamma}\right)-(\Gamma e-y,y-\gamma e)=\|y\|^2-|(y,e)|^2. \tag{2.2.6}$$

Taking the real part in both the above equalities, we can write

$$\operatorname{Re}\left[(\Phi-(x,e))\left(\overline{(x,e)}-\overline{\phi}\right)\right]-\operatorname{Re}(\Phi e-x,x-\phi e)=\|x\|^2-|(x,e)|^2, \tag{2.2.7}$$

and

$$\operatorname{Re}\left[(\Gamma-(y,e))\left(\overline{(y,e)}-\overline{\gamma}\right)\right]-\operatorname{Re}(\Gamma e-y,y-\gamma e)=\|y\|^2-|(y,e)|^2. \tag{2.2.8}$$

From the condition (2.2.1), we deduce

$$\|x\|^2-|(x,e)|^2\leqslant \operatorname{Re}\left[(\Phi-(x,e))\left(\overline{(x,e)}-\overline{\phi}\right)\right], \tag{2.2.9}$$

and

$$\|y\|^2-|(y,e)|^2\leqslant \operatorname{Re}\left[(\Gamma-(y,e))\left(\overline{(y,e)}-\overline{\gamma}\right)\right]. \tag{2.2.10}$$

Using the elementary inequality $4\operatorname{Re}\left(a\overline{b}\right)\leqslant |a+b|^2$ holding for real or complex numbers a,b, for $a:=\Phi-(x,e)$ and $b:=(x,e)-\phi$, we get

$$\operatorname{Re}\left[(\Phi-(x,e))\left(\overline{(x,e)}-\overline{\phi}\right)\right]\leqslant \frac{1}{4}|\Phi-\phi|^2, \tag{2.2.11}$$

and, similarly

$$\operatorname{Re}\left[(\Gamma-(y,e))\left(\overline{(y,e)}-\overline{\gamma}\right)\right]\leqslant \frac{1}{4}|\Gamma-\gamma|^2. \tag{2.2.12}$$

Consequently, using (2.2.3)–(2.2.12), we have successively

$$\begin{aligned}|(x,y)-(x,e)(e,y)|^2 &\leqslant \left(\|x\|^2-|(x,e)|^2\right)\left(\|y\|^2-|(y,e)|^2\right)\\ &\leqslant \operatorname{Re}\left[(\Phi-(x,e))\left(\overline{(x,e)}-\overline{\phi}\right)\right]\operatorname{Re}\left[(\Gamma-(y,e))\left(\overline{(y,e)}-\overline{\gamma}\right)\right]\end{aligned}$$

$$\leqslant \frac{1}{16}|\Phi-\phi|^2|\Gamma-\gamma|^2,$$

from which we get the desired inequality (2.2.2).

To prove that the constant $\frac{1}{4}$ is sharp, we can restrict ourselves to the real case. Let $e,m \in H$ with $\|e\| = \|m\| = 1$, $e \perp m$ and assume that $\phi, \gamma, \Phi, \Gamma$ are real numbers. Define the vectors

$$x := \frac{\Phi+\phi}{2}e + \frac{\Phi-\phi}{2}m, y := \frac{\gamma+\Gamma}{2}e + \frac{\Gamma-\gamma}{2}m.$$

Then

$$(\Phi e - x, x - \phi e) = \left(\frac{\Phi-\phi}{2}\right)^2 (e-m, e+m) = 0,$$

and similarly $(\Gamma e - y, y - \gamma e) = 0$, i.e., the condition (2.2.1) holds. Now, observe that

$$(x,y) = \left(\frac{\Phi+\phi}{2}\right)\left(\frac{\gamma+\Gamma}{2}\right) + \left(\frac{\Phi-\phi}{2}\right)\left(\frac{\Gamma-\gamma}{2}\right),$$

and

$$(x,e)(e,y) = \left(\frac{\Phi+\phi}{2}\right)\left(\frac{\gamma+\Gamma}{2}\right).$$

Consequently,

$$|(x,y)-(x,e)(e,y)| = \frac{1}{4}|\Phi-\phi||\Gamma-\gamma|,$$

which shows that the constant $\frac{1}{4}$ is sharp.

In [53], the author gave an alternative proof of (2.2.2) by using the following lemmas.

Lemma 2.2.1. Let a,x,A be vectors in the inner product space $(H,(\cdot,\cdot))$ over $\mathbb{K}$ ($\mathbb{K} = \mathbb{R},\mathbb{C}$) with $a \neq A$. Then

$$\operatorname{Re}(A-x, x-a) \geqslant 0,$$

if and only if

$$\left\| x - \frac{a+A}{2} \right\| \leqslant \frac{1}{2}\|A-a\|.$$

Proof. Define

$$I_1 := \operatorname{Re}(A-x,x-a), \quad I_2 := \frac{1}{4}\|A-a\|^2 - \left\| x - \frac{a+A}{2} \right\|^2.$$

A simple calculation shows that

$$I_1 = I_2 = \operatorname{Re}[(x,a)+(A,x)] - \operatorname{Re}(A,a) - \|x\|^2,$$

and thus obviously, $I_1 \geqslant 0$ if and only if $I_2 \geqslant 0$, showing the required equivalence.

The following corollary is obvious.

Corollary 2.2.1. Let $x,\ e \in H$ with $\|e\| = 1$ and δ, $Delta \in \mathbb{K}$ with $\delta \neq \Delta$. Then

$$\operatorname{Re}(\Delta e - x, x - \delta e) \geqslant 0,$$

if and only if

$$\left\| x - \frac{\delta + \Delta}{2} e \right\| \leqslant \frac{1}{2} \|\Delta - \delta\| .$$

Lemma 2.2.2. Let $x, e \in H$ with $\|e\| = 1$. Then one has the following representation

$$0 \leqslant \|x\|^2 - |(x,e)|^2 = \inf_{\lambda \in \mathbb{K}} \|x - \lambda e\|^2, \tag{2.2.13}$$

where $\mathbb{K}$ is as in Lemma 2.2.1.

Proof. For any $\lambda \in \mathbb{K}$ observe that

$$(x - \lambda e, x - (x,e)) = \|x\|^2 - |(x,e)|^2 - \lambda\left[(e,x) - (e,x)\|e\|^2\right] = \|x\|^2 - |(x,e)|^2.$$

Using Schwarz inequality, we have

$$\left[\|x\|^2 - |(x,e)|^2\right]^2 = |(x - \lambda e, x - (x,e)e)|^2$$

$$\leqslant \|x - \lambda e\|^2 \|x - (x,e)e\|^2 = \|x - \lambda e\|^2 \left[\|x\|^2 - |(x,e)|^2\right],$$

giving the bound

$$\|x\|^2 - |(x,e)|^2 \leqslant \|x - \lambda e\|^2. \tag{2.2.14}$$

Taking the infimum in (2.2.14) over $\lambda \in \mathbb{K}$, we deduce

$$\|x\|^2 - |(x,e)|^2 \leqslant \inf_{\lambda \in \mathbb{K}} \|x - \lambda e\|^2.$$

Since, for $\lambda_0 = (x,e)$, we get $\|x - \lambda_0 e\|^2 = \|x\|^2 - |(x,e)|^2$, then the representation (2.2.13) is proved.

The following result is proved in [53].

Theorem 2.2.2. Let $(H, (\cdot,\cdot))$ be an inner product space over K ($\mathbb{K} = \mathbb{R},\ \mathbb{C}$) and $e \in H$, $\|e\| = 1$. If $\phi,\ \gamma,\ \Phi,\ \Gamma$ are real or complex numbers and $x,\ y$ are vectors in H such that the conditions (2.2.1) hold, or, equivalently, the following assumptions

$$\left\| x - \frac{\phi + \Phi}{2} e \right\| \leqslant \frac{1}{2} |\Phi - \phi|, \quad \left\| y - \frac{\gamma + \Gamma}{2} e \right\| \leqslant \frac{1}{2} |\Gamma - \gamma|, \tag{2.2.15}$$

are valid, then one has the inequality

$$|(x,y) - (x,e)(e,y)| \leqslant \frac{1}{4} |\Phi - \phi| |\Gamma - \gamma|. \tag{2.2.16}$$

The constant $\frac{1}{4}$ is the best possible.

Proof. As in the proof of Theorem 2.2.1, we have (2.2.3) and (2.2.4). Using Lemma 2.2.2 and conditions (2.2.15), we obviously have

$$\left[\|x\|^2 - |(x,e)|^2\right]^{\frac{1}{2}} = \inf_{\lambda\in\mathbb{K}} \|x-\lambda e\| \leqslant \left\| x - \frac{\phi+\Phi}{2} e \right\| \leqslant \frac{1}{2}|\Phi-\phi|, \tag{2.2.17}$$

and

$$\left[\|y\|^2 - |(y,e)|^2\right]^{\frac{1}{2}} = \inf_{\lambda\in\mathbb{K}} \|y-\lambda e\| \leqslant \left\| y - \frac{\gamma+\Gamma}{2} e \right\| \leqslant \frac{1}{2}|\Gamma-\gamma|. \tag{2.2.18}$$

Using (2.2.17), (2.2.18) in (2.2.4), the desired inequality in (2.2.16) follows. The fact that $\frac{1}{4}$ is the best possible constant, has been shown in the proof of Theorem 2.2.1 and hence we omit the details.

The refinement of the inequality (2.2.2) proved in [53] is embodied in the following theorem.

Theorem 2.2.3. Let $(H,(\cdot,\cdot))$ be an inner product space over K ($\mathbb{K}=\mathbb{R}$, $\mathbb{C}$) and $e\in H$, $\|e\|=1$. If ϕ, γ, Φ, Γ are real or complex numbers and x, y are vectors in H such that the condition (2.2.1) or equivalently, (2.2.15) hold, then we have the inequality

$$|(x,y)-(x,e)(e,y)|$$
$$\leqslant \frac{1}{4}|\Phi-\phi||\Gamma-\gamma| - \left[\mathrm{Re}(\Phi e - x, x-\phi e)\right]^{\frac{1}{2}}\left[\mathrm{Re}(\Gamma e - y, y-\gamma e)\right]^{\frac{1}{2}}. \tag{2.2.19}$$

Proof. Following the proof of Theorem 2.2.1, we have (2.2.3), (2.2.7), (2.2.8), (2.2.11), (2.2.12) and consequently, we observe that

$$|(x,y)-(x,e)(e,y)|^2 \leqslant \left[\frac{1}{4}|\Phi-\phi|^2 - \left(\left[\mathrm{Re}(\Phi e - x, x-\phi e)\right]^{\frac{1}{2}}\right)^2\right]$$
$$\times\left[\frac{1}{4}|\Gamma-\gamma|^2 - \left(\left[\mathrm{Re}(\Gamma e - y, y-\gamma e)\right]^{\frac{1}{2}}\right)^2\right]. \tag{2.2.20}$$

By a suitable application of the elementary inequality

$$\left(m^2-n^2\right)\left(p^2-q^2\right) \leqslant (mp-nq)^2, \tag{2.2.21}$$

for $m,n,p,q\in\mathbb{R}$, to the right hand side of (2.2.20), we have

$$|(x,y)-(x,e)(e,y)|^2$$
$$\leqslant \left[\frac{1}{4}|\Phi-\phi||\Gamma-\gamma| - \left(\left[\mathrm{Re}(\Phi e - x, x-\phi e)\right]^{\frac{1}{2}}\left[\mathrm{Re}(\Gamma e - y, y-\gamma e)\right]^{\frac{1}{2}}\right)^2\right]^2,$$

from which the desired inequality in (2.2.19) follows.

The following Theorem given in [56] deals with the inequalities of the pre-Grüss-type in inner product spaces.

Theorem 2.2.4. Let $(H,(\cdot,\cdot))$ be an inner product space over $\mathbb{K}$ ($\mathbb{K}=\mathbb{R}$, $\mathbb{C}$) and $e\in H$, $\|e\|=1$. If ϕ, Φ are real or complex numbers and x,y are vectors in H such that either the condition

$$\operatorname{Re}(\Phi e-x,x-\phi e)\geqslant 0,$$

or equivalently,

$$\left\|x-\frac{\phi+\Phi}{2}e\right\|\leqslant\frac{1}{2}|\Phi-\phi|, \tag{2.2.22}$$

holds true, then we have the inequalities

$$|(x,y)-(x,e)(e,y)|\leqslant\frac{1}{2}|\Phi-\phi|\left(\|y\|^2-|(y,e)|^2\right)^{\frac{1}{2}}, \tag{2.2.23}$$

and

$$|(x,y)-(x,e)(e,y)|\leqslant\frac{1}{2}|\Phi-\phi|\|y\|-\left(\operatorname{Re}(\Phi e-x,x-\phi e)\right)^{\frac{1}{2}}|(y,e)|. \tag{2.2.24}$$

Proof. As in the proof of Theorem 2.2.1, we have (2.2.3) and (2.2.4). Now, the inequality (2.2.23) is a simple consequence of (2.2.2) for $x=y$ or of Lemma 2.2.2 and (2.2.22). Furthermore, from the proof of Theorem 2.2.1, we have (2.2.7) and (2.2.11). Using (2.2.7) and (2.2.11), we have

$$\|x\|^2-|(x,e)|^2\leqslant\left(\frac{1}{2}|\Phi-\phi|\right)^2-\left(\left(\operatorname{Re}(\Phi e-x,x-\phi e)\right)^{\frac{1}{2}}\right)^2. \tag{2.2.25}$$

From (2.2.4) and (2.2.25), we get

$$\begin{aligned}&|(x-(x,e)e,y-(y,e)e)|^2\\ \leqslant&\left(\left(\frac{1}{2}|\Phi-\phi|\right)^2-\left(\left(\operatorname{Re}(\Phi e-x,x-\phi e)\right)^{\frac{1}{2}}\right)^2\right)\left(\|y\|^2-|(y,e)|^2\right).\end{aligned} \tag{2.2.26}$$

Now, by a suitable application of the elementary inequality (2.2.21) to the right hand side of (2.2.26) and rewriting, we get the desired inequality in (2.2.24). The proof is complete. Before closing this section, we present a Grüss-type inequality for sequences of vectors in inner product spaces given in [43].

The following lemma given in [43] is of interest in itself.

Lemma 2.2.3. Let $(H,(\cdot,\cdot))$ be an inner product space over the real or complex number field $\mathbb{K}$, $x_i\in H$ and $p_i\geqslant 0$ $(i=1,\ldots,n)$ such that $\sum_{i=1}^n p_i=1$ $(n\geqslant 2)$. If x, $X\in H$ are such that

$$\operatorname{Re}(X-x_i,x_i-x)\geqslant 0, \tag{2.2.27}$$

for all $i\in\{1,\ldots,n\}$, then we have the inequality

$$0\leqslant\sum_{i=1}^n p_i\|x_i\|^2-\left\|\sum_{i=1}^n p_ix_i\right\|^2\leqslant\frac{1}{4}\|X-x\|^2. \tag{2.2.28}$$

The constant $\frac{1}{4}$ is sharp.

Proof. Define

$$I_1 := \left(X - \sum_{i=1}^{n} p_i x_i, \sum_{i=1}^{n} p_i x_i - x \right),$$

and

$$I_2 := \sum_{i=1}^{n} p_i (X - x_i, x_i - x).$$

Then

$$I_1 = \sum_{i=1}^{n} p_i (X, x_i) - (X, x) - \left\| \sum_{i=1}^{n} p_i x_i \right\|^2 + \sum_{i=1}^{n} p_i (x_i, x),$$

and

$$I_2 = \sum_{i=1}^{n} p_i (X, x_i) - (X, x) - \sum_{i=1}^{n} p_i \|x_i\|^2 + \sum_{i=1}^{n} p_i (x_i, x).$$

Consequently,

$$I_1 - I_2 = \sum_{i=1}^{n} p_i \|x_i\|^2 - \left\| \sum_{i=1}^{n} p_i x_i \right\|^2. \tag{2.2.29}$$

Taking the real value in (2.2.29), we can state

$$\sum_{i=1}^{n} p_i \|x_i\|^2 - \left\| \sum_{i=1}^{n} p_i x_i \right\|^2 = \operatorname{Re} \left(X - \sum_{i=1}^{n} p_i x_i, \sum_{i=1}^{n} p_i x_i - x \right)$$

$$- \sum_{i=1}^{n} p_i \operatorname{Re}(X - x_i, x_i - x), \tag{2.2.30}$$

which is an identity of interest in itself.

Using the assumption (2.2.27), we can conclude by (2.2.30), that

$$\sum_{i=1}^{n} p_i \|x_i\|^2 - \left\| \sum_{i=1}^{n} p_i x_i \right\|^2 \leqslant \operatorname{Re} \left(X - \sum_{i=1}^{n} p_i x_i, \sum_{i=1}^{n} p_i x_i - x \right). \tag{2.2.31}$$

It is known that if $y,\ z \in H$, then

$$4 \operatorname{Re}(z, y) \leqslant \|z + y\|^2, \tag{2.2.32}$$

with equality if and only if $z = y$. Now, by (2.2.32), we can state that

$$\operatorname{Re} \left(X - \sum_{i=1}^{n} p_i x_i, \sum_{i=1}^{n} p_i x_i - x \right) \leqslant \frac{1}{4} \left\| X - \sum_{i=1}^{n} p_i x_i + \sum_{i=1}^{n} p_i x_i - x \right\|^2 = \frac{1}{4} \|X - x\|^2.$$

Using (2.2.31), we can easily deduce (2.2.28).

To prove the sharpness of the constant $\frac{1}{4}$, let us assume that the inequality (2.2.28) holds with a constant $c > 0$, i.e.,

$$0 \leqslant \sum_{i=1}^{n} p_i \|x_i\|^2 - \left\| \sum_{i=1}^{n} p_i x_i \right\|^2 \leqslant c\|X - x\|^2, \tag{2.2.33}$$

for all p_i, x_i and x, X as in the hypotheses. Assume that $n = 2$, $p_1 = p_2 = \frac{1}{2}$, $x_1 = x$ and $x_2 = X$ with x, $X \in H$ and $x \neq X$. Then, obviously,

$$(X - x_1, x_1 - x) = (X - x_2, x_2 - x) = 0,$$

which shows that the condition (2.2.27) holds. If we replace n, p_1, p_2, x_1, x_2 in (2.2.33) as above, we obtain

$$\sum_{i=1}^{2} p_i \|x_i\|^2 - \left\| \sum_{i=1}^{2} p_i x_i \right\|^2 = \frac{1}{2} \left(\|x\|^2 + \|X\|^2 - \left\| \frac{x+X}{2} \right\|^2 \right) = \frac{1}{4}\|X - x\|^2 \leqslant c\|X - x\|^2,$$

from where we deduce $c \geqslant \frac{1}{4}$ which proves the sharpness of the constant factor $\frac{1}{4}$.

The following Grüss-type inequality holds (see [43]).

Theorem 2.2.5. Let $(H, (\cdot, \cdot))$ be an inner product space over $\mathbb{K}$ ($\mathbb{K} = \mathbb{R}$, $\mathbb{C}$) and x_i, $y_i \in H$, $p_i \geqslant 0$, $(i = 1, \dots, n)$ $(n \geqslant 2)$ with $\sum_{i=1}^{n} p_i = 1$. If x, X, y, $Y \in H$ are such that

$$\operatorname{Re}(X - x_i, x_i - x) \geqslant 0, \quad \operatorname{Re}(Y - y_i, y_i - y) \geqslant 0,$$

for all $i \in \{1, \dots, n\}$, then we have the inequality

$$\left| \sum_{i=1}^{n} p_i (x_i, y_i) - \left(\sum_{i=1}^{n} p_i x_i, \sum_{i=1}^{n} p_i y_i \right) \right| \leqslant \frac{1}{4} \|X - x\| \|Y - y\|. \tag{2.2.34}$$

The constant $\frac{1}{4}$ is sharp.

Proof. A simple calculation shows that

$$\sum_{i=1}^{n} p_i (x_i, y_i) - \left(\sum_{i=1}^{n} p_i x_i, \sum_{i=1}^{n} p_i y_i \right) = \frac{1}{2} \sum_{i,j=1}^{n} p_i p_j (x_i - x_j, y_i - y_j). \tag{2.2.35}$$

Taking modulus in both parts of (2.2.35), and using the generalized triangle inequality, we obtain

$$\left| \sum_{i=1}^{n} p_i (x_i, y_i) - \left(\sum_{i=1}^{n} p_i x_i, \sum_{i=1}^{n} p_i y_i \right) \right| \leqslant \frac{1}{2} \sum_{i,j=1}^{n} p_i p_j |(x_i - x_j, y_i - y_j)|. \tag{2.2.36}$$

By using Schwarz's inequality in inner product spaces we have

$$|(x_i - x_j, y_i - y_j)| \leqslant \|x_i - x_j\| \|y_i - y_j\|, \tag{2.2.37}$$

for $i,\ j \in \{1,\ldots,n\}$, and therefore

$$\left|\sum_{i=1}^{n} p_i(x_i,y_i) - \left(\sum_{i=1}^{n} p_i x_i, \sum_{i=1}^{n} p_i y_i\right)\right| \leqslant \frac{1}{2}\sum_{i,j=1}^{n} p_i p_j \|x_i - x_j\| \|y_i - y_j\|. \tag{2.2.38}$$

Using the Cauchy-Schwarz inequality for double sums, we can state that

$$\frac{1}{2}\sum_{i,j=1}^{n} p_i p_j \|x_i - x_j\| \|y_i - y_j\|$$

$$\leqslant \left(\frac{1}{2}\sum_{i,j=1}^{n} p_i p_j \|x_i - x_j\|^2\right)^{\frac{1}{2}} \left(\frac{1}{2}\sum_{i,j=1}^{n} p_i p_j \|y_i - y_j\|^2\right)^{\frac{1}{2}}, \tag{2.2.39}$$

and a simple calculation shows that

$$\frac{1}{2}\sum_{i,j=1}^{n} p_i p_j \|x_i - x_j\|^2 = \sum_{i=1}^{n} p_i \|x_i\|^2 - \left\|\sum_{i=1}^{n} p_i x_i\right\|^2,$$

and

$$\frac{1}{2}\sum_{i,j=1}^{n} p_i p_j \|y_i - y_j\|^2 = \sum_{i=1}^{n} p_i \|y_i\|^2 - \left\|\sum_{i=1}^{n} p_i y_i\right\|^2.$$

We obtain

$$\left|\sum_{i=1}^{n} p_i(x_i,y_i) - \left(\sum_{i=1}^{n} p_i x_i, \sum_{i=1}^{n} p_i y_i\right)\right|$$

$$\leqslant \left(\sum_{i=1}^{n} p_i \|x_i\|^2 - \left\|\sum_{i=1}^{n} p_i x_i\right\|^2\right)^{\frac{1}{2}} \left(\sum_{i=1}^{n} p_i \|y_i\|^2 - \left\|\sum_{i=1}^{n} p_i y_i\right\|^2\right)^{\frac{1}{2}}. \tag{2.2.40}$$

Using Lemma 2.2.3, we know that

$$\left(\sum_{i=1}^{n} p_i \|x_i\|^2 - \left\|\sum_{i=1}^{n} p_i x_i\right\|^2\right)^{\frac{1}{2}} \leqslant \frac{1}{2}\|X - x\|, \tag{2.2.41}$$

and

$$\left(\sum_{i=1}^{n} p_i \|y_i\|^2 - \left\|\sum_{i=1}^{n} p_i y_i\right\|^2\right)^{\frac{1}{2}} \leqslant \frac{1}{2}\|Y - y\|. \tag{2.2.42}$$

Using (2.2.41) and (2.2.42) in (2.2.40), we get the desired inequality in (2.2.34).

To prove the sharpness of the constant $\frac{1}{4}$, let us assume that (2.2.34) holds with a constant $c > 0$, i.e.,

$$\left|\sum_{i=1}^{n} p_i(x_i,y_i) - \left(\sum_{i=1}^{n} p_i x_i, \sum_{i=1}^{n} p_i y_i\right)\right| \leqslant c\|X - x\| \|Y - y\|, \tag{2.2.43}$$

under the above assumptions p_i, x_i, y_i, x, X, y, Y and $n \geqslant 2$. If we choose $n = 2$, $x_1 = x$, $x_2 = X$, $y_1 = y$, $y_2 = Y$ $(x \neq X,\ y \neq Y)$ and $p_1 = p_2 = \frac{1}{2}$, then

$$\sum_{i=1}^{2} p_i(x_i, y_i) - \left(\sum_{i=1}^{2} p_i x_i, \sum_{i=1}^{2} p_i y_i\right) = \frac{1}{2} \sum_{i,j=1}^{2} p_i p_j (x_i - x_j, y_i - y_j)$$

$$= \sum_{1 \leqslant i < j \leqslant 2} p_i p_j (x_i - x_j, y_i - y_j) = \frac{1}{4}(x - X, y - Y),$$

and then

$$\left| \sum_{i=1}^{2} p_i(x_i, y_i) - \left(\sum_{i=1}^{2} p_i x_i, \sum_{i=1}^{2} p_i y_i\right) \right| = \frac{1}{4} |(x - X, y - Y)|.$$

Choose $X - x = z$, $Y - y = z$, $z \neq 0$. Then using (2.2.43), we derive

$$\frac{1}{4}\|z\|^2 \leqslant c\|z\|^2, z \neq 0,$$

which implies that $c \geqslant \frac{1}{4}$, and the Theorem is proved.

2.3 Grüss-and Čebyšev-type inequalities in two and three variables

This section deals with some Grüss and Čebyšev-type inequalities established by Pachpatte in [89,91,122,129], involving functions of two and three independent variables.

Let $\Delta = [a,b] \times [c,d]$, a, b, c, $d \in \mathbb{R}$. The partial derivatives of a function $h(x,y)$ defined on Δ are denoted by $D_1 h(x,y) = \frac{\partial}{\partial x} h(x,y)$, $D_2 h(x,y) = \frac{\partial}{\partial y} h(x,y)$, $D_2 D_1 h(x,y) = \frac{\partial^2}{\partial y \partial x} h(x,y)$. We denote by $C(\Delta)$ the class of continuous functions $h : \Delta \to \mathbb{R}$ for which $D_1 h(x,y)$, $D_2 h(x,y)$, $D_2 D_1 h(x,y)$ exist and are continuous on Δ and belong to $L_\infty(\Delta)$. For any function $h(x,y) \in L_\infty(\Delta)$, we define $\|h\|_\infty = \sup_{(x,y) \in \Delta} |h(x,y)| < \infty$. For convenience, we introduce the following notation to simplify the details of presentation:

$$k = (b-a)(d-c),$$

$$H_1(x) = \left[\frac{1}{4}(b-a)^2 + \left(x - \frac{a+b}{2}\right)^2\right],$$

$$H_2(y) = \left[\frac{1}{4}(d-c)^2 + \left(y - \frac{c+d}{2}\right)^2\right],$$

$$F(x,y) = \left[(d-c)\int_a^b f(t,y)dt + (b-a)\int_c^d f(x,s)ds\right],$$

$$G(x,y) = \left[(d-c)\int_a^b g(t,y)dt + (b-a)\int_c^d g(x,s)ds\right],$$

$$A_0(x,y) = g(x,y)\int_a^b\int_c^d f(t,s)dsdt + f(x,y)\int_a^b\int_c^d g(t,s)dsdt,$$

$$A_1(x,y) = g(x,y)\int_a^b\int_c^d p(x,t)D_1 f(t,s)dsdt + f(x,y)\int_a^b\int_c^d p(x,t)D_1 g(t,s)dsdt,$$

$$A_2(x,y) = g(x,y)\int_a^b\int_c^d q(y,s)D_2 f(t,s)dsdt + f(x,y)\int_a^b\int_c^d q(y,s)D_2 g(t,s)dsdt,$$

$$A_3(x,y) = g(x,y)\int_a^b\int_c^d p(x,t)q(y,s)D_2D_1 f(t,s)dsdt$$

$$+f(x,y)\int_a^b\int_c^d p(x,t)q(y,s)D_2D_1 g(t,s)dsdt,$$

$$M_1(x,y) = |g(x,y)|\|D_1 f\|_\infty + |f(x,y)|\|D_1 g\|_\infty,$$

$$M_2(x,y) = |g(x,y)|\|D_2 f\|_\infty + |f(x,y)|\|D_2 g\|_\infty,$$

$$M_3(x,y) = |g(x,y)|\|D_2D_1 f\|_\infty + |f(x,y)|\|D_2D_1 g\|_\infty,$$

$$A(x,y) = \|D_1 f\|_\infty(d-c)H_1(x) + \|D_2 f\|_\infty(b-a)H_2(y) + \|D_2D_1 f\|_\infty H_1(x)H_2(y),$$

$$B(x,y) = \|D_1 g\|_\infty(d-c)H_1(x) + \|D_2 g\|_\infty(b-a)H_2(y) + \|D_2D_1 g\|_\infty H_1(x)H_2(y),$$

for some suitable functions f, g defined on Δ, and $p:[a,b]^2 \to \mathbb{R}$, $q:[c,d]^2 \to \mathbb{R}$ are given by

$$p(x,t) = \begin{cases} t-a, & t \in [a,x] \\ t-b, & t \in (x,b] \end{cases}$$

$$q(y,s) = \begin{cases} s-c, & s \in [c,y] \\ s-d, & s \in (y,d] \end{cases}$$

and set

$$L[h(x,y)] = \int_a^b\int_c^d p(x,t)D_1 h(t,s)dsdt$$

$$+\int_a^b\int_c^d q(y,s)D_2 h(t,s)dsdt + \int_a^b\int_c^d p(x,t)q(y,s)D_2D_1 h(t,s)dsdt,$$

$$M[h(x,y)] = \int_a^b\int_c^d p(x,t)q(y,s)D_2D_1 h(t,s)dsdt,$$

for some suitable function h defined on Δ.

We begin with proving some auxiliary results.

Lemma 2.3.1 (see [37]). Let $h : \Delta \to \mathbb{R}$ be such that the partial derivatives $D_1h(x,y)$, $D_2h(x,y)$, $D_2D_1h(x,y)$ exist and are continuous on Δ. Then for all $(x,y) \in \Delta$, we have the representation

$$kh(x,y) - \int_a^b \int_c^d h(t,s)dsdt = L[h(x,y)]. \tag{2.3.1}$$

Proof. We use the following identity, which can be easily proved by integration by parts,

$$g(u) = \frac{1}{\beta-\alpha}\int_\alpha^\beta g(z)dz + \frac{1}{\beta-\alpha}\int_\alpha^\beta e(u,z)g'(z)dz, \tag{2.3.2}$$

where $e : [\alpha,\beta]^2 \to \mathbb{R}$ is given by

$$e(u,z) = \begin{cases} z-\alpha, & z \in [\alpha,u] \\ z-\beta, & z \in (u,\beta] \end{cases}$$

and g is absolutely continuous on $[\alpha,\beta]$. Now, write the identity (2.3.2) for the partial map $h(\cdot,y)$, $y \in [c,d]$, to obtain

$$h(x,y) = \frac{1}{b-a}\int_a^b h(t,y)dt + \frac{1}{b-a}\int_a^b p(x,t)D_1h(t,y)dt, \tag{2.3.3}$$

for all $(x,y) \in \Delta$. Also, if we write (2.3.2) for the map $h(t,\cdot)$, we get

$$h(t,y) = \frac{1}{d-c}\int_c^d h(t,s)ds + \frac{1}{d-c}\int_c^d q(y,s)D_2h(t,s)ds, \tag{2.3.4}$$

for all $(t,y) \in \Delta$. The same formula (2.3.2) applied for the partial derivative $D_1h(\cdot,y)$ will produce

$$D_1h(t,y) = \frac{1}{d-c}\int_c^d D_1h(t,s)ds + \frac{1}{d-c}\int_c^d q(y,s)D_2D_1h(t,s)ds, \tag{2.3.5}$$

for all $(t,y) \in \Delta$. Substituting (2.3.4) and (2.3.5) in (2.3.3), and using the Fubini's theorem, we have

$$h(x,y) = \frac{1}{b-a}\int_a^b \left[\frac{1}{d-c}\int_c^d h(t,s)ds + \frac{1}{d-c}\int_c^d q(y,s)D_2h(t,s)ds\right]dt$$

$$+\frac{1}{b-a}\int_a^b p(x,t)\left[\frac{1}{d-c}\int_c^d D_1h(t,s)ds + \frac{1}{d-c}\int_c^d q(y,s)D_2D_1h(t,s)ds\right]dt$$

$$= \frac{1}{(b-a)(d-c)}\left[\int_a^b\int_c^d h(t,s)dsdt + \int_a^b\int_c^d q(y,s)D_2h(t,s)dsdt\right.$$

$$\left.+\int_a^b\int_c^d p(x,t)D_1h(t,s)dsdt + \int_a^b\int_c^d p(x,t)q(y,s)D_2D_1h(t,s)dsdt\right]. \tag{2.3.6}$$

Rewriting (2.3.6), we get the required identity in (2.3.1).

Lemma 2.3.2 (see [8]). Let $h : \Delta \to \mathbb{R}$ be a continuous mapping on Δ and $D_2D_1h(x,y)$ exists on $(a,b)\times(c,d)$. Then, we have the identity

$$kh(x,y) - H(x,y) = M\left[h(x,y)\right], \tag{2.3.7}$$

where

$$H(x,y) = (d-c)\int_a^b h(t,y)dt + (b-a)\int_c^d h(x,s)ds - \int_a^b\int_c^d h(t,s)dsdt.$$

Proof. Integrating by parts twice, we can state:

$$\int_a^x\int_c^y (t-a)(s-c)D_2D_1h(t,s)dsdt = (x-a)(y-c)h(x,y)$$
$$-(y-c)\int_a^x h(t,y)dt - (x-a)\int_c^y h(x,s)ds + \int_a^x\int_c^y h(t,s)dsdt, \tag{2.3.8}$$
$$\int_a^x\int_y^d (t-a)(s-d)D_2D_1h(t,s)dsdt = (x-a)(d-y)h(x,y)$$
$$-(d-y)\int_a^x h(t,y)dt - (x-a)\int_y^d h(x,s)ds + \int_a^x\int_y^d h(t,s)dsdt, \tag{2.3.9}$$
$$\int_x^b\int_y^d (t-b)(s-d)D_2D_1h(t,s)dsdt = (b-x)(d-y)h(x,y)$$
$$-(d-y)\int_x^b h(t,y)dt - (b-x)\int_y^d h(x,s)ds + \int_x^b\int_y^d h(t,s)dsdt, \tag{2.3.10}$$
$$\int_x^b\int_c^y (t-b)(s-c)D_2D_1h(t,s)dsdt = (b-x)(y-c)h(x,y)$$
$$-(y-c)\int_x^b h(t,y)dt - (b-x)\int_c^y h(x,s)ds + \int_x^b\int_c^y h(t,s)dsdt. \tag{2.3.11}$$

Adding (2.3.8)–(2.3.11) and rewriting, we easily deduce (2.3.7).

In the following theorems, we present the inequalities investigated in [89,122].

Theorem 2.3.1. Let $f,\ g \in C(\Delta)$. Then

$$|E(f,g)| \leqslant \frac{1}{2k^2}\int_a^b\int_c^d \left[M_1(x,y)(d-c)H_1(x) + M_2(x,y)(b-a)H_2(y)\right.$$
$$\left. + M_3(x,y)H_1(x)H_2(y)\right]dydx, \tag{2.3.12}$$

and

$$|E(f,g)| \leqslant \frac{1}{k^3}\int_a^b\int_c^d A(x,y)B(x,y)dydx, \tag{2.3.13}$$

where

$$E(f,g) = \frac{1}{k}\int_a^b\int_c^d f(x,y)g(x,y)dydx$$
$$-\left(\frac{1}{k}\int_a^b\int_c^d f(x,y)dydx\right)\left(\frac{1}{k}\int_a^b\int_c^d g(x,y)dydx\right). \tag{2.3.14}$$

Proof. From the hypotheses, we have the following identities (see, Lemma 2.3.1):

$$kf(x,y) = \int_a^b \int_c^d f(t,s)dsdt + \int_a^b \int_c^d p(x,t)D_1 f(t,s)dsdt$$

$$+ \int_a^b \int_c^d q(y,s)D_2 f(t,s)dsdt + \int_a^b \int_c^d p(x,t)q(y,s)D_2 D_1 f(t,s)dsdt, \tag{2.3.15}$$

and

$$kg(x,y) = \int_a^b \int_c^d g(t,s)dsdt + \int_a^b \int_c^d p(x,t)D_1 g(t,s)dsdt$$

$$+ \int_a^b \int_c^d q(y,s)D_2 g(t,s)dsdt + \int_a^b \int_c^d p(x,t)q(y,s)D_2 D_1 g(t,s)dsdt, \tag{2.3.16}$$

for $(x,y) \in \Delta$. Multiplying (2.3.15) by $g(x,y)$, (2.3.16) by $f(x,y)$, and adding the resulting identities, we get

$$2kf(x,y)g(x,y) = A_0(x,y) + A_1(x,y) + A_2(x,y) + A_3(x,y). \tag{2.3.17}$$

Integrating (2.3.17) over Δ and rewriting, we get

$$E(f,g) = \frac{1}{2k^2} \int_a^b \int_c^d [A_1(x,y) + A_2(x,y) + A_3(x,y)]\, dydx. \tag{2.3.18}$$

It is easy to observe that

$$|A_1(x,y)| \leqslant M_1(x,y)(d-c)H_1(x), \tag{2.3.19}$$

$$|A_2(x,y)| \leqslant M_2(x,y)(b-a)H_2(y), \tag{2.3.20}$$

$$|A_3(x,y)| \leqslant M_3(x,y)H_1(x)H_2(y), \tag{2.3.21}$$

for $(x,y) \in \Delta$. From (2.3.18)–(2.3.21), we get

$$|E(f,g)| \leqslant \frac{1}{2k^2} \int_a^b \int_c^d [|A_1(x,y)| + |A_2(x,y)| + |A_3(x,y)|]\, dydx$$

$$\leqslant \frac{1}{2k^2} \int_a^b \int_c^d [M_1(x,y)(d-c)H_1(x) + M_2(x,y)(b-a)H_2(y) + M_3(x,y)H_1(x)H_2(y)]\, dydx.$$

This is the required inequality in (2.3.12).

The identities (2.3.15) and (2.3.16) can be rewritten as

$$kf(x,y) - \int_a^b \int_c^d f(t,s)dsdt = L[f(x,y)], \tag{2.3.22}$$

and

$$kg(x,y) - \int_a^b \int_c^d g(t,s)dsdt = L[g(x,y)], \tag{2.3.23}$$

for $(x,y) \in \Delta$. Multiplying the left hand sides and right hand sides of (2.3.22) and (2.3.23), we have

$$k^2 f(x,y)g(x,y) - kf(x,y)\int_a^b\int_c^d g(t,s)dsdt - kg(x,y)\int_a^b\int_c^d f(t,s)dsdt$$

$$+\left(\int_a^b\int_c^d f(t,s)dsdt\right)\left(\int_a^b\int_c^d g(t,s)dsdt\right) = L[f(x,y)]L[g(x,y)]. \tag{2.3.24}$$

Integrating (2.3.24) over Δ and rewriting, we get

$$E(f,g) = \frac{1}{k^3}\int_a^b\int_c^d L[f(x,y)]L[g(x,y)]dydx. \tag{2.3.25}$$

From (2.3.25) and using the properties of modulus, we get

$$|E(f,g)| \leqslant \frac{1}{k^3}\int_a^b\int_c^d |L[f(x,y)]||L[g(x,y)]|dydx. \tag{2.3.26}$$

It is easy to observe that

$$|L[f(x,y)]| \leqslant \int_a^b\int_c^d |p(x,t)||D_1 f(t,s)|dsdt$$

$$+\int_a^b\int_c^d |q(y,s)||D_2 f(t,s)|dsdt + \int_a^b\int_c^d |p(x,t)||q(y,s)||D_2D_1 f(t,s)|dsdt$$

$$\leqslant \|D_1 f\|_\infty \int_a^b\int_c^d |p(x,t)|dsdt$$

$$+\|D_2 f\|_\infty \int_a^b\int_c^d |q(y,s)|dsdt + \|D_2D_1 f\|_\infty \int_a^b\int_c^d |p(x,t)||q(y,s)|dsdt$$

$$= \|D_1 f\|_\infty (d-c)\int_a^b |p(x,t)|dt$$

$$+\|D_2 f\|_\infty (b-a)\int_c^d |q(y,s)|ds + \|D_2D_1 f\|_\infty \int_a^b\int_c^d |p(x,t)||q(y,s)|dsdt$$

$$= \|D_1 f\|_\infty (d-c)H_1(x) + \|D_2 f\|_\infty (b-a)H_2(y) + \|D_2D_1 f\|_\infty H_1(x)H_2(y)$$

$$= A(x,y). \tag{2.3.27}$$

Similarly, we have

$$|L[g(x,y)]| \leqslant B(x,y). \tag{2.3.28}$$

Using (2.3.27) and (2.3.28) in (2.3.26), we get the desired inequality in (2.3.13). The proof is complete.

Remark 2.3.1. From (2.3.17), (2.3.19)–(2.3.21), it is easy to obtain the inequality

$$|2kf(x,y)g(x,y)-A_0(x,y)| \leqslant M_1(x,y)(d-c)H_1(x)$$

$$+M_2(x,y)(b-a)H_2(y)+M_3(x,y)H_1(x)H_2(y), \tag{2.3.29}$$

for $(x,y) \in \Delta$ and from (2.3.24), (2.3.27), (2.3.28), it is easy to see that the following inequality

$$\left| f(x,y)g(x,y)-\frac{1}{k}\left[f(x,y)\int_a^b\int_c^d g(t,s)dsdt + g(x,y)\int_a^b\int_c^d f(t,s)dsdt \right.\right.$$

$$\left.\left. -\frac{1}{k}\left(\int_a^b\int_c^d f(t,s)dsdt\right)\left(\int_a^b\int_c^d g(t,s)dsdt\right)\right]\right|$$

$$\leqslant \frac{1}{k^2}A(x,y)B(x,y), \tag{2.3.30}$$

holds for $(x,y) \in \Delta$.

Theorem 2.3.2. Let $f,\ g \in C(\Delta)$. Then

$$\left| \frac{1}{k}\int_a^b\int_c^d f(x,y)g(x,y)dydx + \left(\frac{1}{k}\int_a^b\int_c^d f(x,y)dydx\right)\left(\frac{1}{k}\int_a^b\int_c^d g(x,y)dydx\right) \right.$$

$$\left. -\frac{1}{2k^2}\int_a^b\int_c^d [g(x,y)F(x,y)+f(x,y)G(x,y)]\,dydx \right|$$

$$\leqslant \frac{1}{2k^2}\int_a^b\int_c^d M_3(x,y)H_1(x)H_2(y)dydx, \tag{2.3.31}$$

and

$$\left| \frac{1}{k}\int_a^b\int_c^d f(x,y)g(x,y)dydx \right.$$

$$\left. -\frac{1}{k^2}\int_a^b\int_c^d \left[f(x,y)\overline{G}(x,y)+g(x,y)\overline{F}(x,y)-\frac{1}{k}\overline{F}(x,y)\overline{G}(x,y)\right] dydx \right|$$

$$\leqslant \frac{1}{k^3}\|D_2D_1f\|_\infty\|D_2D_1g\|_\infty\int_a^b\int_c^d [H_1(x)H_2(y)]^2\,dydx, \tag{2.3.32}$$

where

$$\overline{F}(x,y)=F(x,y)-\int_a^b\int_c^d f(t,s)dsdt, \overline{G}(x,y)=G(x,y)-\int_a^b\int_c^d g(t,s)dsdt.$$

Proof. From the hypotheses, we have the following identities (see, Lemma 2.3.2):

$$kf(x,y) = F(x,y) - \int_a^b \int_c^d f(t,s)dsdt + \int_a^b \int_c^d p(x,t)q(y,s)D_2D_1 f(t,s)dsdt, \quad (2.3.33)$$

and

$$kg(x,y) = G(x,y) - \int_a^b \int_c^d g(t,s)dsdt + \int_a^b \int_c^d p(x,t)q(y,s)D_2D_1 g(t,s)dsdt, \quad (2.3.34)$$

for $(x,y) \in \Delta$. Multiplying (2.3.33) by $g(x,y)$, (2.3.34) by $f(x,y)$, and adding the resulting identities, we get

$$2kf(x,y)g(x,y) = g(x,y)F(x,y) + f(x,y)G(x,y) - A_0(x,y) + A_3(x,y). \quad (2.3.35)$$

Integrating (2.3.35) over Δ and rewriting, we have

$$\int_a^b \int_c^d f(x,y)g(x,y)dydx = \frac{1}{2k}\int_a^b \int_c^d [g(x,y)F(x,y) + f(x,y)G(x,y)]\, dydx$$

$$-\frac{1}{k}\left(\int_a^b \int_c^d f(x,y)dydx\right)\left(\int_a^b \int_c^d g(x,y)dydx\right) + \frac{1}{2k}\int_a^b \int_c^d A_3(x,y)dydx. \quad (2.3.36)$$

We note that, here (2.3.21) holds for $(x,y) \in \Delta$. From (2.3.36) and (2.3.21), we observe that

$$\left|\frac{1}{k}\int_a^b \int_c^d f(x,y)g(x,y)dydx + \left(\frac{1}{k}\int_a^b \int_c^d f(x,y)dydx\right)\left(\frac{1}{k}\int_a^b \int_c^d g(x,y)dydx\right)\right.$$

$$\left. -\frac{1}{2k^2}\int_a^b \int_c^d [g(x,y)F(x,y) + f(x,y)G(x,y)]\, dydx\right|$$

$$\leqslant \frac{1}{2k^2}\int_a^b \int_c^d |A_3(x,y)|\, dydx$$

$$\leqslant \frac{1}{2k^2}\int_a^b \int_c^d M_3(x,y)H_1(x)H_2(y)dydx.$$

This is the required inequality in (2.3.31).

The identities (2.3.33) and (2.3.34) can be rewritten as

$$kf(x,y) - \overline{F}(x,y) = M[f(x,y)], \quad (2.3.37)$$

and

$$kg(x,y) - \overline{G}(x,y) = M[g(x,y)], \quad (2.3.38)$$

for $(x,y) \in \Delta$. Multiplying the left hand sides and right hand sides of (2.3.37) and (2.3.38), we have

$$k^2 f(x,y)g(x,y) - kf(x,y)\overline{G}(x,y) - kg(x,y)\overline{F}(x,y)$$

$$+\overline{F}(x,y)\overline{G}(x,y) = M[f(x,y)]M[g(x,y)]. \quad (2.3.39)$$

Rewriting (2.3.39) and integrating over Δ and using the properties of modulus, we have

$$\left|\frac{1}{k}\int_a^b\int_c^d f(x,y)g(x,y)dydx\right.$$

$$\left.-\frac{1}{k^2}\int_a^b\int_c^d\left[f(x,y)\overline{G}(x,y)+g(x,y)\overline{F}(x,y)-\frac{1}{k}\overline{F}(x,y)\overline{G}(x,y)\right]dydx\right|$$

$$\leqslant \frac{1}{k^3}\int_a^b\int_c^d |M[f(x,y)]||M[g(x,y)]|dydx. \tag{2.3.40}$$

It is easy to observe that

$$|M[f(x,y)]| \leqslant \|D_2D_1f\|_\infty \int_a^b\int_c^d |p(x,t)||q(y,s)|dsdt = \|D_2D_1f\|_\infty H_1(x)H_2(y). \tag{2.3.41}$$

Similarly, we get

$$|M[g(x,y)]| \leqslant \|D_2D_1g\|_\infty H_1(x)H_2(y). \tag{2.3.42}$$

Using (2.3.41) and (2.3.42) in (2.3.40), we get the desired inequality in (2.3.32). The proof is complete.

Remark 2.3.2. From (2.3.35) and (2.3.21), it is easy to observe that the following inequality holds,

$$|2kf(x,y)g(x,y)+A_0(x,y)-[g(x,y)F(x,y)+f(x,y)G(x,y)]|$$

$$\leqslant M_3(x,y)H_1(x)H_2(y), \tag{2.3.43}$$

for $(x,y)\in\Delta$ and from (2.3.39), (2.3.41), (2.3.42), one can very easily obtain the following inequality

$$\left|f(x,y)g(x,y)-\frac{1}{k}\left[f(x,y)\overline{G}(x,y)+g(x,y)\overline{F}(x,y)-\frac{1}{k}\overline{F}(x,y)\overline{G}(x,y)\right]\right|$$

$$\leqslant \frac{1}{k^2}\|D_2D_1f\|_\infty\|D_2D_1g\|_\infty\left[H_1(x)H_2(y)\right]^2, \tag{2.3.44}$$

for $(x,y)\in\Delta$.

In our further discussion, the following notation will also be used to simplify the details of presentation.

Let $\Omega=[a,k]\times[b,m]\times[c,n], a,b,c,k,m,n\in\mathbb{R}$. The partial derivative $\frac{\partial^3}{\partial z\partial y\partial x}e(x,y,z)$ of a function e defined on Ω is denoted by $D_3D_2D_1e(x,y,z)$ and the function e is said to be bounded if $\|e\|_\infty=\sup_{(x,y,z)\in\Omega}|e(x,y,z)|<\infty$. For some suitable functions $h:\Delta\to\mathbb{R}$, $e:\Omega\to\mathbb{R}$, we set

$$A(D_2D_1h(x,y)) = A[a,c;x,y;b,d;D_2D_1h(s,t)]$$

$$= \int_a^x \int_c^y D_2D_1h(t,s)dsdt - \int_a^x \int_y^d D_2D_1h(t,s)dsdt$$

$$- \int_x^b \int_c^y D_2D_1h(t,s)dsdt + \int_x^b \int_y^d D_2D_1h(t,s)dsdt,$$

$$E(h(x,y)) = E[a,c;x,y;b,d;h]$$

$$= \frac{1}{2}[h(x,c)+h(x,d)+h(a,y)+h(b,y)] - \frac{1}{4}[h(a,c)+h(a,d)+h(b,c)+h(b,d)],$$

$$B(D_3D_2D_1e(r,s,t)) = B[a,b,c;r,s,t;k,m,n;D_3D_2D_1e(u,v,w)]$$

$$= \int_a^r \int_b^s \int_c^t D_3D_2D_1e(u,v,w)dwdvdu - \int_a^r \int_b^s \int_t^n D_3D_2D_1e(u,v,w)dwdvdu$$

$$- \int_a^r \int_s^m \int_c^t D_3D_2D_1e(u,v,w)dwdvdu - \int_r^k \int_b^s \int_c^t D_3D_2D_1e(u,v,w)dwdvdu$$

$$+ \int_a^r \int_s^m \int_t^n D_3D_2D_1e(u,v,w)dwdvdu + \int_r^k \int_s^m \int_c^t D_3D_2D_1e(u,v,w)dwdvdu$$

$$+ \int_r^k \int_b^s \int_t^n D_3D_2D_1e(u,v,w)dwdvdu - \int_r^k \int_s^m \int_t^n D_3D_2D_1e(u,v,w)dwdvdu,$$

$$L(e(r,s,t)) = L[a,b,c;r,s,t;k,m,n;e]$$

$$= \frac{1}{8}[e(a,b,c)+e(k,m,n)]$$

$$-\frac{1}{4}[e(r,b,c)+e(r,m,n)+e(r,m,c)+e(r,b,n)]$$

$$-\frac{1}{4}[e(a,s,c)+e(k,s,n)+e(a,s,n)+e(k,s,c)]$$

$$-\frac{1}{4}[e(a,b,t)+e(k,m,t)+e(k,b,t)+e(a,m,t)]$$

$$+\frac{1}{2}[e(a,s,t)+e(k,s,t)] + \frac{1}{2}[e(r,b,t)+e(r,m,t)]$$

$$+\frac{1}{2}[e(r,s,c)+e(r,s,n)].$$

The Grüss- and Čebyšev-type inequalities established in [91,129] are given in the following theorems.

Theorem 2.3.3. Let $f,\ g:\Delta\to\mathbb{R}$ be continuous functions on Δ and $D_2D_1f(x,y)$, $D_2D_1g(x,y)$ exist, continuous and bounded on Δ. Then

$$\left|\int_a^b\int_c^d\left[f(x,y)g(x,y)-\frac{1}{2}\left[E(f(x,y))g(x,y)+E(g(x,y))f(x,y)\right]\right]dydx\right|$$

$$\leqslant\frac{1}{8}(b-a)(d-c)\int_a^b\int_c^d\left[|g(x,y)|\|D_2D_1f\|_\infty+|f(x,y)|\|D_2D_1g\|_\infty\right]dydx,\qquad(2.3.45)$$

and

$$\left|\int_a^b\int_c^d\left[f(x,y)g(x,y)-\left[E(f(x,y))g(x,y)+E(g(x,y))f(x,y)\right.\right.\right.$$

$$\left.\left.\left.-E(f(x,y))E(g(x,y))\right]\right]dydx\right|$$

$$\leqslant\frac{1}{16}\{(b-a)(d-c)\}^2\|D_2D_1f\|_\infty\|D_2D_1g\|_\infty.\qquad(2.3.46)$$

Proof. From the hypotheses, it is easy to observe that the following identities hold for $(x,y)\in\Delta$ (see [86,91]).

$$f(x,y)=-f(a,c)+f(x,c)+f(a,y)+\int_a^x\int_c^y D_2D_1f(t,s)dsdt,$$

$$f(x,y)=-f(a,d)+f(x,d)+f(a,y)-\int_a^x\int_y^d D_2D_1f(t,s)dsdt,$$

$$f(x,y)=-f(b,c)+f(x,c)+f(b,y)-\int_x^b\int_c^y D_2D_1f(t,s)dsdt,$$

$$f(x,y)=-f(b,d)+f(x,d)+f(b,y)+\int_x^b\int_y^d D_2D_1f(t,s)dsdt.$$

Adding the above identities and rewriting, we have

$$f(x,y)-E(f(x,y))=\frac{1}{4}A(D_2D_1f(x,y)),\qquad(2.3.47)$$

for $(x,y)\in\Delta$. Similarly, we have

$$g(x,y)-E(g(x,y))=\frac{1}{4}A\left(D_2D_1g(x,y)\right),\qquad(2.3.48)$$

for $(x,y)\in\Delta$. Multiplying (2.3.47) by $g(x,y)$ and (2.3.48) by $f(x,y)$ and adding the resulting identities, rewriting and then integrating over Δ, we have

$$\int_a^b\int_c^d\left[f(x,y)g(x,y)-\frac{1}{2}\left[E(f(x,y))g(x,y)+E(g(x,y))f(x,y)\right]\right]dydx$$

$$=\frac{1}{8}\int_a^b\int_c^d\left[A(D_2D_1f(x,y))g(x,y)+A\left(D_2D_1g(x,y)\right)f(x,y)\right]dydx.\qquad(2.3.49)$$

From the properties of modulus and integrals, it is easy to see that

$$|A(D_2D_1f(x,y))| \leqslant \int_a^b \int_c^d |D_2D_1f(t,s)|dsdt \leqslant \|D_2D_1f\|_\infty(b-a)(d-c), \tag{2.3.50}$$

$$|A(D_2D_1g(x,y))| \leqslant \int_a^b \int_c^d |D_2D_1g(t,s)|dsdt \leqslant \|D_2D_1g\|_\infty(b-a)(d-c). \tag{2.3.51}$$

From (2.3.49)–(2.3.51), we observe that

$$\left|\int_a^b \int_c^d \left[f(x,y)g(x,y) - \frac{1}{2}\left[E(f(x,y))g(x,y) + E(g(x,y))f(x,y)\right]\right] dydx\right|$$
$$\leqslant \frac{1}{8}\int_a^b \int_c^d \left[|g(x,y)|\,|A(D_2D_1f(x,y))| + |f(x,y)|\,|A(D_2D_1g(x,y))|\right] dydx$$
$$\leqslant \frac{1}{8}\int_a^b \int_c^d \left[|g(x,y)|\int_a^b \int_c^d |D_2D_1f(t,s)|dsdt + |f(x,y)|\int_a^b \int_c^d |D_2D_1g(t,s)|dsdt\right] dydx$$
$$\leqslant \frac{1}{8}(b-a)(d-c)\int_a^b \int_c^d \left[|g(x,y)|\,\|D_2D_1f\|_\infty + |f(x,y)|\,\|D_2D_1g\|_\infty\right] dydx,$$

which is the required inequality in (2.3.45).

Multiplying the left hand sides and right hand sides of (2.3.47) and (2.3.48), we get

$$f(x,y)g(x,y) - [f(x,y)E(g(x,y)) + g(x,y)E(f(x,y)) - E(f(x,y))E(g(x,y))]$$
$$= \frac{1}{16}A(D_2D_1f(x,y))A(D_2D_1g(x,y)). \tag{2.3.52}$$

Integrating (2.3.52) over Δ and using the properties of modulus, we have

$$\left|\int_a^b \int_c^d [f(x,y)g(x,y) - [f(x,y)E(g(x,y)) + g(x,y)E(f(x,y)) - E(f(x,y))E(g(x,y))]]dydx\right|$$
$$\leqslant \frac{1}{16}\int_a^b \int_c^d |A(D_2D_1f(x,y))||A(D_2D_1g(x,y))|dydx. \tag{2.3.53}$$

Now, using (2.3.50) and (2.3.51) in (2.3.53), we get (2.3.46). The proof is complete.

Theorem 2.3.4. Let $f,\ g : \Omega \to \mathbb{R}$ be continuous functions on Ω and $D_3D_2D_1f(r,s,t)$, $D_3D_2D_1g(r,s,t)$, exist, continuous and bounded on Ω. Then

$$\left|\int_a^k \int_b^m \int_c^n \left[f(r,s,t)g(r,s,t) - \frac{1}{2}[L(f(r,s,t))g(r,s,t) + L(g(r,s,t))f(r,s,t)]\right] dtdsdr\right|$$
$$\leqslant \frac{1}{16}(k-a)(m-b)(n-c)\int_a^k \int_b^m \int_c^n [|g(r,s,t)|\,\|D_3D_2D_1f\|_\infty$$
$$+|f(r,s,t)|\,\|D_3D_2D_1g\|_\infty]\, dtdsdr, \tag{2.3.54}$$

and

$$\left|\int_a^k \int_b^m \int_c^n [f(r,s,t)g(r,s,t) - [L(f(r,s,t))g(r,s,t)\right.$$
$$\left. + L(g(r,s,t))f(r,s,t) - L(f(r,s,t))L(g(r,s,t))]]\, dtdsdr\right|$$
$$\leqslant \frac{1}{64}\{(k-a)(m-b)(n-c)\}^2 \|D_3D_2D_1f\|_\infty \|D_3D_2D_1g\|_\infty. \tag{2.3.55}$$

Proof. From the hypotheses, it is easy to observe that the following identities hold (see [83,90]):

$$f(r,s,t) = f(a,b,c) + f(a,s,t) + f(r,s,c) + f(r,b,t)$$

$$-f(a,b,t) - f(a,s,c) - f(r,b,c) + \int_a^r \int_b^s \int_c^t D_3D_2D_1 f(u,v,w)\,dw\,dv\,du,$$

$$f(r,s,t) = f(r,s,n) + f(a,s,t) + f(r,b,t) + f(a,b,n)$$

$$-f(a,b,t) - f(a,s,n) - f(r,b,n) - \int_a^r \int_b^s \int_t^n D_3D_2D_1 f(u,v,w)\,dw\,dv\,du,$$

$$f(r,s,t) = f(r,m,t) + f(r,s,c) + f(a,m,c) + f(a,s,t)$$

$$-f(r,m,c) - f(a,m,t) - f(a,s,c) - \int_a^r \int_s^m \int_c^t D_3D_2D_1 f(u,v,w)\,dw\,dv\,du,$$

$$f(r,s,t) = f(k,s,t) + f(k,b,c) + f(r,s,c) + f(r,b,t)$$

$$-f(k,s,c) - f(k,b,t) - f(r,b,c) - \int_r^k \int_b^s \int_c^t D_3D_2D_1 f(u,v,w)\,dw\,dv\,du,$$

$$f(r,s,t) = f(r,m,t) + f(r,s,n) + f(a,m,n) + f(a,s,t)$$

$$-f(r,m,n) - f(a,m,t) - f(a,s,n) + \int_a^r \int_s^m \int_t^n D_3D_2D_1 f(u,v,w)\,dw\,dv\,du,$$

$$f(r,s,t) = f(r,m,t) + f(r,s,c) + f(k,s,t) + f(k,m,c)$$

$$-f(k,m,t) - f(k,s,c) - f(r,m,c) + \int_r^k \int_s^m \int_c^t D_3D_2D_1 f(u,v,w)\,dw\,dv\,du,$$

$$f(r,s,t) = f(k,s,t) + f(k,b,n) + f(r,s,n) + f(r,b,t)$$

$$-f(k,s,n) - f(k,b,t) - f(r,b,n) + \int_r^k \int_b^s \int_t^n D_3D_2D_1 f(u,v,w)\,dw\,dv\,du,$$

$$f(r,s,t) = f(k,m,n) + f(k,s,t) + f(r,m,t) + f(r,s,n)$$

$$-f(k,m,t) - f(k,s,n) - f(r,m,n) - \int_r^k \int_s^m \int_t^n D_3D_2D_1 f(u,v,w)\,dw\,dv\,du.$$

Adding the above identities and rewriting, we have

$$f(r,s,t) - L(f(r,s,t)) = \frac{1}{8} B(D_3D_2D_1 f(r,s,t)), \tag{2.3.56}$$

for $(r,s,t) \in \Omega$. Similarly, we have

$$g(r,s,t) - L(g(r,s,t)) = \frac{1}{8} B(D_3D_2D_1 g(r,s,t)), \tag{2.3.57}$$

for $(r,s,t) \in \Omega$. Multiplying (2.3.56) by $g(r,s,t)$ and (2.3.57) by $f(r,s,t)$ and adding the resulting identities, then integrating over Ω and rewriting, we have

$$\int_a^k \int_b^m \int_c^n \left[f(r,s,t)g(r,s,t) - \frac{1}{2}\left[f(r,s,t)L(g(r,s,t)) + g(r,s,t)L(f(r,s,t))\right] \right] dtdsdr$$

$$= \frac{1}{16} \int_a^k \int_b^m \int_c^n \left[g(r,s,t)B(D_3D_2D_1f(r,s,t)) + f(r,s,t)B(D_3D_2D_1g(r,s,t))\right] dtdsdr. \tag{2.3.58}$$

From the properties of modulus and integrals, we observe that

$$|B(D_3D_2D_1f(r,s,t))| \leqslant \int_a^k \int_b^m \int_c^n |D_3D_2D_1f(r,s,t)| dwdvdu$$

$$\leqslant \|D_3D_2D_1f\|_\infty (k-a)(m-b)(n-c). \tag{2.3.59}$$

Similarly, we get

$$|B(D_3D_2D_1g(r,s,t))| \leqslant \|D_3D_2D_1g\|_\infty (k-a)(m-b)(n-c). \tag{2.3.60}$$

Now, from (2.3.58)-(2.3.60) and following the same arguments as in the proof of inequality (2.3.45) with suitable changes, we get the required inequality in (2.5.54).
Multiplying the left hand sides and right hand sides of (2.3.56) and (2.3.57) and integrating over Ω, we get

$$\int_a^k \int_b^m \int_c^n \left[f(r,s,t)g(r,s,t) - \left[f(r,s,t)L(g(r,s,t))\right.\right.$$

$$\left.\left. + g(r,s,t)L(f(r,s,t)) - L(f(r,s,t))L(g(r,s,t))\right]\right] dtdsdr$$

$$= \frac{1}{64} \int_a^k \int_b^m \int_c^n B(D_3D_2D_1f(r,s,t))B(D_3D_2D_1g(r,s,t)) dtdsdr. \tag{2.3.61}$$

From (2.3.61), using the properties of modulus and (2.3.59) and (2.3.60), we get the desired inequality in (2.3.55). The proof is complete.

2.4 Trapezoid-type inequalities in two variables

In this section we present some Trapezoid-type inequalities involving functions of two independent variables, recently established by Dragomir, Barnett and Pearce [39], Barnett and Dragomir [6] and Pachpatte [86]. In our subsequent discussion, we make use of some of the notations and definitions given in Section 2.3 without further mention.
We start with the Trapezoid-type inequality established in [86].

Theorem 2.4.1. Let $f : \Delta \to \mathbb{R}$ be a continuous function on Δ, $D_2D_1f(x,y)$ exists and continuous on Δ. Then

$$\left|\int_a^b\int_c^d f(t,s)dsdt - \frac{1}{2}\left[(d-c)\int_a^b[f(t,c)+f(t,d)]dt + (b-a)\int_c^d[f(a,s)+f(b,s)]ds\right]\right.$$

$$\left.+\frac{1}{4}(b-a)(d-c)[f(a,c)+f(a,d)+f(b,c)+f(b,d)]\right|$$

$$\leqslant \frac{1}{4}(b-a)(d-c)\int_a^b\int_c^d |D_2D_1f(t,s)|dsdt. \tag{2.4.1}$$

Proof. By following the proof of Theorem 2.3.3, we have the following identity

$$f(x,y) - \frac{1}{2}[f(x,c)+f(x,d)+f(a,y)+f(b,y)] + \frac{1}{4}[f(a,c)+f(a,d)+f(b,c)+f(b,d)]$$

$$= \frac{1}{4}A(D_2D_1f(x,y)), \tag{2.4.2}$$

for $(x,y) \in \Delta$. Integrating both sides of (2.4.2) over Δ, we get

$$\int_a^b\int_c^d f(t,s)dsdt - \frac{1}{2}\left[(d-c)\int_a^b[f(t,c)+f(t,d)]dt + (b-a)\int_c^d[f(a,s)+f(b,s)]ds\right]$$

$$+\frac{1}{4}(b-a)(d-c)[f(a,c)+f(a,d)+f(b,c)+f(b,d)]$$

$$= \frac{1}{4}\int_a^b\int_c^d A(D_2D_1f(t,s))dsdt. \tag{2.4.3}$$

Using the properties of modulus and integrals, we observe that

$$|A(D_2D_1f(x,y))| \leqslant \int_a^b\int_c^d |D_2D_1f(t,s)|dsdt. \tag{2.4.4}$$

From (2.4.3) and (2.4.4), we have

$$\left|\int_a^b\int_c^d f(t,s)dsdt - \frac{1}{2}\left[(d-c)\int_a^b[f(t,c)+f(t,d)]dt + (b-a)\int_c^d[f(a,s)+f(b,s)]ds\right]\right.$$

$$\left.+\frac{1}{4}(b-a)(d-c)[f(a,c)+f(a,d)+f(b,c)+f(b,d)]\right|$$

$$\leqslant \frac{1}{4}\int_a^b\int_c^d |A(D_2D_1f(t,s))|\,dsdt$$

$$\leqslant \frac{1}{4}(b-a)(d-c)\int_a^b\int_c^d |D_2D_1f(t,s)|dsdt,$$

which is the required inequality in (2.4.1) and the proof is complete.

Remark 2.4.1. From (2.4.2) it is easy to observe that the following inequality holds

$$\left| f(x,y) - \frac{1}{2}[f(x,c)+f(x,d)+f(a,y)+f(b,y)] + \frac{1}{4}[f(a,c)+f(a,d)+f(b,c)+f(b,d)] \right|$$

$$\leqslant \frac{1}{4}\int_a^b \int_c^d |D_2D_1 f(t,s)|dsdt, \tag{2.4.5}$$

for $(x,y) \in \Delta$.

The next Theorem deals with the Trapezoid-type inequality investigated in [39].

Theorem 2.4.2. Let $f : \Delta \to \mathbb{R}$ be a continuous mapping on Δ, $D_2D_1 f(x,y)$ exists on $(a,b)\times(c,d)$ and is bounded, then

$$\left| \int_a^b \int_c^d f(t,s)dsdt - \frac{1}{2}\left[(d-c)\int_a^b [f(t,c)+f(t,d)]dt \right.\right.$$

$$\left. + (b-a)\int_c^d [f(a,s)+f(b,s)]ds \right]$$

$$\left. + \frac{1}{4}(b-a)(d-c)[f(a,c)+f(a,d)+f(b,c)+f(b,d)] \right|$$

$$\leqslant \frac{1}{16}\{(b-a)(d-c)\}^2 \|D_2D_1 f\|_\infty. \tag{2.4.6}$$

Proof. From the hypotheses, the following identity holds (see, Lemma 2.3.2):

$$\int_a^b \int_c^d p(x,t)q(y,s)D_2D_1 f(t,s)dsdt$$

$$= (d-c)(b-a)f(x,y) - (d-c)\int_a^b f(t,y)dt$$

$$-(b-a)\int_c^d f(x,s)ds + \int_a^b \int_c^d f(t,s)dsdt, \tag{2.4.7}$$

for all $(x,y) \in \Delta$, where $p(x,t)$, $q(y,s)$ are as given in Section 2.3. In (2.4.7) choose (i) $x = a,\ y = c$; (ii) $x = b,\ y = c$; (iii) $x = a,\ y = d$; and (iv) $x = b,\ y = d$ to obtain the following identities:

$$\int_a^b \int_c^d p(a,t)q(c,s)D_2D_1 f(t,s)dsdt$$

$$= (d-c)(b-a)f(a,c) - (d-c)\int_a^b f(t,c)dt$$

$$-(b-a)\int_c^d f(a,s)ds + \int_a^b \int_c^d f(t,s)dsdt, \tag{2.4.8}$$

$$\int_a^b \int_c^d p(b,t)q(c,s)D_2D_1 f(t,s)dsdt$$

$$= (d-c)(b-a)f(b,c) - (d-c)\int_a^b f(t,c)dt$$

$$-(b-a)\int_c^d f(b,s)ds + \int_a^b \int_c^d f(t,s)dsdt, \tag{2.4.9}$$

$$\int_a^b \int_c^d p(a,t)q(d,s)D_2D_1 f(t,s)dsdt$$

$$= (d-c)(b-a)f(a,d) - (d-c)\int_a^b f(t,d)dt$$

$$-(b-a)\int_c^d f(a,s)ds + \int_a^b \int_c^d f(t,s)dsdt, \tag{2.4.10}$$

and

$$\int_a^b \int_c^d p(b,t)q(d,s)D_2D_1 f(t,s)dsdt$$

$$= (d-c)(b-a)f(b,d) - (d-c)\int_a^b f(t,d)dt$$

$$-(b-a)\int_c^d f(b,s)ds + \int_a^b \int_c^d f(t,s)dsdt, \tag{2.4.11}$$

Adding (2.4.8)–(2.4.11) and dividing by 4, we have

$$\frac{1}{4}\int_a^b \int_c^d (p(a,t)+p(b,t))\,(q(c,s)+q(d,s))\,D_2D_1 f(t,s)dsdt$$

$$= \int_a^b \int_c^d f(t,s)dsdt + \frac{1}{4}(b-a)(d-c)[f(a,c)+f(a,d)+f(b,c)+f(b,d)]$$

$$-\frac{1}{2}\left[(d-c)\int_a^b [f(t,c)+f(t,d)]dt + (b-a)\int_c^d [f(a,s)+f(b,s)]ds\right],$$

and as

$$p(a,t)+p(b,t) = 2t-(a+b), q(c,s)+q(d,s) = 2s-(c+d),$$

then we get the identity:

$$\int_a^b \int_c^d f(t,s)dsdt - \frac{1}{2}\left[(d-c)\int_a^b [f(t,c)+f(t,d)]dt + (b-a)\int_c^d [f(a,s)+f(b,s)]ds\right]$$

$$+\frac{1}{4}(b-a)(d-c)[f(a,c)+f(a,d)+f(b,c)+f(b,d)]$$

$$= \int_a^b \int_c^d \left(t - \frac{a+b}{2}\right)\left(s - \frac{c+d}{2}\right) D_2 D_1 f(t,s) ds dt. \tag{2.4.12}$$

Now, using the identity (2.4.12) and the properties of the integral, we get

$$\left| \int_a^b \int_c^d f(t,s) ds dt - \frac{1}{2}\left[(d-c)\int_a^b [f(t,c)+f(t,d)]dt + (b-a)\int_c^d [f(a,s)+f(b,s)]ds\right]\right.$$
$$\left. + \frac{1}{4}(b-a)(d-c)[f(a,c)+f(a,d)+f(b,c)+f(b,d)]\right|$$
$$\leqslant \int_a^b \int_c^d \left|t - \frac{a+b}{2}\right| \left|s - \frac{c+d}{2}\right| |D_2 D_1 f(t,s)| ds dt$$
$$\leqslant \frac{1}{16}\{(b-a)(d-c)\}^2 \|D_2 D_1 f\|_\infty.$$

Since a simple calculation gives,

$$\int_a^b \left|t - \frac{a+b}{2}\right| dt = \frac{(b-a)^2}{4}, \int_c^d \left|s - \frac{c+d}{2}\right| dt = \frac{(d-c)^2}{4},$$

the inequality (2.4.6) is thus obtained. The prof is complete.

In order to prove the next two Theorems we need the following integral identity proved in [6].

Lemma 2.4.1. Let $f : \Delta \to \mathbb{R}$ be a continuous mapping on Δ such that $D_2 f(a,\cdot), D_2 f(b,\cdot)$ are continuous on $[c,d]$, $D_1 f(\cdot,c), D_1 f(\cdot,d)$ are continuous on $[a,b]$ and $D_2 D_1 f(\cdot,\cdot)$ is continuous on Δ. Then we have the identity:

$$\int_a^b \int_c^d f(x,y) dy dx + (b-a)\int_c^d \left(y - \frac{c+d}{2}\right) f_2(y) dy + (d-c)\int_a^b \left(x - \frac{a+b}{2}\right) f_1(x) dx$$
$$= \frac{1}{4}(b-a)(d-c)[f(a,c)+f(a,d)+f(b,c)+f(b,d)]$$
$$+ \int_a^b \int_c^d \left(x - \frac{a+b}{2}\right)\left(y - \frac{c+d}{2}\right) D_2 D_1 f(x,y) dy dx, \tag{2.4.13}$$

where

$$f_1(x) = \frac{1}{2}[D_1 f(x,c) + D_1 f(x,d)], \tag{2.4.14}$$

for $x \in [a,b]$ and

$$f_2(y) = \frac{1}{2}[D_2 f(a,y) + D_2 f(b,y)], \tag{2.4.15}$$

for $y \in [c,d]$.

Proof. A simple integration by parts gives

$$\int_\alpha^\beta h(x)dx = \frac{h(\alpha)+h(\beta)}{2}(\beta-\alpha) - \int_\alpha^\beta \left(x-\frac{\alpha+\beta}{2}\right)h'(x)dx, \tag{2.4.16}$$

provided that $h:[\alpha,\beta]\to\mathbb{R}$ is absolutely continuous on $[\alpha,\beta]$. Using (2.4.15), we can write:

$$\int_a^b f(x,y)dx = (b-a)\frac{f(a,y)+f(b,y)}{2} - \int_a^b \left(x-\frac{a+b}{2}\right)D_1f(x,y)dx, \tag{2.4.17}$$

for all $y\in[c,d]$. Integrating (2.4.17) over the interval $[c,d]$, we obtain

$$\int_c^d\left(\int_a^b f(x,y)dx\right)dy = \frac{1}{2}(b-a)\left[\int_c^d f(a,y)dy + \int_c^d f(b,y)dy\right]$$
$$-\int_c^d\left(\int_a^b\left(x-\frac{a+b}{2}\right)D_1f(x,y)dx\right)dy.$$

Using x's theorem, we can state:

$$\int_a^b\int_c^d f(x,y)dydx = \frac{1}{2}(b-a)\left[\int_c^d f(a,y)dy + \int_c^d f(b,y)dy\right]$$
$$-\int_a^b\left(x-\frac{a+b}{2}\right)\left(\int_c^d D_1f(x,y)dy\right)dx. \tag{2.4.18}$$

By the identity (2.4.16), we can also state:

$$\int_c^d f(a,y)dy = \frac{1}{2}[f(a,c)+f(a,d)](d-c) - \int_c^d\left(y-\frac{c+d}{2}\right)D_2f(a,y)dy, \tag{2.4.19}$$

$$\int_c^d f(b,y)dy = \frac{1}{2}[f(b,c)+f(b,d)](d-c) - \int_c^d\left(y-\frac{c+d}{2}\right)D_2f(b,y)dy, \tag{2.4.20}$$

and

$$\int_c^d D_1f(x,y)dy = \frac{1}{2}[D_1f(x,c)+D_1f(x,d)](d-c)$$
$$-\int_c^d\left(y-\frac{c+d}{2}\right)D_2D_1f(x,y)dy. \tag{2.4.21}$$

Now, using (2.4.18) and (2.4.19)–(2.4.21), we have successively

$$\int_a^b\int_c^d f(x,y)dydx = \frac{1}{2}(b-a)\left[\frac{1}{2}[f(a,c)+f(a,d)](d-c) - \int_c^d\left(y-\frac{c+d}{2}\right)D_2f(a,y)dy\right.$$
$$\left.+\frac{1}{2}[f(b,c)+f(b,d)](d-c) - \int_c^d\left(y-\frac{c+d}{2}\right)D_2f(b,y)dy\right]$$
$$-\int_a^b\left(x-\frac{a+b}{2}\right)\left[\frac{1}{2}[D_1f(x,c)+D_1f(x,d)](d-c)\right.$$

$$-\int_c^d \left(y-\frac{c+d}{2}\right) D_2D_1f(x,y)dy\Bigg] dx$$

$$= \frac{1}{4}[f(a,c)+f(a,d)+f(b,c)+f(b,d)](b-a)(d-c)$$

$$-\frac{1}{2}(b-a)\int_c^d \left(y-\frac{c+d}{2}\right) D_2f(a,y)dy - \frac{1}{2}(b-a)\int_c^d \left(y-\frac{c+d}{2}\right) D_2f(b,y)dy$$

$$-\frac{1}{2}(d-c)\int_a^b \left(x-\frac{a+b}{2}\right) D_1f(x,c)dx - \frac{1}{2}(d-c)\int_a^b \left(x-\frac{a+b}{2}\right) D_1f(x,d)dx$$

$$+\int_a^b\int_c^d \left(x-\frac{a+b}{2}\right)\left(y-\frac{c+d}{2}\right) D_2D_1f(x,y)dydx. \tag{2.4.22}$$

Rewriting (2.4.22), we get the desired identity in (2.4.13).

The Trapezoid-type inequalities given in [6] are embodied in the following theorems.

Theorem 2.4.3. Let f, f_1, f_2 be as in Lemma 2.4.1 and assume that $D_2D_1f(x,y)$ is bounded. Then

$$\Bigg|\int_a^b\int_c^d f(x,y)dydx + (b-a)\int_c^d \left(y-\frac{c+d}{2}\right) f_2(y)dy + (d-c)\int_a^b \left(x-\frac{a+b}{2}\right) f_1(x)dx$$

$$-\frac{1}{4}\left[f(a,c)+f(a,d)+f(b,c)+f(b,d)\right](b-a)(b-c)\Bigg|$$

$$\leqslant \frac{1}{16}\{(b-a)(d-c)\}^2\|D_2D_1f\|_\infty, \tag{2.4.23}$$

where $f_1(x)$ and $f_2(y)$ are given by (2.4.14) and (2.4.15).

Proof. Using the identity (2.4.13) and the properties of integral, we have

$$\Bigg|\int_a^b\int_c^d f(x,y)dydx + (b-a)\int_c^d \left(y-\frac{c+d}{2}\right) f_2(y)dy + (d-c)\int_a^b \left(x-\frac{a+b}{2}\right) f_1(x)dx$$

$$-\frac{1}{4}\left[f(a,c)+f(a,d)+f(b,c)+f(b,d)\right](b-a)(b-c)\Bigg|$$

$$\leqslant \int_a^b\int_c^d \left|x-\frac{a+b}{2}\right|\left|y-\frac{c+d}{2}\right| |D_2D_1f(x,y)|\,dydx$$

$$\leqslant \|D_2D_1f\|_\infty \int_a^b \left|x-\frac{a+b}{2}\right| dx \int_c^d \left|y-\frac{c+d}{2}\right| dy$$

$$= \|D_2D_1f\|_\infty \frac{(b-a)^2}{4}\frac{(d-c)^2}{4},$$

and the inequality (2.4.23) is proved.

Theorem 2.4.4. Let f, f_1, f_2 be as in Theorem 2.4.3 and assume that

$$\|f_1\|_\infty = \sup_{x\in[a,b]} |f_1(x)| < \infty, \quad \|f_2\|_\infty = \sup_{y\in[c,d]} |f_2(y)| < \infty.$$

Then

$$\left|\int_a^b\int_c^d f(x,y)dydx - \frac{1}{4}[f(a,c)+f(a,d)+f(b,c)+f(b,d)](b-a)(b-c)\right|$$
$$\leqslant \frac{1}{4}(b-a)(d-c)\left[(b-a)\|f_1\|_\infty + (d-c)\|f_2\|_\infty + \frac{1}{4}(b-a)(d-c)\|D_2D_1f\|_\infty\right]. \quad (2.4.24)$$

Proof. As in the proof of Theorem 2.4.3, we have, by the identity (2.4.13) that

$$\left|\int_a^b\int_c^d f(x,y)dydx - \frac{1}{4}[f(a,c)+f(a,d)+f(b,c)+f(b,d)](b-a)(b-c)\right|$$
$$\leqslant (b-a)\int_c^d \left|y-\frac{c+d}{2}\right| |f_2(y)|dy + (d-c)\int_a^b \left|x-\frac{a+b}{2}\right| |f_1(x)|dx$$
$$+\int_a^b\int_c^d \left|x-\frac{a+b}{2}\right|\left|y-\frac{c+d}{2}\right| |D_2D_1f(x,y)|\,dydx$$
$$\leqslant (b-a)\|f_2\|_\infty \frac{(d-c)^2}{4} + (d-c)\|f_1\|_\infty \frac{(b-a)^2}{4} + \|D_2D_1f\|_\infty \frac{(b-a)^2}{4}\frac{(d-c)^2}{4}$$
$$= \frac{1}{4}(b-a)(d-c)\left[(b-a)\|f_1\|_\infty + (d-c)\|f_2\|_\infty + \frac{1}{4}(b-a)(d-c)\|D_2D_1f\|_\infty\right].$$

Hence the proof is completed.

Remark 2.4.2. We note that, one can very easily obtain bounds on the right hand sides in (2.4.23) and (2.4.24) for $\|\cdot\|_p$ norm, $p\in[1,\infty)$. Here, we do not discuss the details.

2.5 Some multivariate Grüss-type integral inequalities

Our main goal in this section is to present some multivariate Grüss-type integral inequalities recently investigated by Pachpatte in [94,130].

Let $B=\prod_{i=1}^n[a_i,b_i]$ be a bounded domain in $\mathbb{R}^n$, the n-dimensional Euclidean space. For $x_i\in\mathbb{R}$, $x=(x_1,\ldots,x_n)$ is a variable point in B and $dx=dx_1\cdots dx_n$. For any integrable function $u(x):B\to\mathbb{R}$ we denote by $\int_B u(x)dx$ the n-fold integral $\int_{a_n}^{b_n}\cdots\int_{a_1}^{b_1}u(x_1,\ldots,x_n)dx_1\cdots dx_n$. For integrable functions $f,\ g:B\to\mathbb{R}$ on B and $p:B\to\mathbb{R}_+$ an integrable function on B such that $P=\int_B p(x)dx>0$, we set

$$T(P,p,f,g;B) = \frac{1}{P}\int_B p(x)f(x)g(x)dx - \left(\frac{1}{P}\int_B p(x)f(x)dx\right)\left(\frac{1}{P}\int_B p(x)g(x)dx\right), \quad (2.5.1)$$

and assume that the integrals involved in (2.5.1) exist.

In the following theorems, we present the integral inequalities investigated in [130].

Theorem 2.5.1. Let $f,\ g : B \to \mathbb{R}$ be integrable functions on B and $p : B \to \mathbb{R}_+$ an integrable function on B such that $P = \int_B p(x)dx > 0$. Then

$$|T(P,p,f,g;B)| \leqslant \sqrt{T(P,p,f,f;B)}\sqrt{T(P,p,g,g;B)}, \tag{2.5.2}$$

and in addition if $\phi \leqslant f(x) \leqslant \Phi$, $\gamma \leqslant g(x) \leqslant \Gamma$ for each $x \in B$, where $\phi,\ \Phi,\ \gamma,\ \Gamma$ are given real constants, then

$$|T(P,p,f,g;B)| \leqslant \frac{1}{4}(\Phi - \phi)(\Gamma - \gamma). \tag{2.5.3}$$

Proof. By direct computation, it is easy to observe that the following version of the Korkine's identity [79, p. 242] holds:

$$T(P,p,f,g;B) = \frac{1}{2P^2}\int_B\int_B p(x)p(y)(f(x)-f(y))(g(x)-g(y))dydx. \tag{2.5.4}$$

From (2.5.4), it is easy to observe that

$$T(P,p,f,f;B) = \frac{1}{P}\int_B p(x)f^2(x)dx - \left(\frac{1}{P}\int_B p(x)f(x)dx\right)^2. \tag{2.5.5}$$

Furthermore, by using the multivariate version of the Schwarz integral inequality, it is easy to observe that $T(P,p,f,f;B) \geqslant 0$. Similarly, we have

$$T(P,p,g,g;B) = \frac{1}{P}\int_B p(x)g^2(x)dx - \left(\frac{1}{P}\int_B p(x)g(x)dx\right)^2, \tag{2.5.6}$$

and $T(P,p,g,g;B) \geqslant 0$. From (2.5.4) and using the multivariate version of the Schwarz integral inequality, we have

$$|T(P,p,f,g;B)|^2 = \left\{\frac{1}{2P^2}\int_B\int_B p(x)p(y)(f(x)-f(y))(g(x)-g(y))dydx\right\}^2$$

$$\leqslant \left\{\frac{1}{2P^2}\int_B\int_B p(x)p(y)(f(x)-f(y))^2dydx\right\}\left\{\frac{1}{2P^2}\int_B\int_B p(x)p(y)(g(x)-g(y))^2dydx\right\}$$

$$= \left\{\frac{1}{P}\int_B p(x)f^2(x)dx - \left(\frac{1}{P}\int_B p(x)f(x)dx\right)^2\right\}$$

$$\times \left\{\frac{1}{P}\int_B p(x)g^2(x)dx - \left(\frac{1}{P}\int_B p(x)g(x)dx\right)^2\right\}$$

$$= T(P,p,f,f;B)T(P,p,g,g;B). \tag{2.5.7}$$

The desired inequality in (2.5.2) follows from (2.5.7).

It is easy to observe that the following identity also holds:

$$T(P,p,f,f;B) = \left(\Phi - \frac{1}{P}\int_B p(x)f(x)dx\right)\left(\frac{1}{P}\int_B p(x)f(x)dx - \phi\right)$$

$$-\frac{1}{P}\int_B p(x)\,(\Phi - f(x))\,(f(x)-\phi)\,dx. \tag{2.5.8}$$

Using the fact that $(\Phi - f(x))\,(f(x)-\phi) \geqslant 0$ in (2.5.8), we have

$$T(P,p,f,f;B) \leqslant \left(\Phi - \frac{1}{P}\int_B p(x)f(x)dx\right)\left(\frac{1}{P}\int_B p(x)g(x)dx - \phi\right). \tag{2.5.9}$$

Similarly, we have

$$T(P,p,g,g;B) \leqslant \left(\Gamma - \frac{1}{P}\int_B p(x)g(x)dx\right)\left(\frac{1}{P}\int_B p(x)g(x)dx - \gamma\right). \tag{2.5.10}$$

Using (2.5.9) and (2.5.10) in (2.5.7), we get

$$|T(P,p,f,g;B)|^2 \leqslant \left(\Phi - \frac{1}{P}\int_B p(x)f(x)dx\right)\left(\frac{1}{P}\int_B p(x)f(x)dx - \phi\right)$$

$$\times\left(\Gamma - \frac{1}{P}\int_B p(x)g(x)dx\right)\left(\frac{1}{P}\int_B p(x)g(x)dx - \gamma\right). \tag{2.5.11}$$

By using the elementary inequality $cd \leqslant \left(\frac{c+d}{2}\right)^2$; $c,\ d \in \mathbb{R}$, we observe that

$$\left(\Phi - \frac{1}{P}\int_B p(x)f(x)dx\right)\left(\frac{1}{P}\int_B p(x)f(x)dx - \phi\right) \leqslant \left(\frac{\Phi-\phi}{2}\right)^2, \tag{2.5.12}$$

$$\left(\Gamma - \frac{1}{P}\int_B p(x)g(x)dx\right)\left(\frac{1}{P}\int_B p(x)g(x)dx - \gamma\right) \leqslant \left(\frac{\Gamma-\gamma}{2}\right)^2. \tag{2.5.13}$$

The required inequality in (2.5.3) follows from (2.5.11)–(2.5.13).

Remark 2.5.1. We note that the inequality established in (2.5.3) can be considered as a weighted multivariate generalization of the Grüss inequality. In the special case when $n=1$, from (2.5.3), we get the generalization of the Grüss inequality (3) given by Dragomir in [42].

Theorem 2.5.2. Let $f,\ g : B \to \mathbb{R}$ be integrable functions on B and $p : B \to \mathbb{R}_+$ an integrable function on B such that $P = \int_B p(x)dx > 0$. Then

$$|T(P,p,f,g;B)|$$

$$\leqslant \frac{1}{P}\int_B p(x)\left|\left(f(x) - \frac{1}{P}\int_B p(y)f(y)dy\right)\left(g(x) - \frac{1}{P}\int_B p(y)g(y)dy\right)\right|dx, \tag{2.5.14}$$

and in addition if $\phi \leqslant f(x) \leqslant \Phi$ for each $x \in B$, where $\phi,\ \Phi$ are real constants, then

$$|T(P,p,f,g;B)| \leqslant \frac{\Phi-\phi}{2}\sqrt{T(P,p,g,g;B)}. \tag{2.5.15}$$

Proof. In order to establish the inequality (2.5.14), we observe that

$$\frac{1}{P}\int_B p(x)\left(f(x)-\frac{1}{P}\int_B p(y)f(y)dy\right)\left(g(x)-\frac{1}{P}\int_B p(y)g(y)dy\right)dx$$

$$=\frac{1}{P}\int_B p(x)\left\{f(x)g(x)-\frac{1}{P}f(x)\int_B p(y)g(y)dy\right.$$

$$\left.-\frac{1}{P}g(x)\int_B p(y)f(y)dy+\frac{1}{P^2}\left(\int_B p(y)f(y)dy\right)\left(\int_B p(y)g(y)dy\right)\right\}dx$$

$$=\frac{1}{P}\int_B p(x)f(x)g(x)dx-\left(\frac{1}{P}\int_B p(x)f(x)dx\right)\left(\frac{1}{P}\int_B p(x)g(x)dx\right)$$

$$-\left(\frac{1}{P}\int_B p(x)f(x)dx\right)\left(\frac{1}{P}\int_B p(x)g(x)dx\right)+\frac{1}{P}\frac{1}{P^2}P\left(\int_B p(x)f(x)dx\right)\left(\int_B p(x)g(x)dx\right)$$

$$=\frac{1}{P}\int_B p(x)f(x)g(x)dx-\left(\frac{1}{P}\int_B p(x)f(x)dx\right)\left(\frac{1}{P}\int_B p(x)g(x)dx\right)$$

$$=T(P,p,f,g;B). \tag{2.5.16}$$

From (2.5.16) and using the properties of modulus, we get the desired inequality in (2.5.14). Following the proof of Theorem 2.5.1, we have $T(P,p,f,f;B)\geqslant 0$, $T(P,p,g,g;B)\geqslant 0$ and (2.5.7), (2.5.9), (2.5.12) hold. From (2.5.9) and (2.5.12), we get

$$T(P,p,f,f;B)\leqslant\frac{1}{4}(\Phi-\phi)^2. \tag{2.5.17}$$

The required inequality in (2.5.15) follows from (2.5.7) and (2.5.17). The proof is complete.

Remark 2.5.2. By taking $p(x)=1$ and hence $P=\prod_{i=1}^n(b_i-a_i)$ in (2.5.14), we get

$$|T(P,1,f,g;B)|\leqslant\frac{1}{P}\int_B\left|\left(f(x)-\frac{1}{P}\int_B f(y)dy\right)\left(g(x)-\frac{1}{P}\int_B g(y)dy\right)\right|dx. \tag{2.5.18}$$

The inequality (2.5.18) can be considered as the multivariate version of the integral inequality of the Grüss-type given by Dragomir and McAndrew in [34]. We note that the inequality (2.5.15) can be considered as a multivariate version of the Pre-Grüss inequality given by Matić, Pečarić and Ujević in [72].

Before giving the next result, we introduce the following notation used to simplify the details of presentation.

Let $D_i[a_i,b_i]=\{x_i : a_i<x_i<b_i\}$ for $i=1,\ldots,n$; $a_i,\ b_i\in\mathbb{R}$, $D=\prod_{i=1}^n D_i[a_i,b_i]$ and $\overline{D}$ be the closure of D. For any function $u:\overline{D}\to\mathbb{R}$, differentiable on D, we denote the first order partial derivatives by $\frac{\partial u(x)}{\partial x_i}$ and by $\int_D u(x)dx$ the n-fold integral. If

$$\left\|\frac{\partial u}{\partial x_i}\right\|_\infty=\sup_{x\in D}\left|\frac{\partial u(x)}{\partial x_i}\right|<\infty,$$

then we say that the partial derivatives $\frac{\partial u(x)}{\partial x_i}$ are bounded. For continuous functions p, q defined on $\overline{D}$ and differentiable on D and $w(x)$ a real-valued, nonnegative and integrable function for $x \in D$ and $\int_D w(x)dx > 0$ and $x_i,\ y_i \in D_i[a_i,b_i]$, we set

$$A(w,p,q;D) = \int_D w(x)p(x)q(x)dx$$

$$-\frac{1}{\int_D w(x)dx}\left(\int_D w(x)p(x)dx\right)\left(\int_D w(x)q(x)dx\right), \tag{2.5.19}$$

$$H(p,x_i,y_i) = \sum_{i=1}^{n}\left\|\frac{\partial p}{\partial x_i}\right\|_\infty |x_i - y_i|, \tag{2.5.20}$$

and assume that the integrals involved in (2.5.19) exist.

The Grüss-type integral inequalities established in [94] are given in the following theorem.

Theorem 2.5.3. Let $f(x)$, $g(x)$ be real-valued continuous functions on $\overline{D}$ and differentiable on D, with partial derivatives $\frac{\partial f(x)}{\partial x_i}$, $\frac{\partial g(x)}{\partial x_i}$ being bounded. Let $w(x)$ be a real-valued, nonnegative and integrable function for $x \in D$ and $\int_D w(x)dx > 0$. Then

$$|A(w,f,g;D)| \leqslant \frac{1}{2\int_D w(x)dx}\int_D w(x)\left[|g(x)|\int_D H(f,x_i,y_i)w(y)dy\right.$$

$$\left.+|f(x)|\int_D H(g,x_i,y_i)w(y)dy\right]dx, \tag{2.5.21}$$

$$|A(w,f,g;D)| \leqslant \frac{1}{(\int_D w(x)dx)^2}\int_D w(x)\left(\int_D H(f,x_i,y_i)w(y)dy\right)$$

$$\times\left(\int_D H(g,x_i,y_i)w(y)dy\right)dx. \tag{2.5.22}$$

Proof. Let $x = (x_1,\ldots,x_n) \in \overline{D}$, $y = (y_1,\ldots,y_n) \in D$. From the n-dimensional version of the mean value theorem, we have (see [146, p.174])

$$f(x) - f(y) = \sum_{i=1}^{n}\frac{\partial f(c_1)}{\partial x_i}(x_i - y_i), \tag{2.5.23}$$

and

$$g(x) - g(y) = \sum_{i=1}^{n}\frac{\partial g(c_2)}{\partial x_i}(x_i - y_i), \tag{2.5.24}$$

where $c_i = (y_1 + \alpha_i(x_1 - y_1),\ldots,y_n + \alpha_i(x_n - y_n))$ $(0 < \alpha_i < 1)$ for $i = 1,2$. Multiplying both sides of (2.5.23) and (2.5.24) by $g(x)$ and $f(x)$ respectively and adding, we get

$$2f(x)g(x) - g(x)f(y) - f(x)g(y)$$

$$= g(x)\sum_{i=1}^{n}\frac{\partial f(c_1)}{\partial x_i}(x_i - y_i) + f(x)\sum_{i=1}^{n}\frac{\partial g(c_2)}{\partial x_i}(x_i - y_i). \tag{2.5.25}$$

Multiplying both sides of (2.5.25) by $w(y)$ and integrating the resulting identity with respect to y over D, we have

$$2\left(\int_D w(y)dy\right) f(x)g(x) - g(x)\int_D w(y)f(y)dy - f(x)\int_D w(y)g(y)dy$$

$$= g(x)\int_D \sum_{i=1}^{n}\frac{\partial f(c_1)}{\partial x_i}(x_i - y_i)w(y)dy + f(x)\int_D \sum_{i=1}^{n}\frac{\partial g(c_2)}{\partial x_i}(x_i - y_i)w(y)dy. \tag{2.5.26}$$

Next, multiplying both sides of (2.5.26) by $w(x)$ and integrating the resulting identity with respect to x over D, we get

$$2\left(\int_D w(y)dy\right)\int_D w(x)f(x)g(x)dx$$

$$-\left(\int_D w(x)g(x)dx\right)\left(\int_D w(y)f(y)dy\right) - \left(\int_D w(x)f(x)dx\right)\left(\int_D w(y)g(y)dy\right)$$

$$= \int_D w(x)g(x)\left(\int_D \sum_{i=1}^{n}\frac{\partial f(c_1)}{\partial x_i}(x_i - y_i)w(y)dy\right)dx$$

$$+ \int_D w(x)f(x)\left(\int_D \sum_{i=1}^{n}\frac{\partial g(c_2)}{\partial x_i}(x_i - y_i)w(y)dy\right)dx. \tag{2.5.27}$$

Rewriting (2.5.27), using the notation given in (2.5.19) and the properties of modulus, we have

$$|A(w,f,g;D)| \leqslant \frac{1}{2\int_D w(x)dx}\left[\int_D w(x)|g(x)|\left(\int_D \sum_{i=1}^{n}\left|\frac{\partial f(c_1)}{\partial x_i}\right||x_i - y_i|w(y)dy\right)dx\right.$$

$$\left. + \int_D w(x)|f(x)|\left(\int_D \sum_{i=1}^{n}\left|\frac{\partial g(c_2)}{\partial x_i}\right||x_i - y_i|w(y)dy\right)dx\right]$$

$$\leqslant \frac{1}{2\int_D w(x)dx}\left[\int_D w(x)\left[|g(x)|\int_D \sum_{i=1}^{n}\left\|\frac{\partial f}{\partial x_i}\right\|_\infty |x_i - y_i|w(y)dy\right.\right.$$

$$\left.\left. + |f(x)|\int_D \sum_{i=1}^{n}\left\|\frac{\partial g}{\partial x_i}\right\|_\infty |x_i - y_i|w(y)dy\right]dx\right.$$

$$= \frac{1}{2\displaystyle\int_D w(x)dx}\int_D w(x)\left[|g(x)|\int_D H(f,x_i,y_i)w(y)dy + |f(x)|\int_D H(g,x_i,y_i)w(y)dy\right]dx.$$

This is the required inequality in (2.5.21).

Multiplying both sides of (2.5.23) and (2.5.24) by $w(y)$ and integrating the resulting identities with respect to y over D, we get

$$\left(\int_D w(y)dy\right) f(x) - \int_D w(y)f(y)dy = \int_D \sum_{i=1}^{n} \frac{\partial f(c_1)}{\partial x_i}(x_i - y_i)w(y)dy, \qquad (2.5.28)$$

and

$$\left(\int_D w(y)dy\right) g(x) - \int_D w(y)g(y)dy = \int_D \sum_{i=1}^{n} \frac{\partial g(c_2)}{\partial x_i}(x_i - y_i)w(y)dy. \qquad (2.5.29)$$

Multiplying the left hand sides and right hand sides of (2.5.28) and (2.5.29), we get

$$\left(\int_D w(y)dy\right)^2 f(x)g(x) - \left(\int_D w(y)dy\right) f(x) \int_D w(y)g(y)dy$$

$$- \left(\int_D w(y)dy\right) g(x) \int_D w(y)f(y)dy + \left(\int_D w(y)f(y)dy\right)\left(\int_D w(y)g(y)dy\right)$$

$$= \left(\int_D \sum_{i=1}^{n} \frac{\partial f(c_1)}{\partial x_i}(x_i - y_i)w(y)dy\right)\left(\int_D \sum_{i=1}^{n} \frac{\partial g(c_2)}{\partial x_i}(x_i - y_i)w(y)dy\right). \qquad (2.5.30)$$

Multiplying both sides of (2.5.30) by $w(x)$ and integrating the resulting identity with respect to x over D and rewriting, we obtain

$$A(w,f,g;D) = \frac{1}{\left(\int_D w(y)dy\right)^2} \int_D w(x) \left(\int_D \sum_{i=1}^{n} \frac{\partial f(c_1)}{\partial x_i}(x_i - y_i)w(y)dy\right)$$

$$\times \left(\int_D \sum_{i=1}^{n} \frac{\partial g(c_2)}{\partial x_i}(x_i - y_i)w(y)dy\right) dx. \qquad (2.5.31)$$

From (2.5.31) and following the proof of inequality (2.5.21) with suitable modifications, we get the required inequality in (2.5.22). The proof is complete.

Remark 2.5.3. If we take $n = 1$ and $D = I = \{x : a < x < b\}$ in (2.5.21), then we get

$$\left| \int_a^b w(t)f(t)g(t)dt - \frac{1}{\int_a^b w(t)dt}\left(\int_a^b w(t)f(t)dt\right)\left(\int_a^b w(t)g(t)dt\right) \right|$$

$$\leqslant \frac{1}{2\int_a^b w(t)dt} \int_a^b w(t) \left[|g(t)| \int_a^b \|f'\|_\infty |t-s| w(s)ds \right.$$

$$\left. + |f(t)| \int_a^b \|g'\|_\infty |t-s| w(s)ds \right] dt. \qquad (2.5.32)$$

Similarly, one can obtain the special version of (2.5.22). It is easy to see that the bound obtained on the right hand side in (2.5.32) (when $w(t) = 1$) is different from those given by Grüss in [61].

2.6 Multivariate Grüss-and Čebyšev-type discrete inequalities

In this section, we deal with certain multivariate Grüss- and Čebyšev-type discrete inequalities established by Pachpatte in [95,103,129].

Let $N_1 = \{1,2,\dots,k+1\}$, $N_2 = \{1,2,\dots,m+1\}$, $N_3 = \{1,2,\dots,n+1\}$ for $k,m,n \in N$ and denote by $G = N_1 \times N_2$, $H = N_1 \times N_2 \times N_3$. For functions $h(x,y)$ and $e(x,y,z)$ defined respectively on G and H we define the operators $\Delta_1 h(x,y) = h(x+1,y) - h(x,y)$, $\Delta_2 h(x,y) = h(x,y+1) - h(x,y)$, $\Delta_2\Delta_1 h(x,y) = \Delta_2(\Delta_1 h(x,y))$ and $\Delta_1 e(x,y,z) = e(x+1,y,z) - e(x,y,z)$, $\Delta_2 e(x,y,z) = e(x,y+1,z) - e(x,y,z)$, $\Delta_3 e(x,y,z) = e(x,y,z+1) - e(x,y,z)$, $\Delta_2\Delta_1 e(x,y,z) = \Delta_2(\Delta_1 e(x,y,z))$, $\Delta_3\Delta_2\Delta_1 e(x,y,z) = \Delta_3(\Delta_2\Delta_1 e(x,y,z))$. Let $N_i[0,a_i] = \{0,1,2,\dots,a_i\}$, $a_i \in N$, $i = 1,\dots,n$ and $Q = \prod_{i=1}^{n} N_i[0,a_i]$. For a function $u(x): Q \to \mathbb{R}$ we define the first order difference operators as $\Delta_1 u(x) = u(x_1+1,x_2,\dots,x_n) - u(x),\dots,$ $\Delta_n u(x) = u(x_1,\dots,x_{n-1},x_n+1) - u(x)$ and denote the n-fold sum over Q with respect to the variable $y = (y_1,\dots,y_n) \in Q$ by

$$\sum_y u(y) = \sum_{y_1=0}^{a_1-1} \cdots \sum_{y_n=0}^{a_n-1} u(y_1,\dots,y_n).$$

Clearly $\sum_y u(y) = \sum_x u(x)$ for $x,\ y \in Q$. The notation

$$\sum_{t_i=y_i}^{x_i-1} \Delta_i u(y_1,\dots,y_{i-1},t_i,x_{i+1},\dots,x_n)$$

for $x_i,\ y_i \in \mathrm{N}_i[0,a_i]$; $i = 1,\dots,n$, we mean for $i=1$, it is $\sum_{t_1=y_1}^{x_1-1} \Delta_1 u(t_1,x_2,\dots,x_n)$ and so on and for $i=n$, it is $\sum_{t_n=y_n}^{x_n-1} \Delta_n u(y_1,\dots,y_{n-1},t_n)$. The functions $h(x,y)$, $e(x,y,z)$ and $u(x)$ defined on G,H and Q respectively are said to be bounded if $\|h\|_\infty = \sup_{(x,y)\in G} |h(x,y)| < \infty$, $\|e\|_\infty = \sup_{(x,y,z)\in H} |e(x,y,z)| < \infty$ and $\|u\|_\infty = \sup_{x\in Q} |u(x)| < \infty$. We use the usual convention that the empty sum is taken to be zero. We give the following notation used to simplify the details of presentation:

$$A(\Delta_2\Delta_1 h(x,y)) = A[1,1;x,y;k,m;\Delta_2\Delta_1 h(s,t)]$$

$$= \sum_{s=1}^{x-1}\sum_{t=1}^{y-1} \Delta_2\Delta_1 h(s,t) - \sum_{s=1}^{x-1}\sum_{t=y}^{m} \Delta_2\Delta_1 h(s,t) - \sum_{s=x}^{k}\sum_{t=1}^{y-1} \Delta_2\Delta_1 h(s,t) + \sum_{s=x}^{k}\sum_{t=y}^{m} \Delta_2\Delta_1 h(s,t),$$

$$E(h(x,y)) = E[1,1;x,y;k+1,m+1;h]$$

$$= \frac{1}{2}[h(x,1) + h(x,m+1) + h(1,y) + h(k+1,y)]$$

$$-\frac{1}{4}[h(1,1) + h(1,m+1) + h(k+1,1) + h(k+1,m+1)],$$

$$B(\Delta_3\Delta_2\Delta_1 e(r,s,t)) = B[1,1,1;r,s,t;k,m,n;\Delta_3\Delta_2\Delta_1 e(u,v,w)]$$

$$= \sum_{u=1}^{r-1}\sum_{v=1}^{s-1}\sum_{w=1}^{t-1} \Delta_3\Delta_2\Delta_1 e(u,v,w) - \sum_{u=1}^{r-1}\sum_{v=1}^{s-1}\sum_{w=t}^{n} \Delta_3\Delta_2\Delta_1 e(u,v,w)$$

$$- \sum_{u=1}^{r-1}\sum_{v=s}^{m}\sum_{w=1}^{t-1} \Delta_3\Delta_2\Delta_1 e(u,v,w) - \sum_{u=r}^{k}\sum_{v=1}^{s-1}\sum_{w=1}^{t-1} \Delta_3\Delta_2\Delta_1 e(u,v,w)$$

$$+ \sum_{u=1}^{r-1}\sum_{v=s}^{m}\sum_{w=t}^{n} \Delta_3\Delta_2\Delta_1 e(u,v,w) + \sum_{u=r}^{k}\sum_{v=s}^{m}\sum_{w=1}^{t-1} \Delta_3\Delta_2\Delta_1 e(u,v,w)$$

$$+ \sum_{u=r}^{k}\sum_{v=1}^{s-1}\sum_{w=t}^{n} \Delta_3\Delta_2\Delta_1 e(u,v,w) - \sum_{u=r}^{k}\sum_{v=s}^{m}\sum_{w=t}^{n} \Delta_3\Delta_2\Delta_1 e(u,v,w),$$

$$L(e(r,s,t)) = L[1,1,1;r,s,t;k+1,m+1,n+1;e]$$

$$= \frac{1}{8}[e(1,1,1) + e(k+1,m+1,n+1)]$$

$$-\frac{1}{4}[e(r,1,1) + e(r,1,n+1) + e(r,m+1,1) + e(r,m+1,n+1)]$$

$$-\frac{1}{4}[e(k+1,s,n+1) + e(k+1,s,1) + e(1,s,n+1) + e(1,s,1)]$$

$$-\frac{1}{4}[e(k+1,m+1,t) + e(k+1,1,t) + e(1,m+1,t) + e(1,1,t)]$$

$$+\frac{1}{2}[e(1,s,t) + e(k+1,s,t)] + \frac{1}{2}[e(r,1,t) + e(r,m+1,t)]$$

$$+\frac{1}{2}[e(r,s,1) + e(r,s,n+1)],$$

and for $x \in Q$,

$$S(f,g,M;Q) = \frac{1}{M}\sum_{x} f(x)g(x) - \left(\frac{1}{M}\sum_{x} f(x)\right)\left(\frac{1}{M}\sum_{x} g(x)\right),$$

for some suitable functions f, g and a constant M.

In the following Theorems we present the inequalities of the Grüss-and Čebyšev-type established in [103,129].

Theorem 2.6.1. Let $f,\ g : G \to \mathbb{R}$ be functions such that $\Delta_2\Delta_1 f(x,y)$, $\Delta_2\Delta_1 g(x,y)$ exist and bounded on G. Then

$$\left|\sum_{x=1}^{k}\sum_{y=1}^{m}\left[f(x,y)g(x,y)-\frac{1}{2}\{g(x,y)E(f(x,y))+f(x,y)E(g(x,y))\}\right]\right|$$

$$\leqslant \frac{1}{8}km\sum_{x=1}^{k}\sum_{y=1}^{m}[|g(x,y)|\,\|\Delta_2\Delta_1 f\|_\infty+|f(x,y)|\,\|\Delta_2\Delta_1 g\|_\infty], \tag{2.6.1}$$

$$\left|\sum_{x=1}^{k}\sum_{y=1}^{m}[f(x,y)g(x,y)-\ \{g(x,y)E(f(x,y))+f(x,y)E(g(x,y))\ \ -E(f(x,y))E(g(x,y))\}]\right|$$

$$\leqslant \frac{1}{16}(km)^2\,\|\Delta_2\Delta_1 f\|_\infty\,\|\Delta_2\Delta_1 g\|_\infty. \tag{2.6.2}$$

Proof. For $(x,y)\in G$ it is easy to observe that the following identities hold (see [86]):

$$f(x,y)=f(x,1)+f(1,y)-f(1,1)+\sum_{s=1}^{x-1}\sum_{t=1}^{y-1}\Delta_2\Delta_1 f(s,t),$$

$$f(x,y)=f(x,m+1)+f(1,y)-f(1,m+1)-\sum_{s=1}^{x-1}\sum_{t=y}^{m}\Delta_2\Delta_1 f(s,t),$$

$$f(x,y)=f(x,1)+f(k+1,y)-f(k+1,1)-\sum_{s=x}^{k}\sum_{t=1}^{y-1}\Delta_2\Delta_1 f(s,t),$$

$$f(x,y)=f(x,m+1)+f(k+1,y)-f(k+1,m+1)+\sum_{s=x}^{k}\sum_{t=y}^{m}\Delta_2\Delta_1 f(s,t).$$

Adding the above identities and rewriting, we have

$$f(x,y)-E(f(x,y))=\frac{1}{4}A\left(\Delta_2\Delta_1 f(x,y)\right), \tag{2.6.3}$$

for $(x,y)\in G$. Similarly, we have

$$g(x,y)-E(g(x,y))=\frac{1}{4}A(\Delta_2\Delta_1 g(x,y)), \tag{2.6.4}$$

for $(x,y)\in G$. Multiplying (2.6.3) by $g(x,y)$ and (2.6.4) by $f(x,y)$, adding the resulting identities, rewriting and then summing over G, we have

$$\sum_{x=1}^{k}\sum_{y=1}^{m}\left[f(x,y)g(x,y)-\frac{1}{2}\{g(x,y)E(f(x,y))+f(x,y)E(g(x,y))\}\right]$$

$$=\frac{1}{8}\sum_{x=1}^{k}\sum_{y=1}^{m}[g(x,y)A\left(\Delta_2\Delta_1 f(x,y)\right)+f(x,y)A(\Delta_2\Delta_1 g(x,y))]. \tag{2.6.5}$$

From the properties of modulus and sums, it is easy to see that

$$|A(\Delta_2\Delta_1 f(x,y))| \leqslant \sum_{s=1}^{k}\sum_{t=1}^{m} |\Delta_2\Delta_1 f(s,t)| \leqslant \|\Delta_2\Delta_1 f\|_\infty (km), \tag{2.6.6}$$

$$|A(\Delta_2\Delta_1 g(x,y))| \leqslant \sum_{s=1}^{k}\sum_{t=1}^{m} |\Delta_2\Delta_1 g(s,t)| \leqslant \|\Delta_2\Delta_1 g\|_\infty (km). \tag{2.6.7}$$

From (2.6.5)–(2.6.7), we observe that

$$\left|\sum_{x=1}^{k}\sum_{y=1}^{m}\left[f(x,y)g(x,y)-\frac{1}{2}\{g(x,y)E(f(x,y))+f(x,y)E(g(x,y))\}\right]\right|$$
$$\leqslant \frac{1}{8}\sum_{x=1}^{k}\sum_{y=1}^{m}\left[|g(x,y)|\,|A(\Delta_2\Delta_1 f(x,y))|+|f(x,y)|\,|A(\Delta_2\Delta_1 g(x,y))|\right]$$
$$\leqslant \frac{1}{8}km\sum_{x=1}^{k}\sum_{y=1}^{m}\left[|g(x,y)|\,\|\Delta_2\Delta_1 f\|_\infty+|f(x,y)|\,\|\Delta_2\Delta_1 g\|_\infty\right],$$

which is the required inequality in (2.6.1).

Multiplying the left hand sides and right hand sides of (2.6.3) and (2.6.4), we get

$$f(x,y)g(x,y)-\{g(x,y)E(f(x,y))+f(x,y)E(g(x,y))-E(f(x,y))E(g(x,y))\}$$
$$=\frac{1}{16}A(\Delta_2\Delta_1 f(x,y))A(\Delta_2\Delta_1 g(x,y)). \tag{2.6.8}$$

Summing both sides of (2.6.8) over G and using the properties of modulus, we have

$$\left|\sum_{x=1}^{k}\sum_{y=1}^{m}[f(x,y)g(x,y)-\{g(x,y)E(f(x,y))+f(x,y)E(g(x,y))-E(f(x,y))E(g(x,y))\}]\right|$$
$$\leqslant \frac{1}{16}\sum_{x=1}^{k}\sum_{y=1}^{m}|A(\Delta_2\Delta_1 f(x,y))|\,|A(\Delta_2\Delta_1 g(x,y))|. \tag{2.6.9}$$

Now, using (2.6.6) and (2.6.7) in (2.6.9), we get (2.6.2). The proof is complete.

Theorem 2.6.2. Let $f,\ g : H \to \mathbb{R}$ be functions such that $\Delta_3\Delta_2\Delta_1 f(r,s,t)$, $\Delta_3\Delta_2\Delta_1 g(r,s,t)$ exist and bounded on H. Then

$$\left|\sum_{r=1}^{k}\sum_{s=1}^{m}\sum_{t=1}^{n}\left[f(r,s,t)g(r,s,t)-\frac{1}{2}\{f(r,s,t)L(g(r,s,t))+g(r,s,t)L(f(r,s,t))\}\right]\right|$$
$$\leqslant \frac{1}{16}kmn\sum_{r=1}^{k}\sum_{s=1}^{m}\sum_{t=1}^{n}\left[|g(r,s,t)|\,\|\Delta_3\Delta_2\Delta_1 f\|_\infty+|f(r,s,t)|\,\|\Delta_3\Delta_2\Delta_1 g\|_\infty\right], \tag{2.6.10}$$

$$\left|\sum_{r=1}^{k}\sum_{s=1}^{m}\sum_{t=1}^{n}[f(r,s,t)g(r,s,t)-\{f(r,s,t)L(g(r,s,t))\right.$$
$$\left.+g(r,s,t)L(f(r,s,t))-L(f(r,s,t))L(g(r,s,t))\}]\right|$$
$$\leqslant \frac{1}{64}(kmn)^2\|\Delta_3\Delta_2\Delta_1 f\|_\infty\,\|\Delta_3\Delta_2\Delta_1 g\|_\infty. \tag{2.6.11}$$

Proof. For $(r,s,t) \in H$, it is easy to observe that the following identities hold (see [83]):

$$f(r,s,t) = f(1,1,1) + f(1,s,t) + f(r,1,t) + f(r,s,1)$$
$$-f(1,1,t) - f(1,s,1) - f(r,1,1) + \sum_{u=1}^{r-1}\sum_{v=1}^{s-1}\sum_{w=1}^{t-1} \Delta_3\Delta_2\Delta_1 f(u,v,w),$$
$$f(r,s,t) = f(1,1,n+1) + f(r,s,n+1) + f(1,s,t) + f(r,1,t)$$
$$-f(1,s,n+1) - f(r,1,n+1) - f(1,1,t) - \sum_{u=1}^{r-1}\sum_{v=1}^{s-1}\sum_{w=t}^{n} \Delta_3\Delta_2\Delta_1 f(u,v,w),$$
$$f(r,s,t) = f(r,m+1,t) + f(1,s,t) + f(1,m+1,1) + f(r,s,1)$$
$$-f(1,m+1,t) - f(r,m+1,1) - f(1,s,1) - \sum_{u=1}^{r-1}\sum_{v=s}^{m}\sum_{w=1}^{t-1} \Delta_3\Delta_2\Delta_1 f(u,v,w),$$
$$f(r,s,t) = f(k+1,s,t) + f(r,1,t) + f(r,s,1) + f(k+1,1,1)$$
$$-f(k+1,1,t) - f(k+1,s,t) - f(r,1,1) - \sum_{u=r}^{k}\sum_{v=1}^{s-1}\sum_{w=1}^{t-1} \Delta_3\Delta_2\Delta_1 f(u,v,w),$$
$$f(r,s,t) = f(r,m+1,t) + f(r,s,n+1) + f(r,m+1,n+1) + f(1,s,t)$$
$$-f(r,m+1,n+1) - f(1,m+1,t) - f(1,s,n+1) + \sum_{u=1}^{r-1}\sum_{v=s}^{m}\sum_{w=t}^{n} \Delta_3\Delta_2\Delta_1 f(u,v,w),$$
$$f(r,s,t) = f(r,m+1,t) + f(r,s,1) + f(k+1,s,t) + f(k+1,m,1)$$
$$-f(k+1,m+1,t) - f(k+1,s,1) - f(r,m+1,1) + \sum_{u=r}^{k}\sum_{v=s}^{m}\sum_{w=1}^{t-1} \Delta_3\Delta_2\Delta_1 f(u,v,w),$$
$$f(r,s,t) = f(k+1,s,t) + f(k+1,1,n+1) + f(r,s,n+1) + f(r,1,t)$$
$$-f(k+1,1,n+1) - f(k+1,1,t) - f(r,1,n+1) + \sum_{u=r}^{k}\sum_{v=1}^{s-1}\sum_{w=t}^{n} \Delta_3\Delta_2\Delta_1 f(u,v,w),$$
$$f(r,s,t) = f(k+1,m+1,n+1) + f(k+1,s,t) + f(r,m+1,t) + f(r,s,n+1)$$
$$-f(k+1,m+1,t) - f(k+1,s,n+1) - f(r,m+1,n+1) - \sum_{u=r}^{k}\sum_{v=s}^{m}\sum_{w=t}^{n} \Delta_3\Delta_2\Delta_1 f(u,v,w).$$

Adding the above identities and rewriting, we have

$$f(r,s,t) - L(f(r,s,t)) = \frac{1}{8}B\left(\Delta_3\Delta_2\Delta_1 f(r,s,t)\right), \tag{2.6.12}$$

for $(r,s,t) \in H$. Similarly, we have

$$g(r,s,t) - L(g(r,s,t)) = \frac{1}{8}B\left(\Delta_3\Delta_2\Delta_1 g(r,s,t)\right), \tag{2.6.13}$$

for $(r,s,t) \in H$. From the properties of modulus and sums, we observe that

$$\left|B\left(\Delta_3\Delta_2\Delta_1 f(r,s,t)\right)\right| \leqslant \sum_{u=1}^{k}\sum_{v=1}^{m}\sum_{w=1}^{n} \left|\Delta_3\Delta_2\Delta_1 f(u,v,w)\right| \leqslant \|\Delta_3\Delta_2\Delta_1 f\|_\infty (kmn), \tag{2.6.14}$$

$$\left|B\left(\Delta_3\Delta_2\Delta_1 g(r,s,t)\right)\right| \leqslant \sum_{u=1}^{k}\sum_{v=1}^{m}\sum_{w=1}^{n} \left|\Delta_3\Delta_2\Delta_1 g(u,v,w)\right| \leqslant \|\Delta_3\Delta_2\Delta_1 g\|_\infty (kmn). \tag{2.6.15}$$

The rest of the proofs of (2.6.10) and (2.6.11) can be completed by closely looking at the proofs of (2.6.1) and (2.6.2) given in Theorem 2.6.1 with suitable modifications. We omit the further details.

The inequalities in the following Theorem are proved in [95].

Theorem 2.6.3. Let $f,\ g: Q \to \mathbb{R}$ be functions such that $\Delta_i f(x)$, $\Delta_i g(x)$ for $i = 1,\dots,n$ exist and bounded on Q. Then

$$|S(f,g,M;Q)| \leqslant \frac{1}{2M^2}\sum_x \left[\sum_{i=1}^n [|g(x)|\|\Delta_i f\|_\infty + |f(x)|\|\Delta_i g\|_\infty] H_i(x)\right], \tag{2.6.16}$$

$$|S(f,g,M;Q)| \leqslant \frac{1}{2M^2}\sum_x \left[\sum_y \left[\sum_{i=1}^n [\|\Delta_i f\|_\infty]\,|x_i - y_i|\right]\left[\sum_{i=1}^n [\|\Delta_i g\|_\infty]\,|x_i - y_i|\right]\right], \tag{2.6.17}$$

where $M = \prod_{i=1}^n a_i$ and $H_i(x) = \sum_y |x_i - y_i|$.

Proof. For $x = (x_1,\dots,x_n)$, $y = (y_1,\dots,y_n)$ in Q, it is easy to observe that the following identities hold:

$$f(x) - f(y) = \sum_{i=1}^n \left\{\sum_{t_i=y_i}^{x_i-1} \Delta_i f(y_1,\dots,y_{i-1},t_i,x_{i+1},\dots,x_n)\right\}, \tag{2.6.18}$$

$$g(x) - g(y) = \sum_{i=1}^n \left\{\sum_{t_i=y_i}^{x_i-1} \Delta_i g(y_1,\dots,y_{i-1},t_i,x_{i+1},\dots,x_n)\right\}. \tag{2.6.19}$$

Multiplying both sides of (2.6.18) and (2.6.19) by $g(x)$ and $f(x)$ respectively and adding, we get

$$\begin{aligned} &2f(x)g(x) - g(x)f(y) - f(x)g(y) \\ &= g(x)\sum_{i=1}^n \left\{\sum_{t_i=y_i}^{x_i-1} \Delta_i f(y_1,\dots,y_{i-1},t_i,x_{i+1},\dots,x_n)\right\} \\ &\quad + f(x)\sum_{i=1}^n \left\{\sum_{t_i=y_i}^{x_i-1} \Delta_i g(y_1,\dots,y_{i-1},t_i,x_{i+1},\dots,x_n)\right\}. \end{aligned} \tag{2.6.20}$$

Summing both sides of (2.6.20) with respect to y over Q, using the fact that $M > 0$ and rewriting, we have

$$\begin{aligned} &f(x)g(x) - \frac{1}{2M}g(x)\sum_y f(y) - \frac{1}{2M}f(x)\sum_y g(y) \\ &= \frac{1}{2M}\left[g(x)\sum_y \left[\sum_{i=1}^n \left\{\sum_{t_i=y_i}^{x_i-1} \Delta_i f(y_1,\dots,y_{i-1},t_i,x_{i+1},\dots,x_n)\right\}\right]\right. \\ &\quad \left. + f(x)\sum_y \left[\sum_{i=1}^n \left\{\sum_{t_i=y_i}^{x_i-1} \Delta_i g(y_1,\dots,y_{i-1},t_i,x_{i+1},\dots,x_n)\right\}\right]\right]. \end{aligned} \tag{2.6.21}$$

Summing both sides of (2.6.21) with respect to x over Q, rewriting and using the properties of modulus, we have

$$|S(f,g,M;Q)| \leqslant \frac{1}{2M^2}\sum_x \left[|g(x)| \sum_y \left[\sum_{i=1}^n \left| \left\{ \sum_{t_i=y_i}^{x_i-1} |\Delta_i f(y_1,\ldots,y_{i-1},t_i,x_{i+1},\ldots,x_n)| \right\} \right| \right] \right.$$
$$\left. +|f(x)| \sum_y \left[\sum_{i=1}^n \left| \left\{ \sum_{t_i=y_i}^{x_i-1} |\Delta_i g(y_1,\ldots,y_{i-1},t_i,x_{i+1},\ldots,x_n)| \right\} \right| \right] \right].$$
$$\leqslant \frac{1}{2M^2}\sum_x \left[|g(x)| \sum_y \left[\sum_{i=1}^n \|\Delta_i f\|_\infty \left| \sum_{t_i=y_i}^{x_i-1} 1 \right| \right] + |f(x)| \sum_y \left[\sum_{i=1}^n \|\Delta_i g\|_\infty \left| \sum_{t_i=y_i}^{x_i-1} 1 \right| \right] \right]$$
$$= \frac{1}{2M^2}\sum_x \left[|g(x)| \sum_{i=1}^n \|\Delta_i f\|_\infty \sum_y |x_i-y_i| + |f(x)| \sum_{i=1}^n \|\Delta_i g\|_\infty \sum_y |x_i-y_i| \right]$$
$$= \frac{1}{2M^2}\sum_x \left[\sum_{i=1}^n [|g(x)|\|\Delta_i f\|_\infty + |f(x)|\|\Delta_i g\|_\infty] H_i(x) \right],$$

which proves the inequality (2.6.16).

Multiplying the left hand sides and right hand sides of (2.6.18) and (2.6.19), we get

$$f(x)g(x) - g(x)f(y) - f(x)g(y) + f(y)g(y)$$
$$= \left[\sum_{i=1}^n \left\{ \sum_{t_i=y_i}^{x_i-1} \Delta_i f(y_1,\ldots,y_{i-1},t_i,x_{i+1},\ldots,x_n) \right\} \right]$$
$$\times \left[\sum_{i=1}^n \left\{ \sum_{t_i=y_i}^{x_i-1} \Delta_i g(y_1,\ldots,y_{i-1},t_i,x_{i+1},\ldots,x_n) \right\} \right]. \tag{2.6.22}$$

Summing both sides of (2.6.22) with respect to y over Q and rewriting, we have

$$f(x)g(x) - \frac{1}{M}g(x)\sum_y f(y) - \frac{1}{M}f(x)\sum_y g(y) + \frac{1}{M}\sum_y f(y)g(y)$$
$$= \frac{1}{M}\sum_y \left[\sum_{i=1}^n \left\{ \sum_{t_i=y_i}^{x_i-1} \Delta_i f(y_1,\ldots,y_{i-1},t_i,x_{i+1},\ldots,x_n) \right\} \right]$$
$$\times \left[\sum_{i=1}^n \left\{ \sum_{t_i=y_i}^{x_i-1} \Delta_i g(y_1,\ldots,y_{i-1},t_i,x_{i+1},\ldots,x_n) \right\} \right]. \tag{2.6.23}$$

Summing both sides of (2.6.23) with respect to x over Q, rewriting and using the properties of modulus, we have

$$|S(f,g,M;Q)| \leqslant \frac{1}{2M^2}\sum_x \left[\sum_y \left[\sum_{i=1}^n \left| \left\{ \sum_{t_i=y_i}^{x_i-1} |\Delta_i f(y_1,\ldots,y_{i-1},t_i,x_{i+1},\ldots,x_n)| \right\} \right| \right] \right.$$
$$\left. \times \left[\sum_{i=1}^n \left| \left\{ \sum_{t_i=y_i}^{x_i-1} |\Delta_i g(y_1,\ldots,y_{i-1},t_i,x_{i+1},\ldots,x_n)| \right\} \right| \right] \right]$$
$$\leqslant \frac{1}{2M^2}\sum_x \left[\sum_y \left[\sum_{i=1}^n [\|\Delta_i f\|_\infty] |x_i-y_i| \right] \left[\sum_{i=1}^n [\|\Delta_i g\|_\infty] |x_i-y_i| \right] \right],$$

which is the required inequality in (2.6.17). The proof is complete.

2.7 Applications

One of the main motivations for investigating different types of inequalities given in earlier sections was to apply them as tools in various applications. In this section we give applications of some of the inequalities and it is hoped that these inequalities will provide a fruitful source for future research.

2.7.1 *Some integral inequalities via Grüss inequality in inner product spaces*

In this subsection, we present some integral versions of Theorem 2.2.1 given by Dragomir and Gomm [54], that have potential for applications.

Let (Ω, Σ, μ) be a measure space consisting of a set Ω, a σ-algebra Σ of subsets of Ω and a countably additive and positive measure μ on Σ with values in $\mathbb{R} \cup \{\infty\}$. Denote $L^2_\rho(\Omega, \mathbb{K})$ $(\mathbb{K} = \mathbb{R},\ \mathbb{C})$ the Hilbert space of all measurable functions $f : \Omega \to \mathbb{K}$ that are $2-\rho$-integrable on Ω, i.e., $\int_\Omega \rho(s)|f(s)|^2 d\mu(s) < \infty$, where $\rho : \Omega \to [0, \infty)$ is a given measurable function on Ω. The inner product $(\cdot, \cdot) : L^2_\rho(\Omega, \mathbb{K}) \times L^2_\rho(\Omega, \mathbb{K}) \to \mathbb{K}$ that generates the norm of $L^2_\rho(\Omega, \mathbb{K})$ is

$$(f, g)_\rho := \int_\Omega f(s)\overline{g(s)}\rho(s)d\mu(s). \tag{2.7.1}$$

The following proposition holds.

Proposition 2.7.1. Let $\phi,\ \gamma,\ \Phi,\ \Gamma \in \mathbb{K}$ and $h,\ f,\ g \in L^2_\rho(\Omega, \mathbb{K})$ be such that

$$\operatorname{Re}\left[(\Phi h(x) - f(x))\left(\overline{f(x)} - \overline{\phi}\overline{h(x)}\right)\right] \geqslant 0, \tag{2.7.2(a)}$$

$$\operatorname{Re}\left[(\Gamma h(x) - g(x))\left(\overline{g(x)} - \overline{\gamma}\overline{h(x)}\right)\right] \geqslant 0, \tag{2.7.2(b)}$$

for a.e. $x \in \Omega$ and

$$\int_\Omega |h(x)|^2 \rho(x)d\mu(x) = 1. \tag{2.7.3}$$

Then one has the inequality

$$\left|\int_\Omega \rho(x)f(x)\overline{g(x)}d\mu(x) - \left(\int_\Omega \rho(x)f(x)\overline{h(x)}d\mu(x)\right)\left(\int_\Omega \rho(x)h(x)\overline{g(x)}d\mu(x)\right)\right|$$

$$\leqslant \frac{1}{4}|\Phi - \phi||\Gamma - \gamma|, \tag{2.7.4}$$

and the constant $\frac{1}{4}$ in (2.7.4) is sharp.

Proof. Follows from Theorem 2.2.1 applied for the inner product (2.7.1), on taking into account that

$$\mathrm{Re}(\Phi h - f, f - \phi h)_\rho = \int_\Omega \rho(x)\,\mathrm{Re}\left[(\Phi h(x) - f(x))\left(\overline{f(x)} - \overline{\phi}\overline{h(x)}\right)\right] d\mu(x) \geqslant 0$$

and

$$\mathrm{Re}\,(\Gamma h - g, g - \gamma h)_\rho = \int_\Omega \rho(x)\,\mathrm{Re}\left[(\Gamma h(x) - g(x))\left(\overline{g(x)} - \overline{\gamma}\overline{h(x)}\right)\right] d\mu(x) \geqslant 0.$$

The details are omitted.

The following result may be stated as well:

Corollary 2.7.1. If $z,\ Z,\ t,\ T \in \mathbb{K}$, $\rho \in L(\Omega,\mathbb{R})$ with $\int_\Omega \rho(x)d\mu(x) > 0$ and $f,\ g \in L^2_\rho(\Omega,\mathbb{K})$ are such that

$$\mathrm{Re}\left[(Z - f(x))\left(\overline{f(x)} - \overline{z}\right)\right] \geqslant 0, \tag{2.7.5(a)}$$

$$\mathrm{Re}\left[(T - g(x))\left(\overline{g(x)} - \overline{t}\right)\right] \geqslant 0, \tag{2.7.5(b)}$$

for a.e. $x \in \Omega$, then

$$\left|\frac{1}{\int_\Omega \rho(x)d\mu(x)}\int_\Omega \rho(x)f(x)\overline{g(x)}d\mu(x)\right.$$

$$\left. - \left(\frac{1}{\int_\Omega \rho(x)d\mu(x)}\int_\Omega \rho(x)f(x)d\mu(x)\right)\left(\frac{1}{\int_\Omega \rho(x)d\mu(x)}\int_\Omega \rho(x)\overline{g(x)}d\mu(x)\right)\right|$$

$$\leqslant \frac{1}{4}|Z - z||T - t|. \tag{2.7.6}$$

The constant $\frac{1}{4}$ in (2.7.6) is sharp.

Proof. Follows by Proposition 2.7.1 on choosing

$$h = \frac{1}{[\int_\Omega \rho(x)d\mu(x)]^{\frac{1}{2}}},$$

$$\Phi = Z\left[\int_\Omega \rho(x)d\mu(x)\right]^{\frac{1}{2}}, \quad \phi = z\left[\int_\Omega \rho(x)d\mu(x)\right]^{\frac{1}{2}}$$

$$\Gamma = T\left[\int_\Omega \rho(x)d\mu(x)\right]^{\frac{1}{2}}, \quad \gamma = t\left[\int_\Omega \rho(x)d\mu(x)\right]^{\frac{1}{2}}.$$

We omit the details.

As mentioned in [32], if $\rho : \Omega \subseteq \mathbb{R} \to [0,\infty)$ is a probability density function, i.e., $\int_\Omega \rho(t)dt = 1$, then $\rho^{\frac{1}{2}} \in L^2(\Omega,\mathbb{R})$ and obviously $\left\|\rho^{\frac{1}{2}}\right\|_2 = 1$. Consequently, if we assume that $f,\ g \in L^2(\Omega,\mathbb{R})$ and

$$a\rho^{\frac{1}{2}} \leqslant f \leqslant A\rho^{\frac{1}{2}}, b\rho^{\frac{1}{2}} \leqslant g \leqslant B\rho^{\frac{1}{2}},$$

a.e. on Ω, where a, A, b, B are given real numbers, then by Proposition 2.7.1, one has the Grüss-type inequality

$$\left|\int_\Omega f(t)g(t)dt - \left(\int_\Omega f(t)\rho^{\frac{1}{2}}(t)dt\right)\left(\int_\Omega g(t)\rho^{\frac{1}{2}}(t)dt\right)\right| \leqslant \frac{1}{4}(A-a)(B-b). \tag{2.7.7}$$

The following particular inequalities are of interest.

1. If $f,\ g \in L^2(\mathbb{R},\mathbb{R})$ are such that

$$\frac{a}{\sqrt{\sigma\sqrt[4]{2\pi}}}e^{-\frac{1}{4}\left(\frac{x-m}{\sigma}\right)^2} \leqslant f(x) \leqslant \frac{A}{\sqrt{\sigma\sqrt[4]{2\pi}}}e^{-\frac{1}{4}\left(\frac{x-m}{\sigma}\right)^2},$$

$$\frac{b}{\sqrt{\sigma\sqrt[4]{2\pi}}}e^{-\frac{1}{4}\left(\frac{x-m}{\sigma}\right)^2} \leqslant g(x) \leqslant \frac{B}{\sqrt{\sigma\sqrt[4]{2\pi}}}e^{-\frac{1}{4}\left(\frac{x-m}{\sigma}\right)^2},$$

for a.e. $x \in \mathbb{R}$, where a, A, b, $B \in \mathbb{R}$, $m \in \mathbb{R}$, $\sigma > 0$, then one has the following Normal-Grüss inequality

$$\left|\int_{-\infty}^{\infty} f(x)g(x)dx - \frac{1}{\sigma\sqrt{2\pi}}\left(\int_{-\infty}^{\infty} f(x)e^{-\frac{1}{4}\left(\frac{x-m}{\sigma}\right)^2}dx\right)\left(\int_{-\infty}^{\infty} g(x)e^{-\frac{1}{4}\left(\frac{x-m}{\sigma}\right)^2}dx\right)\right|$$

$$\leqslant \frac{1}{4}(A-a)(B-b). \tag{2.7.8}$$

2. If $f,\ g \in L^2(\mathbb{R},\mathbb{R})$ are such that

$$\frac{a}{\sqrt{2\beta}}e^{-\left|\frac{x-\alpha}{2\beta}\right|} \leqslant f(x) \leqslant \frac{A}{\sqrt{2\beta}}e^{-\left|\frac{x-\alpha}{2\beta}\right|},$$

$$\frac{b}{\sqrt{2\beta}}e^{-\left|\frac{x-\alpha}{2\beta}\right|} \leqslant g(x) \leqslant \frac{B}{\sqrt{2\beta}}e^{-\left|\frac{x-\alpha}{2\beta}\right|},$$

for a.e. $x \in \mathbb{R}$, where a, A, b, $B \in \mathbb{R}$, $\alpha \in \mathbb{R}$, $\beta > 0$, then one has the following Laplace-Grüss inequality

$$\left|\int_{-\infty}^{\infty} f(x)g(x)dx - \frac{1}{2\beta}\left(\int_{-\infty}^{\infty} f(x)e^{-\left|\frac{x-\alpha}{2\beta}\right|}dx\right)\left(\int_{-\infty}^{\infty} g(x)e^{-\left|\frac{x-\alpha}{2\beta}\right|}dx\right)\right|$$

$$\leqslant \frac{1}{4}(A-a)(B-b). \tag{2.7.9}$$

3. If $f,\ g \in L^2([0,\infty),R)$ are such that

$$\frac{a}{\sqrt{\Gamma(p)}}x^{\frac{p-1}{2}}e^{-\frac{x}{2}} \leqslant f(x) \leqslant \frac{A}{\sqrt{\Gamma(p)}}x^{\frac{p-1}{2}}e^{-\frac{x}{2}},$$

$$\frac{b}{\sqrt{\Gamma(p)}}x^{\frac{p-1}{2}}e^{-\frac{x}{2}} \leqslant g(x) \leqslant \frac{B}{\sqrt{\Gamma(p)}}x^{\frac{p-1}{2}}e^{-\frac{x}{2}},$$

for a.e. $x \in [0,\infty)$, where a, A, b, $B \in \mathbb{R}$, $p > 0$, then one has the following Gamma-Grüss inequality

$$\left|\int_0^\infty f(x)g(x)dx - \frac{1}{\sqrt{\Gamma(p)}}\left(\int_0^\infty f(x)x^{\frac{p-1}{2}}e^{-\frac{x}{2}}dx\right)\left(\int_0^\infty g(x)x^{\frac{p-1}{2}}e^{-\frac{x}{2}}dx\right)\right| \leqslant \frac{1}{4}(A-a)(B-b). \tag{2.7.10}$$

4. If f, $g \in L^2(x \in [0,1], R)$ are such that

$$\frac{a}{\sqrt{\mathrm{B}(p,q)}}x^{\frac{p-1}{2}}(1-x)^{\frac{q-1}{2}} \leqslant f(x) \leqslant \frac{A}{\sqrt{\mathrm{B}(p,q)}}x^{\frac{p-1}{2}}(1-x)^{\frac{q-1}{2}},$$

$$\frac{\overline{b}}{\sqrt{\mathrm{B}(p,q)}}x^{\frac{p-1}{2}}(1-x)^{\frac{q-1}{2}} \leqslant g(x) \leqslant \frac{\overline{B}}{\sqrt{\mathrm{B}(p,q)}}x^{\frac{p-1}{2}}(1-x)^{\frac{q-1}{2}},$$

for a.e. $x \in [0,1]$ where a, A, $\overline{b}$, $\overline{B} \in \mathbb{R}$, p, $q \in [1,\infty)$, then one has the following Beta-Grüss inequality

$$\left|\int_0^1 f(x)g(x)dx - \frac{1}{\sqrt{B(p,q)}}\left(\int_0^1 f(x)x^{\frac{p-1}{2}}(1-x)^{\frac{q-1}{2}}dx\right)\left(\int_0^1 g(x)x^{\frac{p-1}{2}}(1-x)^{\frac{q-1}{2}}dx\right)\right| \leqslant \frac{1}{4}(A-a)(\overline{B}-\overline{b}). \tag{2.7.11}$$

Finally, we note that Theorem 2.2.1 allows us to state some discrete versions of the Grüss-type inequalities for real and complex sequences, see [54]. Here we omit the details.

2.7.2 *Application to numerical integration*

In this section, we consider an application of Theorem 2.4.2 to numerical integration in connection with a general cubature formula given by Dragomir, Barnett and Pearce in [39]. First, by employing the identity (2.4.12), we present the perturbed version of Grüss inequality proved in [39], which may be useful in certain applications.

Theorem 2.7.1. Let f, $g : [a,b] \to \mathbb{R}$ be continuous on $[a,b]$ and differentiable on (a,b) with derivatives being bounded. Then

$$\left|\frac{1}{b-a}\int_a^b f(x)g(x)dx - \left(\frac{1}{b-a}\int_a^b f(x)dx\right)\left(\frac{1}{b-a}\int_a^b g(x)dx\right) + \frac{1}{4}(f(b)-f(a))(g(b)-g(a))\right|$$
$$\leqslant \frac{1}{2}\left[\|f-f(a)\|_\infty\|g-g(a)\|_\infty + \|f(b)-f\|_\infty\|g(b)-g\|_\infty\right] + \frac{1}{16}(b-a)^2\|f'\|_\infty\|g'\|_\infty. \tag{2.7.12}$$

Proof. Define the mapping $h : [a,b]^2 \to \mathbb{R}$ given by

$$h(x,y) = (f(x) - f(y))(g(x) - g(y)),$$

and write the identity (2.4.12) for h, to get

$$\int_a^b \int_a^b h(x,y)dydx + \frac{1}{4}(b-a)^2[h(a,a) + h(a,b) + h(b,a) + h(b,b)]$$

$$= \frac{1}{2}(b-a)\int_a^b [h(s,a) + h(s,b)]ds + \frac{1}{2}(b-a)\int_a^b [h(a,s) + h(b,s)]ds$$

$$+ \int_a^b \int_a^b \left(s - \frac{a+b}{2}\right)\left(t - \frac{a+b}{2}\right) D_2 D_1 h(s,t)dtds. \tag{2.7.13}$$

We observe that

$$\frac{1}{4}[h(a,a) + h(a,b) + h(b,a) + h(b,b)] = \frac{1}{2}(f(b) - f(a))(g(b) - g(a)),$$

$$\frac{1}{2}\int_a^b [h(s,a) + h(s,b)]\, ds = \frac{1}{2}\int_a^b [h(a,s) + h(b,s)]ds$$

$$= \frac{1}{2}\int_a^b [(f(s) - f(a))(g(s) - g(a)) + (f(b) - f(s))(g(b) - g(s))]\, ds,$$

$$D_2 D_1 h(x,y) = -f'(x)g'(y) - f'(y)g'(x),$$

and

$$\int_a^b \int_a^b \left(s - \frac{a+b}{2}\right)\left(t - \frac{a+b}{2}\right) D_2 D_1 h(s,t)dtds$$

$$= -\int_a^b \int_a^b \left(s - \frac{a+b}{2}\right)\left(t - \frac{a+b}{2}\right)[f'(s)g'(t) + f'(t)g'(s)]dtds$$

$$= -2\int_a^b \int_a^b \left(s - \frac{a+b}{2}\right)\left(t - \frac{a+b}{2}\right) f'(s)g'(t)dtds$$

$$= -2\int_a^b \left(s - \frac{a+b}{2}\right) f'(s)ds \int_a^b \left(t - \frac{a+b}{2}\right) g'(t)dt.$$

Consequently, by (2.7.13), we get

$$\int_a^b \int_a^b (f(x) - f(y))(g(x) - g(y))dydx + \frac{1}{2}(b-a)^2(f(b) - f(a))(g(b) - g(a))$$

$$= (b-a)\int_a^b [(f(s) - f(a))(g(s) - g(a)) + (f(b) - f(s))(g(b) - g(s))]\, ds$$

$$-2\int_a^b \left(s - \frac{a+b}{2}\right) f'(s)ds \int_a^b \left(t - \frac{a+b}{2}\right) g'(t)dt. \tag{2.7.14}$$

Now, dividing by 2 and taking into account the fact that

$$\frac{1}{2}\int_a^b\int_a^b (f(x)-f(y))(g(x)-g(y))dydx = (b-a)\int_a^b f(x)g(x)dx - \int_a^b f(x)dx\int_a^b g(x)dx,$$

the identity (2.7.14) becomes

$$(b-a)\int_a^b f(x)g(x)dx - \int_a^b f(x)dx\int_a^b g(x)dx + \frac{1}{4}(b-a)^2(f(b)-f(a))(g(b)-g(a))$$

$$= \frac{1}{2}(b-a)\int_a^b [(f(s)-f(a))(g(s)-g(a)) + (f(b)-f(s))(g(b)-g(s))]ds$$

$$-\int_a^b \left(s-\frac{a+b}{2}\right) f'(s)ds \int_a^b \left(t-\frac{a+b}{2}\right) g'(t)dt. \tag{2.7.15}$$

Rewriting (2.7.15) and using the properties of modulus, we get

$$\left|\frac{1}{b-a}\int_a^b f(x)g(x)dx - \left(\frac{1}{b-a}\int_a^b f(x)dx\right)\left(\frac{1}{b-a}\int_a^b g(x)dx\right)\right.$$
$$\left. + \frac{1}{4}(f(b)-f(a))(g(b)-g(a))\right|$$

$$\leqslant \frac{1}{2(b-a)}\left[(b-a)\|f-f(a)\|_\infty\|g-g(a)\|_\infty + (b-a)\|f(b)-f\|_\infty\|g(b)-g\|_\infty\right]$$

$$+\|f'\|_\infty\int_a^b \left|s-\frac{a+b}{2}\right| ds\|g'\|_\infty\int_a^b \left|s-\frac{a+b}{2}\right| ds. \tag{2.7.16}$$

A simple calculation gives

$$\int_a^b \left|s-\frac{a+b}{2}\right| ds = \frac{(b-a)^2}{4}. \tag{2.7.17}$$

Using (2.7.17) in (2.7.16), we deduce the desired inequality in (2.7.12). The proof is complete.

Consider the arbitrary division $I_n = a = x_0 < x_1 < \cdots < x_{n-1} < x_n = b$ of $[a,b]$ and $J_m = c = y_0 < y_1 < \cdots < y_{m-1} < y_m = d$ of $[c,d]$, put $h_i := x_{i+1} - x_i$, $l_j := y_{j+1} - y_j$, $i = 0,1,\ldots,n-1$; $j = 0,1,\ldots,m-1$. Define the sum given by

$$C_T(f,I_n,J_m) := \sum_{i=0}^{n-1}\sum_{j=0}^{m-1}\frac{1}{2}h_i\int_{y_j}^{y_{j+1}}[f(x_i,t)+f(x_{i+1},t)]dt$$

$$+\sum_{i=0}^{n-1}\sum_{j=0}^{m-1}\frac{1}{2}l_j\int_{x_j}^{x_{j+1}}\left[f(s,y_j)+f(s,y_{j+1})\right]ds$$

$$-\sum_{i=0}^{n-1}\sum_{j=0}^{m-1}\frac{1}{4}h_il_j\left[f(x_i,y_j)+f(x_i,y_{j+1})+f(x_{i+1},y_j)+f(x_{i+1},y_{j+1})\right]. \tag{2.7.18}$$

As an application of Theorem 2.4.2, in [39], the authors proved the following theorem.

Theorem 2.7.2. Let $f : [a,b] \times [c,d]$ be as in Theorem 2.4.2 and I_n, J_m as above. Then we have the cubature formula

$$\int_a^b \int_c^d f(s,t)dtds = C_T(f,I_n,J_m) + R_T(f,I_n,J_m), \tag{2.7.19}$$

where the remainder term $R_T(f,I_n,J_m)$ satisfies the estimation

$$|R_T(f,I_n,J_m)| \leqslant \frac{1}{16}\|D_2D_1f\|_\infty \sum_{i=0}^{n-1} h_i^2 \sum_{j=0}^{m-1} l_i^2. \tag{2.7.20}$$

Proof. Apply Theorem 2.4.2 on $[x_i,x_{i+1}] \times [y_j,y_{j+1}]$ for $i = 0,1,\ldots,n-1$; $j = 0,1,\ldots,m-1$, to get

$$\Bigg| \int_{x_i}^{x_{i+1}} \int_{yj}^{y_{j+1}} f(s,t)dtds$$

$$- \left[\frac{1}{2}h_i \int_{y_j}^{y_{j+1}} [f(x_i,t)+f(x_{i+1},t)]dt + \frac{1}{2}l_j \int_{x_i}^{x_{i+1}} [f(s,y_j)+f(s,y_{j+1})]\,ds\right.$$

$$\left. -\frac{1}{4}h_il_j\left[f(x_i,y_j)+f(x_i,y_{j+1})+f(x_{i+1},y_j)+f(x_{i+1},y_{j+1})\right]\right]\Bigg|$$

$$\leqslant \frac{1}{16}h_i^2 l_i^2 \|D_2D_1f\|_\infty. \tag{2.7.21}$$

Summing both sides of (2.7.21) over i from 0 to $n-1$ and over j from 0 to $m-1$ and using the generalized triangle inequality, we deduce the desired inequality in (2.7.20).

2.7.3 *Approximation for the finite Fourier transform of two independent variables*

The Fourier transform has applications in a wide variety of fields in science and engineering. In this section, we present the inequality established by Hanna, Dragomir and Roumeliotis [66] for the error, in approximating the finite Fourier transform in two independent variables.

Let $\Delta = [a,b] \times [c,d]$ and $f : \Delta \to \mathbb{R}$ be a continuous mapping defined on Δ and $F(f)$ its finite Fourier transform. That is

$$F(f)(u,v;a,b,c,d) = \int_a^b \int_c^d f(x,y)e^{-2\pi i(ux+vy)}dydx, \tag{2.7.22}$$

$(u,v) \in \Delta$. For a function of one variable we use the notation

$$F(g)(u,a,b) = \int_a^b g(x)e^{-2\pi iux}dx.$$

The following inequality in approximating the finite Fourier transform (2.7.22) in terms of the exponential means was obtained in [66].

Theorem 2.7.3. Let $f : \Delta \to \mathbb{R}$ be an absolutely continuous mapping on Δ and assume that $D_2D_1f(x,y)$ exists on $(a,b)\times(c,d)$, then we have the inequality

$$|F(f)(u,v;a,b,c,d)-I_1-I_2+I_3| \leqslant \begin{cases} J_1; \\ J_2; \\ J_3; \end{cases} \tag{2.7.23}$$

for all $(u,v)\in\Delta$, where

$$\begin{aligned} I_1 &:= I_1(u,v;a,b,c,d) = E(u)\int_a^b F(f(s,\cdot))(v;c,d)ds, \\ I_2 &:= I_2(u,v;a,b,c,d) = E(v)\int_c^d F(f(\cdot,t))(u;a,b)dt, \\ I_3 &:= I_3(u,v;a,b,c,d) = E(u)E(v)\int_a^b\int_c^d f(s,t)dtds, \end{aligned}$$

with

$$\begin{aligned} E(u) &= E(-2\pi iub,-2\pi iua), \\ E(v) &= E(-2\pi ivd,-2\pi ivc), \end{aligned}$$

given that E is the exponential mean of complex numbers, that is

$$E(z,w) := \begin{cases} \dfrac{e^z-e^w}{z-w} & \text{if } z \neq w, \\ e^w & \text{if } z = w, \end{cases}$$

for $z,\ w \in C$, and

$$\begin{aligned} J_1 &:= J_1(a,b,c,d,\|D_2D_1f\|_\infty) = \frac{(b-a)^2(d-c)^2}{9}\|D_2D_1f\|_\infty, \\ &\qquad \text{if } D_2D_1f(x,y)\in L_\infty(\Delta); \\ J_2 &:= J_2(a,b,c,d,\|D_2D_1f\|_p) = \left[\frac{2[(b-a)(d-c)]^{\frac{q+1}{2}}}{(q+1)(q+2)}\right]^{\frac{2}{q}}\|D_2D_1f\|_p, \\ &\qquad \text{if } D_2D_1f(x,y)\in L_p(\Delta),\ \frac{1}{p}+\frac{1}{q}=1,\ p>1; \\ J_3 &:= J_3(a,b,c,d,\|D_2D_1f\|_1) = (b-a)(d-c)\|D_2D_1f\|_1, \\ &\qquad \text{if } D_2D_1f(x,y)\in L_1(\Delta), \end{aligned}$$

where

$$\|D_2D_1f\|_\infty = \sup_{(s,t)\in\Delta}|D_2D_1f(s,t)| < \infty,$$

$$\|D_2D_1f\|_p = \left(\int_a^b\int_c^d |D_2D_1f(s,t)|^p dtds\right)^{\frac{1}{p}} < \infty, \quad 1\leqslant p<\infty,$$

are the usual Lebesgue norms.

Proof. Using the identity obtained by Barnett and Dragomir in [8] (see, Lemma 2.3.2), we have

$$f(x,y)=\frac{1}{b-a}\int_a^b f(s,y)ds+\frac{1}{d-c}\int_c^d f(x,t)dt-\frac{1}{(b-a)(d-c)}\int_a^b\int_c^d f(s,t)dtds$$

$$+\frac{1}{(b-a)(d-c)}\int_a^b\int_c^d P(x,s)Q(y,t)D_2D_1f(s,t)dtds, \tag{2.7.24}$$

provided that f is continuous on Δ and

$$P(x,s)=\begin{cases} s-a, \ a\leqslant s\leqslant x,\\ s-b, \ x<s\leqslant b.\end{cases}$$

$$Q(y,t)=\begin{cases} t-c, & c\leqslant t\leqslant y,\\ t-d, & y<t\leqslant d.\end{cases}$$

If we replace $f(x,y)$ in (2.7.22) by its representation from (2.7.24), we get

$$F(f)(u,v;a,b,c,d)$$

$$=\int_a^b\int_c^d\left(\frac{e^{-2\pi i(ux+vy)}}{b-a}\int_a^b f(s,y)ds\right)dydx+\int_a^b\int_c^d\left(\frac{e^{-2\pi i(ux+vy)}}{d-c}\int_c^d f(x,t)dt\right)dydx$$

$$-\int_a^b\int_c^d\left(\frac{e^{-2\pi i(ux+vy)}}{(b-a)(d-c)}\int_a^b\int_c^d f(s,t)dtds\right)dydx+R(f,u,v;a,b,c,d), \tag{2.7.25}$$

where

$$R(f,u,v;a,b,c,d)=\frac{1}{(b-a)(d-c)}\int_a^b\int_c^d\left(e^{-2\pi i(ux+vy)}\right)$$

$$\times\left(\int_a^b\int_c^d P(x,s)Q(y,t)D_2D_1f(s,t)dtds\right)dydx. \tag{2.7.26}$$

Let

$$I_1=\int_a^b\int_c^d\left(\frac{e^{-2\pi i(ux+vy)}}{b-a}\int_a^b f(s,y)ds\right)dydx,$$

then

$$I_1=\int_a^b\frac{e^{-2\pi iux}}{b-a}dx\left(\int_c^d e^{-2\pi ivy}\left(\int_a^b f(s,y)ds\right)dy\right)$$

$$=\frac{e^{-2\pi iub}-e^{-2\pi iua}}{-2\pi iu(b-a)}\int_a^b\left(\int_c^d e^{-2\pi ivy}f(s,y)dy\right)ds=E(u)\int_a^b F(f(s,\cdot))(v;c,d)ds.$$

In a similar fashion we obtain

$$I_2=\int_a^b\int_c^d\left(\frac{e^{-2\pi i(ux+vy)}}{d-c}\int_c^d f(x,t)dt\right)dydx=E(v)\int_c^d F(f(\cdot,t))(u;a,b)dt$$

and

$$I_3 = \int_a^b \int_c^d \left(\frac{e^{-2\pi i(ux+vy)}}{(b-a)(d-c)} \int_a^b \int_c^d f(s,t)dtds \right) dydx$$

$$= \frac{1}{(b-a)(d-c)} \int_a^b \int_c^d f(s,t)dtds \int_a^b \int_c^d e^{-2\pi iux} e^{-2\pi ivy} dydx$$

$$= E(u)E(v) \int_a^b \int_c^d f(s,t)dtds.$$

From (2.7.25) and using the properties of modulus, we have

$$|F(f)(u,v;a,b,c,d) - I_1 - I_2 + I_3|$$

$$= \left| \int_a^b \int_c^d \left(\int_a^b \int_c^d \frac{e^{-2\pi i(ux+vy)}}{(b-a)(d-c)} P(x,s)Q(y,t)D_2D_1f(s,t)dtds \right) dydx \right|$$

$$\leqslant \int_a^b \int_c^d \int_a^b \int_c^d \left| \frac{e^{-2\pi i(ux+vy)}}{(b-a)(d-c)} \right| |p(x,s)||Q(y,t)||D_2D_1f(s,t)|dtdsdydz$$

$$= \int_a^b \int_c^d \int_a^b \int_c^d \frac{|p(x,s)||Q(y,t)|}{(b-a)(d-c)} |D_2D_1f(s,t)|dtdsdydx. \tag{2.7.27}$$

Now, we observe that

$$\int_a^b \int_c^d \int_a^b \int_c^d |p(x,s)||Q(y,t)||D_2D_1f(s,t)|dtdsdydx$$

$$\leqslant \|D_2D_1f\|_\infty \left[\int_a^b \left(\int_a^b |p(x,s)|ds \right) dx \int_c^d \left(\int_c^d |Q(y,t)|dt \right) dy \right]$$

$$= \|D_2D_1f\|_\infty \left[\int_a^b \left\{ \frac{(s-a)^2}{2}\bigg|_a^x + \frac{(b-s)^2}{2}\bigg|_x^b \right\} dx \int_c^d \left\{ \frac{(t-c)^2}{2}\bigg|_c^y + \frac{(d-t)^2}{2}\bigg|_y^d \right\} dy \right]$$

$$= \|D_2D_1f\|_\infty \left[\left(\int_a^b \frac{(x-a)^2}{2} dx + \int_a^b \frac{(b-x)^2}{2} dx \right) \left(\int_c^d \frac{(y-c)^2}{2} dy + \int_c^d \frac{(d-y)^2}{2} dy \right) \right]$$

$$= \|D_2D_1f\|_\infty \left[\frac{(b-a)^3}{3} \frac{(d-c)^3}{3} \right]. \tag{2.7.28}$$

Substituting (2.7.28) in (2.7.27), we obtain the first inequality in (2.7.28).

Applying Hölder's integral inequality for double integrals, we get

$$\int_a^b \int_c^d \int_a^b \int_c^d |p(x,s)||Q(y,t)||D_2D_1f(s,t)|dtdsdydx$$

$$\leqslant \left(\int_a^b\int_c^d\int_a^b\int_c^d \{|p(x,s)||Q(y,s)|\}^q dtdsdydx\right)^{\frac{1}{q}}$$

$$\times \left(\int_a^b\int_c^d\int_a^b\int_c^d |D_2D_1f(s,t)|^p dtdsdydx\right)^{\frac{1}{p}}$$

$$= \|D_2D_1f\|_p\{(b-a)(d-c)\}^{\frac{1}{p}}$$

$$\times \left(\int_a^b\left(\int_a^b |p(x,s)|^q ds\right)dx\right)^{\frac{1}{q}}\left(\int_c^d\left(\int_c^d |Q(y,t)|^q dt\right)dy\right)^{\frac{1}{q}}$$

$$= \|D_2D_1f\|_p\{(b-a)(d-c)\}^{\frac{1}{p}}$$

$$\times \left(\int_a^b\left(\frac{(x-a)^{q+1}}{q+1}+\frac{(b-x)^{q+1}}{q+1}\right)dx\right)^{\frac{1}{q}}\left(\int_c^d\left(\frac{(y-c)^{q+1}}{q+1}+\frac{(d-y)^{q+1}}{q+1}\right)dy\right)^{\frac{1}{q}}$$

$$= \|D_2D_1f\|_p\left[\frac{2^{\frac{2}{q}}(b-a)^{1+\frac{2}{q}}(d-c)^{1+\frac{2}{q}}}{\{(q+1)(q+2)\}^{\frac{2}{q}}}\right]. \tag{2.7.29}$$

Using (2.7.29) in (2.7.27), we get the second inequality in (2.7.23)

Finally, we obtain that

$$\int_a^b\int_c^d\int_a^b\int_c^d |p(x,s)||Q(y,t)||D_2D_1f(s,t)|dtdsdydx$$

$$\leqslant \sup_{(x,s)\in[a,b]^2}|p(x,s)|\sup_{(y,t)\in[c,d]^2}|Q(y,t)|\int_a^b\int_c^d\int_a^b\int_c^d |D_2D_1f(s,t)|dtdsdydx$$

$$= (b-a)(d-c)\int_a^b\int_c^d\int_a^b\int_c^d |D_2D_1f(s,t)|dtdsdydx$$

$$= \|D_2D_1f\|_1(b-a)^2(d-c)^2. \tag{2.7.30}$$

Using (2.7.30) in (2.7.27), gives the final inequality in (2.7.23), where we have used the fact that

$$\max\{X,Y\} = \frac{x+y}{2}+\left|\frac{y-x}{2}\right|$$

The proof is complete.

2.8 Miscellaneous inequalities

2.8.1 *Dragomir [53]*

Let $(H,(\cdot,\cdot))$ be an inner product space over $\mathbb{K}$ ($\mathbb{K}=\mathbb{R}$, $\mathbb{C}$) and $e \in H$, $\|e\|=1$. If γ, $\Gamma \in \mathbb{K}$ and x, $y \in H$ are such that

$$\operatorname{Re}\left(\Gamma e-\frac{x+y}{2},\frac{x+y}{2}-\gamma e\right)\geqslant 0$$

or, equivalently,

$$\left\|\frac{x+y}{2}-\frac{\gamma+\Gamma}{2}e\right\|\leqslant\frac{1}{2}|\Gamma-\gamma|,$$

then, we have the inequality

$$\operatorname{Re}[(x,y)-(x,e)(e,y)]\leqslant\frac{1}{4}|\Gamma-\gamma|^2.$$

The constant $\frac{1}{4}$ is the best possible in the sense that it cannot be replaced by a smaller constant.

2.8.2 *Ujević [153]*

Let $(X,(\cdot,\cdot))$ be a real inner product space and $\{e_i\}_1^n \subset X$, $(e_i,e_j)=\delta_{ij}$, where

$$\delta_{ij}=\begin{cases}1 \text{ if } i=j,\\ 0 \text{ if } i\neq j.\end{cases}$$

If ϕ_i, γ_i, Φ_i, Γ_i, $i=1,2,\ldots,n$, are real numbers and x, $y \in X$ such that the conditions

$$\left(x-\sum_{i=1}^n\gamma_ie_i,\sum_{i=1}^n\Gamma_ie_i-x\right)\geqslant 0$$

and

$$\left(y-\sum_{i=1}^n\phi_ie_i,\sum_{i=1}^n\Phi_ie_i-y\right)\geqslant 0$$

hold, then we have the inequality

$$\left|(x,y)-\sum_{i=1}^n(x,e_i)(y,e_i)\right|\leqslant\frac{1}{4}\sqrt{\sum_{i=1}^n(\Phi_i-\phi_i)^2\sum_{i=1}^n(\Gamma_i-\gamma_i)^2}.$$

The constant $\frac{1}{4}$ is the best possible.

2.8.3 *Dragomir [55]*

Let $(H,(\cdot,\cdot))$ be an inner product space over $\mathbb{K}$ ($\mathbb{K}=\mathbb{R}$, $\mathbb{C}$). Let $\{e_i\}_{i\in I}$ be a family of orthonormal vectors in H, F a finite part of I, ϕ_i, Φ_i, γ_i, $\Gamma_i \in \mathbb{K}$, $i\in F$ and x, $y\in H$. If

$$\operatorname{Re}\left(\sum_{i=1}^{n}\Phi_i e_i - x, x-\sum_{i=1}^{n}\phi_i e_i\right)\geqslant 0,$$

$$\operatorname{Re}\left(\sum_{i=1}^{n}\Gamma_i e_i - y, y-\sum_{i=1}^{n}\gamma_i e_i\right)\geqslant 0,$$

or, equivalently,

$$\left\|x-\sum_{i\in F}\frac{\Phi_i+\phi_i}{2}e_i\right\|\leqslant\frac{1}{2}\left(\sum_{i\in F}|\Phi_i-\phi_i|^2\right)^{\frac{1}{2}},$$

$$\left\|y-\sum_{i\in F}\frac{\Gamma_i+\gamma_i}{2}e_i\right\|\leqslant\frac{1}{2}\left(\sum_{i\in F}|\Gamma_i-\gamma_i|^2\right)^{\frac{1}{2}},$$

hold, then we have the inequalities

$$\left|(x,y)-\sum_{i\in F}(x,e_i)(e_i,y)\right|\leqslant\frac{1}{4}\left(\sum_{i\in F}|\Phi_i-\phi_i|^2\right)^{\frac{1}{2}}\left(\sum_{i\in F}|\Gamma_i-\gamma_i|^2\right)^{\frac{1}{2}}$$

$$-\left[\operatorname{Re}\left(\sum_{i\in F}\Phi_i e_i - x, x-\sum_{i\in F}\phi_i e_i\right)\right]^{\frac{1}{2}}\times\left[\operatorname{Re}\left(\sum_{i\in F}\Gamma_i e_i - y, y-\sum_{i\in F}\gamma_i e_i\right)\right]^{\frac{1}{2}}$$

$$\leqslant\frac{1}{4}\left(\sum_{i\in F}|\Phi_i-\phi_i|^2\right)^{\frac{1}{2}}\left(\sum_{i\in F}|\Gamma_i-\gamma_i|^2\right)^{\frac{1}{2}}.$$

The constant $\frac{1}{4}$ is the best possible.

2.8.4 *Dragomir [55]*

Let $(H,(\cdot,\cdot))$ be an inner product space over K ($\mathbb{K}=\mathbb{R}$, $\mathbb{C}$). Let $\{e_i\}_{i\in I}$ be a family of orthonormal vectors in H, F a finite part of I and ϕ_i, $\Phi_i\in\mathbb{K}$, $i\in F$, x, $y\in H$ and $\lambda\in(0,1)$, such that either

$$\operatorname{Re}\left(\sum_{i\in F}\Phi_i e_i-(\lambda x+(1-\lambda)y),\lambda x+(1-\lambda)y-\sum_{i\in F}\phi_i e_i\right)\geqslant 0,$$

or, equivalently,

$$\left\|\lambda x+(1-\lambda)y-\sum_{i\in F}\frac{\Phi_i+\phi_i}{2}e_i\right\|\leqslant\frac{1}{2}\left(\sum_{i\in F}|\Phi_i-\phi_i|^2\right)^{\frac{1}{2}},$$

holds. Then we have the inequality

$$\operatorname{Re}\left[(x,y)-\sum_{i\in F}(x,e_i)(e_i,y)\right]\leqslant\frac{1}{16}\frac{1}{\lambda(1-\lambda)}\sum_{i\in F}|\Phi_i-\phi_i|^2$$

$$-\frac{1}{4}\frac{1}{\lambda(1-\lambda)}\sum_{i\in F}\left|\frac{\Phi_i+\phi_i}{2}-(\lambda x+(1-\lambda)y,e_i)\right|^2\leqslant\frac{1}{16}\frac{1}{\lambda(1-\lambda)}\sum_{i\in F}|\Phi_i-\phi_i|^2.$$

The constant $\frac{1}{16}$ is the best possible.

2.8.5 *Dragomir, Pečarić and Tepeš [56]*

Let $(H,(\cdot,\cdot))$ be an inner product space over $\mathbb{K}$ ($\mathbb{K}=\mathbb{R}$, $\mathbb{C}$). Let $\{e_i\}_{i\in I}$ be a family of orthonormal vectors in H, F a finite part of I and ϕ_i, $\Phi_i\in\mathbb{K}$, $i\in F$, x, $y\in H$ such that either the condition

$$\operatorname{Re}\left(\sum_{i\in F}\Phi_i e_i-x,x-\sum_{i\in F}\phi_i e_i\right)\geqslant 0,$$

or, equivalently,

$$\left\|x-\sum_{i\in F}\frac{\Phi_i+\phi_i}{2}e_i\right\|\leqslant\frac{1}{2}\left(\sum_{i\in F}|\Phi_i-\phi_i|^2\right)^{\frac{1}{2}},$$

holds. Then we have

$$\left|(x,y)-\sum_{i\in F}(x,e_i)(e_i,y)\right|\leqslant\frac{1}{2}\left(\sum_{i\in F}|\Phi_i-\phi_i|^2\right)^{\frac{1}{2}}\|y\|-\sum_{i\in F}\left|\frac{\Phi_i+\phi_i}{2}-(x,e_i)\right||(y,e_i)|.$$

2.8.6 *Hanna, Dragomir and Roumeliotis [67]*

Let (Ω,Σ,μ) be a measure space consisting of a set Ω, Σ a σ-algebra of subsets of Ω and μ a countable additive and positive measure with values in $\mathbb{R}\cup\{\infty\}$ and $\rho:\Omega\to[0,\infty)$ be a μ-measurable function on Ω with $\int_\Omega p(s)d\mu(s)=1$. Denote by $L^2_\rho(\Omega,\mathbb{K})$ ($\mathbb{K}=\mathbb{R}$, $\mathbb{C}$) the Hilbert space of all measurable functions $f:\Omega\to K$ that are 2-ρ-integrable on Ω, i.e. $\int_\Omega\rho(s)|f(s)|^2d\mu(s)<\infty$. If f, $g\in L^2_\rho(\Omega,\mathbb{K})$ and there exist constants γ, $\Gamma\in\mathbb{K}$ such that either the condition

$$\operatorname{Re}\left[(\Gamma-f(s))\left(\overline{f(s)}-\overline{\gamma}\right)\right]\geqslant 0,$$

for μ-a.e., $s\in\Omega$ or equivalently

$$\left|f(s)-\frac{\gamma+\Gamma}{2}\right|\leqslant\frac{1}{2}|\Gamma-\gamma|,$$

for μ-a.e., $s \in \Omega$ holds, then

$$\left| \int_{\Omega} \rho(s) f(s) \overline{g(s)} d\mu(s) - \int_{\Omega} \rho(s) f(s) d\mu(s) \int_{\Omega} \rho(s) \overline{g(s)} d\mu(s) \right|$$

$$\leqslant \frac{1}{2} |\Gamma - \gamma| \left[\int_{\Omega} \rho(s) |g(s)|^2 d\mu(s) - \left| \int_{\Omega} \rho(s) g(s) d\mu(s) \right|^2 \right]^{\frac{1}{2}}.$$

2.8.7 *Buşe, Cerone, Dragomir and Roumeliotis [12]*

Let $(H, \langle \cdot, \cdot \rangle)$ be a real or complex Hilbert space, $\Omega \subset \mathbb{R}^n$ be a Lebesgue measurable set and $\rho : \Omega \to [0, \infty)$ a Lebesgue measurable function with $\int_{\Omega} \rho(s) ds = 1$. Denote by $L_{2,\rho}(\Omega, H)$ the set of all Bochner measurable functions f on Ω such that $\|f\|_{2,\rho}^2 := \int_{\Omega} \rho(s) \|f(s)\|^2 ds < \infty$. Assume that $f,\ g \in L_{2,\rho}(\Omega, H)$ and there exist vectors $x,\ X,\ y,\ Y \in H$ such that

$$\int_{\Omega} \rho(t) \operatorname{Re}\langle X - f(t), f(t) - x \rangle dt \geqslant 0,$$
$$\int_{\Omega} \rho(t) \operatorname{Re}\langle Y - g(t), g(t) - y \rangle dt \geqslant 0,$$

or equivalently,

$$\int_{\Omega} \rho(t) \left\| f(t) - \frac{X + x}{2} \right\|^2 dt \leqslant \frac{1}{4} \|X - x\|^2,$$
$$\int_{\Omega} \rho(t) \left\| g(t) - \frac{Y + y}{2} \right\|^2 dt \leqslant \frac{1}{4} \|Y - y\|^2.$$

Then we have the inequality

$$\left| \int_{\Omega} \rho(t) \langle f(t), g(t) \rangle dt - \left\langle \int_{\Omega} \rho(t) f(t) dt, \int_{\Omega} \rho(t) g(t) dt \right\rangle \right| \leqslant \frac{1}{4} \|X - x\| \|Y - y\|$$

$$- \left[\int_{\Omega} \rho(t) \operatorname{Re}\langle X - f(t), f(t) - x \rangle dt \int_{\Omega} \rho(t) \operatorname{Re}\langle Y - g(t), g(t) - y \rangle dt \right]^{\frac{1}{2}} \leqslant \frac{1}{4} \|X - x\| \|Y - y\|.$$

The constant $\frac{1}{4}$ in both inequalities is sharp.

2.8.8 *Hanna, Dragomir and Cerone [62]*

Let $f,\ g : [a,b] \times [c,d] \to \mathbb{R}$ be two mappings such that

$$|f(x,y) - f(u,v)| \leqslant M_1 |x - u|^{\alpha_1} + M_2 |y - v|^{\alpha_2},$$

$$|g(x,y) - g(u,v)| \leqslant N_1 |x - u|^{\beta_1} + N_2 |y - v|^{\beta_2},$$

where M_1, M_2, N_1, N_2 are positive constants and $\alpha_1, \alpha_2, \beta_1, \beta_2$ are constants lying in $(0,1]$. Then we have the inequality:

$$\left|\frac{1}{(b-a)(d-c)}\int_a^b\int_c^d f(x,y)g(x,y)dydx\right.$$
$$\left.-\left(\frac{1}{(b-a)(d-c)}\int_a^b\int_c^d f(x,y)dydx\right)\left(\frac{1}{(b-a)(d-c)}\int_a^b\int_c^d g(x,y)dydx\right)\right|$$
$$\leqslant 4\left[M_1N_1\frac{(b-a)^{\alpha_1+\beta_1}}{(\alpha_1+\beta_1+1)(\alpha_1+\beta_1+2)}+M_1N_2\frac{2(b-a)^{\alpha_1}(d-c)^{\beta_2}}{(\alpha_1+1)(\alpha_1+2)(\beta_2+1)(\beta_2+2)}\right.$$
$$\left.+M_2N_1\frac{2(b-a)^{\beta_1}(d-c)^{\alpha_2}}{(\alpha_2+1)(\alpha_2+2)(\beta_1+1)(\beta_1+2)}+M_2N_2\frac{(d-c)^{\alpha_2+\beta_2}}{(\alpha_2+\beta_2+1)(\alpha_2+\beta_2+2)}\right].$$

2.8.9 *Pachpatte [94]*

Let the assumptions of Theorem 2.5.3 hold. Then

$$|A(w,f,g;D)| \leqslant \frac{1}{2\int_D w(x)dx}\int_D w(x)\left(H(f,x_i,y_i)H(g,x_i,y_i)w(y)dy\right)dx,$$

where A and H are as in Theorem 2.5.3.

2.8.10 *Pachpatte [94]*

Assume that the hypotheses of Theorem 2.6.3 hold. Let $w(x)$ be a real-valued nonnegative function defined on Q and $\sum_x w(x) > 0$. Then

$$|P(w,f,g;Q)| \leqslant \frac{1}{2\sum_x w(x)}\sum_x w(x)\left[|g(x)|\sum_y E(f,x_i,y_i)w(y)+|f(x)|\sum_y E(g,x_i,y_i)w(y)\right],$$

$$|P(w,f,g;Q)| \leqslant \frac{1}{(\sum_x w(x))^2}\sum_x w(x)\left(\sum_y E(f,x_i,y_i)w(y)\right)\left(\sum_y E(g,x_i,y_i)w(y)\right),$$

$$|P(w,f,g;Q)| \leqslant \frac{1}{2\sum_x w(x)}\sum_x w(x)\left(\sum_y E(f,x_i,y_i)E(g,x_i,y_i)w(y)\right),$$

where we have set the notations

$$P(w,p,q;Q) = \sum_x w(x)p(x)q(x) - \frac{1}{\sum_x w(x)}\left(\sum_x w(x)p(x)\right)\left(\sum_x w(x)q(x)\right),$$

$$E(p,x_i,y_i) = \sum_{i=1}^{n} \|\Delta_i p\|_\infty |x_i - y_i|,$$

for some functions $p,\ q : Q \to \mathbb{R}$.

2.8.11 *Pachpatte [130]*

Under the notations and definitions given in section 2.6, let $f,\ g : Q \to \mathbb{R}$ be summable functions on Q and $p : Q \to \mathbb{R}_+$ a summable function on Q such that $\overline{p} = \sum_x p(x) > 0$. Then

$$|F(\overline{P},p,f,g;Q)| \leqslant \sqrt{F(\overline{P},p,f,f;Q)}\sqrt{F(\overline{P},p,g,g;Q)},$$

and in addition if $\phi \leqslant f(x) \leqslant \Phi$, $\gamma \leqslant g(x) \leqslant \Gamma$ for each $x \in Q$, where $\phi,\ \Phi,\ \gamma,\ \Gamma$ are given real constants, then

$$|F(\overline{P},p,f,g;Q)| \leqslant \frac{1}{4}(\Phi-\phi)(\Gamma-\gamma),$$

where

$$F(\overline{P},p,f,g,Q) = \frac{1}{\overline{P}}\sum_x p(x)f(x)g(x) - \left(\frac{1}{\overline{P}}\sum_x p(x)f(x)\right)\left(\frac{1}{\overline{P}}\sum_x p(x)g(x)\right). \qquad \text{(o)}$$

2.8.12 *Pachpatte [130]*

Under the notations and definitions given in Section 2.6, let $f,\ g : Q \to \mathbb{R}$ be summable functions on Q and $p : Q \to \mathbb{R}$ a summable function on Q such that $\overline{P} = \sum_x p(x) > 0$. Then

$$|F(\overline{P},p,f,g;Q)| \leqslant \frac{1}{\overline{P}}\sum_x p(x)\left|\left(f(x) - \frac{1}{\overline{P}}\sum_y p(y)f(y)\right)\left(g(x) - \frac{1}{\overline{P}}\sum_y p(y)g(y)\right)\right|$$

and in addition if $\phi \leqslant f(x) \leqslant \Phi$ for each $x \in Q$, where $\phi,\ \Phi$ are given real constants, then

$$|F(\overline{P},p,f,g;Q)| \leqslant \frac{\Phi-\phi}{2}\sqrt{F(\overline{P},p,g,g;Q)},$$

where $\mathrm{F}\left(\overline{\mathrm{P}},p,f,g;Q\right)$ is defined by (o).

2.9 Notes

The Grüss inequality has been generalized and extended over the last years in a number of ways. In [32], Dragomir investigated the Grüss type inequality in Theorem 2.2.1 in real or Complex inner product spaces. Lemmas 2.2.1 and 2.2.2 are due to Dragomir [53]. Theorem 2.2.2 provides a new proof of Theorem 2.2.1 by replacing the condition (2.2.5) by an equivalent but simpler assumption and is due to Dragomir [53]. Theorem 2.2.3 deals with the refinement of the inequality in Theorem 2.2.1 and is taken from Dragomir [53]. Theorem 2.2.4 is due to Dragomir, Pečarić and Tepeš [56], while Theorem 2.2.5 is due to Dragomir [43].

Lemmas 2.3.1 and 2.3.2 are respectively taken from Dragomir, Corone, Barnett and Rouleliotis [37] and Barnett and Dragomir [8] and the results presented in Theorems 2.3.1–2.3.4 are taken from Pachpatte [89,122,91,129]. Theorems 2.4.1 and 2.4.2 deal with the Trapezoid type inequalities and are taken respectively from Pachpatte [86], Dragomir, Barnett and Pearce [39], while Theorems 2.4.3 and 2.4.4 are due to Barnett and Dragomir [6]. The results presented in Sections 2.5 and 2.6 are due to Pachpatte and taken from [130,94,103,129,95]. The material included in Section 2.7 is devoted to the applications and adapted from Dragomir and Gomm [54], Dragomir, Barnett and Pearce [39] and Hanna, Dragomir and Roumeliotis [66]. Section 2.8 contains a few miscellanous inequalities investigated by various investigators.

Chapter 3

Ostrowski-type inequalities

3.1 Introduction

In [81], A.M. Ostrowski proved the inequality (7), which is now known in the literature as Ostrowski's inequality. Since its apperance in 1938, a good deal of research activity has been concentrated on the investigation of the inequalities of the type (7) and their applications. The books [50,80] contain a considerable amount of results related to Ostrowski's inequality. In the last two decades, the inequalities which claim their origin to the Ostrowski's inequality (7) have renewed interests and several studies dedicated to obtain various generalizations, extensions, variants and applications have appeared in the literature. In this chapter, we present some of the more recent developments related to the Ostrowski's inequality (7), not covered in [50,80]. Applications are discussed to illustrate the usefulness of certain inequalities.

3.2 Inequalities of the Ostrowski-type

In this section, we present some inequalities of the Ostrowski's type recently established by various investigators.
We start with the following generalization of the Ostrowski's inequality for Lipschitzian mappings established by Dragomir in [52].

Theorem 3.2.1. Let $f:[a,b]\to\mathbb{R}$ be an L-lipschitzian mapping on $[a,b]$, i.e.,

$$|f(x)-f(y)|\leqslant L|x-y|,$$

for all $x,\ y\in[a,b]$ and the constant $L\geqslant 0$. Then we have the inequality

$$\left|\int_a^b f(t)dt-f(x)(b-a)\right|\leqslant L(b-a)^2\left[\frac{1}{4}+\frac{\left(x-\frac{a+b}{2}\right)^2}{(b-a)^2}\right], \qquad (3.2.1)$$

for all $x\in[a,b]$. The constant $\frac{1}{4}$ is the best possible.

Proof. Using the integration by parts formula for Riemann-Stieltjes integral, we have

$$\int_a^x (t-a)df(t) = f(x)(x-a) - \int_a^x f(t)dt$$

and

$$\int_x^b (t-b)df(t) = f(x)(b-x) - \int_x^b f(t)dt.$$

If we add the above two equalities, we get

$$f(x)(b-a) - \int_a^b f(t)dt = \int_a^x (t-a)df(t) + \int_x^b (t-b)df(t). \tag{3.2.2}$$

Now, assume that $\Delta_n : a = x_0^{(n)} < x_1^{(n)} < \cdots < x_{(n-1)}^{(n)} < x_n^{(n)} = b$ is a sequence of divisions with $\nu(\Delta_n) \to 0$ as $n \to \infty$ where, $\nu(\Delta_n) := \max_{i\in\{0,\ldots,n-1\}} \left(x_{i+1}^{(n)} - x_i^{(n)}\right)$ and $\xi_i^{(n)} \in \left[x_i^n, x_{i+1}^{(n)}\right]$. If $p : [a,b] \to \mathbb{R}$ is Riemann integrable on $[a,b]$ and $\nu : [a,b] \to \mathbb{R}$ is L-Lispschitzian on $[a,b]$, then

$$\begin{aligned}
\left|\int_a^b p(x)d\nu(x)\right| &= \left|\lim_{\nu(\Delta_n)\to 0} \sum_{i=0}^{n-1} p\left(\xi_i^{(n)}\right)\left[\nu\left(x_{i+1}^{(n)}\right) - \nu\left(x_i^{(n)}\right)\right]\right| \\
&\leqslant \lim_{\nu(\Delta_n)\to 0} \sum_{i=0}^{n-1} \left|p\left(\xi_i^{(n)}\right)\right| \left(x_{i+1}^{(n)} - x_i^{(n)}\right) \left|\frac{\nu\left(x_{i+1}^{(n)}\right) - \nu\left(x_i^{(n)}\right)}{x_{i+1}^{(n)} - x_i^{(n)}}\right| \\
&\leqslant L \lim_{\nu(\Delta_n)\to 0} \sum_{i=0}^{n-1} \left|p\left(\xi_i^{(n)}\right)\right| \left(x_{i+1}^{(n)} - x_i^{(n)}\right) \\
&= L\int_a^b |p(x)|dx.
\end{aligned} \tag{3.2.3}$$

Applying the inequality (3.2.3) on $[a,x]$ and $[x,b]$ successively, we get

$$\begin{aligned}
\left|\int_a^x (t-a)df(t) + \int_x^b (t-b)df(t)\right| &\leqslant \left|\int_a^x (t-a)df(t)\right| + \left|\int_x^b (t-b)df(t)\right| \\
&\leqslant L\left[\int_a^x |t-a|\,dt + \int_x^b |t-b|dt\right] \\
&= \frac{L}{2}\left[(x-a)^2 + (b-x)^2\right] \\
&= L(b-a)^2\left[\frac{1}{4} + \frac{\left(x - \frac{a+b}{2}\right)^2}{(b-a)^2}\right].
\end{aligned} \tag{3.2.4}$$

and then by (3.2.4), via the identity (3.2.2), we deduce the desired inequality (3.2.1).

Now, assume that the inequality (3.2.1) holds with a constant $C > 0$, i.e.,

$$\left|\int_a^b f(t)dt - f(x)(b-a)\right| \leqslant L(b-a)^2\left[C + \frac{\left(x - \frac{a+b}{2}\right)^2}{(b-a)^2}\right] \tag{3.2.5}$$

for all $x \in [a,b]$. Consider the mapping $f : [a,b] \to \mathbb{R}$, $f(x) = x$ in (3.2.5). Then

$$\left| x - \frac{a+b}{2} \right| \leqslant \left[C + \frac{\left(x - \frac{a+b}{2}\right)^2}{(b-a)^2} \right] (b-a),$$

for all $x \in [a,b]$, and then for $x = a$, we get

$$\frac{b-a}{2} \leqslant \left[C + \frac{1}{4} \right] (b-a),$$

which implies that $C \geqslant \frac{1}{4}$, and the proof is complete.

Remark 3.2.1. If the mapping f is differentiable on (a,b) and with derivative f' being bounded on (a,b), i.e., $\|f'\|_\infty = \sup_{x\in(a,b)} |f'(t)| < \infty$, then instead of L in (3.1.1) we can put $\|f'\|_\infty$.

In [26], Dragomir and Wang proved the following Ostrowski-type inequality.

Theorem 3.2.2. Let $f : [a,b] \to \mathbb{R}$ be a differentiable mapping on (a,b) $(a < b)$. Suppose that f' is integrable on $[a,b]$ and $\gamma \leqslant f'(x) \leqslant \Gamma$ for all $x \in [a,b]$ and $\gamma,\ \Gamma \in \mathbb{R}$. Then

$$\left| f(x) - \frac{1}{b-a} \int_a^b f(t)dt - \frac{f(b)-f(a)}{b-a} \left(x - \frac{a+b}{2} \right) \right| \leqslant \frac{1}{4}(b-a)(\Gamma - \gamma), \tag{3.2.6}$$

for all $x \in [a,b]$.

Proof. Define the function

$$p(x,t) = \begin{cases} t-a & \text{if } t \in [a,x], \\ t-b & \text{if } t \in (x,b]. \end{cases} \tag{3.2.7}$$

Integrating by parts (see the proof of Theorem 1.2.3), we have

$$\frac{1}{b-a} \int_a^b p(x,t) f'(t)dt = f(x) - \frac{1}{b-a} \int_a^b f(t)dt, \tag{3.2.8}$$

for all $x \in [a,b]$. It is clear that for all $x \in [a,b]$ and $t \in [a,b]$ from (3.2.7), we have

$$x - b \leqslant p(x,t) \leqslant x - a.$$

Applying Grüss inequality (3), to the mappings $p(x,\cdot)$ and $f'(c)$, we obtain

$$\left| \frac{1}{b-a} \int_a^b p(x,t) f'(t)dt - \left(\frac{1}{b-a} \int_a^b p(x,t)dt \right) \left(\frac{1}{b-a} \int_a^b f'(t)dt \right) \right|$$
$$\leqslant \frac{1}{4}(x-a-x+b)(\Gamma-\gamma) = \frac{1}{4}(b-a)(\Gamma-\gamma). \tag{3.2.9}$$

By simple calculation, we get

$$\frac{1}{b-a} \int_a^b p(x,t)dt = \frac{1}{b-a} \left[\int_a^x (t-a)dt + \int_x^b (t-b)dt \right] = x - \frac{a+b}{2},$$

and

$$\frac{1}{b-a} \int_a^b f'(t)dt = \frac{f(b)-f(a)}{b-a}.$$

The required inequality (3.2.6) follows from (3.2.9), (3.2.8) and the above two equalities. The proof is complete.

Remark 3.2.2. If we choose in (3.2.6), $x = \frac{a+b}{2}$ and $x = b$ respectively, then we get

$$\left| f\left(\frac{a+b}{2}\right) - \frac{1}{(b-a)} \int_a^b f(t)dt \right| \leqslant \frac{1}{4}(b-a)(\Gamma - \gamma), \tag{3.2.10}$$

and

$$\left| \frac{f(a)+f(b)}{2} - \frac{1}{(b-a)} \int_a^b f(t)dt \right| \leqslant \frac{1}{4}(b-a)(\Gamma - \gamma). \tag{3.2.11}$$

An inequality similar to (3.2.6) obtained by Ujević [154], is embodied in the following theorem.

Theorem 3.2.3. Let $f : [a,b] \to \mathbb{R}$ be an absolutely continous function with derivative $f' \in L_2[a,b]$. Then

$$\left| (b-a)f(x) - \left(x - \frac{a+b}{2}\right)[f(b)-f(a)] - \int_a^b f(t)dt \right| \leqslant \frac{(b-a)^{\frac{3}{2}}}{2\sqrt{3}} \sqrt{\sigma(f')}, \tag{3.2.12}$$

for all $x \in [a,b]$, where

$$\sigma(f') = (b-a)\left[\frac{1}{b-a}\|f'\|_2^2 - \frac{1}{(b-a)^2}\left(\int_a^b f'(t)dt\right)^2\right].$$

The constant $\frac{1}{2\sqrt{3}}$ is the best possible.

Proof. Let $p(x,t)$ be the mapping defined by (3.2.7). Integrating by parts, we obtain

$$\int_a^b p(x,t)f'(t)dt = (b-a)f(x) - \int_a^b f(t)dt. \tag{3.2.13}$$

We also have

$$\int_a^b p(x,t)dt = (b-a)\left(x - \frac{a+b}{2}\right), \tag{3.2.14}$$

and

$$\int_a^b f'(t)dt = f(b) - f(a). \tag{3.2.15}$$

From (3.2.13)–(3.5.15), it follows

$$\int_a^b \left[p(x,t) - \frac{1}{b-a}\int_a^b p(x,s)ds\right]\left[f'(t) - \frac{1}{b-a}\int_a^b f'(s)ds\right]dt$$

$$= (b-a)f(x) - \left(x - \frac{a+b}{2}\right)[f(b)-f(a)] - \int_a^b f(t)dt. \tag{3.2.16}$$

On the other hand, we have

$$\left|\int_a^b \left[p(x,t) - \frac{1}{b-a}\int_a^b p(x,s)ds\right]\left[f'(t) - \frac{1}{b-a}\int_a^b f'(s)ds\right]dt\right|$$

$$\leqslant \left\| p(x,\cdot) - \frac{1}{b-a}\int_a^b p(x,s)ds \right\|_2 \left\| f' - \frac{1}{b-a}\int_a^b f'(s)ds \right\|_2 . \tag{3.2.17}$$

We also have

$$\left\| p(x,\cdot) - \frac{1}{b-a}\int_a^b p(x,s)ds \right\|_2^2 = \frac{(b-a)^3}{12}, \tag{3.2.18}$$

and

$$\left\| f' - \frac{1}{b-a}\int_a^b f'(s)ds \right\|_2^2 = \|f'\|_2^2 - \frac{(f(b)-f(a))^2}{b-a}. \tag{3.2.19}$$

From (3.2.16)–(3.2.19), we easily get (3.2.12), since

$$\sqrt{\sigma(f')} = \left[\|f'\|_2^2 - \frac{(f(b)-f(a))^2}{b-a} \right]^{\frac{1}{2}} .$$

We must show that (3.2.12) is sharp. For that purpose,

We define the function

$$f(t) = \begin{cases} \frac{1}{2}t^2, & t \in [0,x], \\ \frac{1}{2}t^2 - t + x, & t \in (x,1], \end{cases} \tag{3.2.20}$$

where $x \in [0,1]$. The function given in (3.2.20) is absolutely continuous since it is a continuous piecewise polynomial function.

We now suppose that (3.2.12) holds with a constant $C > 0$, i.e.,

$$\left| (b-a)f(x) - \left(x - \frac{a+b}{2} \right) [f(b)-f(a)] - \int_a^b f(t)dt \right|$$

$$\leqslant C(b-a)^{\frac{3}{2}} \left[\|f'\|_2^2 - \frac{(f(b)-f(a))^2}{b-a} \right]^{\frac{1}{2}} . \tag{3.2.21}$$

Choosing $a = 0,\ b = 1$ and f defined by (3.2.20), we get

$$\int_0^1 f(t)dt = x - \frac{1}{3} - \frac{x^2}{2}, \quad f(0) = 0, \quad f(1) = x - \frac{1}{2}, \quad f(x) = \frac{x^2}{2},$$

and the left hand side of (3.2.21) becomes $\frac{1}{12}$. We also find that the right hand side of (3.2.21) becomes $\frac{C}{2\sqrt{3}}$ and hence, we find that $C \geqslant \frac{1}{2\sqrt{3}}$, proving that the constant $\frac{1}{2\sqrt{3}}$ is the best possible in (3.2.12). The proof is complete.

The following Ostrowski type inequalities established by Ujević in [151] and [150], enlarge their applicability to obtain better error bounds.

Theorem 3.2.4. Let $f : I \to \mathbb{R}$(where $I \subset \mathbb{R}$ is an interval), be a mapping differentiable in $\mathring{I}$ (the interior of I), and let $a,\ b \in \mathring{I}$, $a < b$. If there exist constants $\gamma,\ \Gamma \in \mathbb{R}$ such that $\gamma \leqslant f'(t) \leqslant \Gamma$, for all $t \in [a,b]$ and f' is integrable on $[a,b]$, then we have

$$\left| f(x) - \left(x - \frac{a+b}{2}\right) \frac{f(b)-f(a)}{b-a} - \frac{1}{b-a}\int_a^b f(t)dt \right| \leqslant \frac{b-a}{2}(S-\gamma), \tag{3.2.22}$$

and

$$\left| f(x) - \left(x - \frac{a+b}{2}\right) \frac{f(b)-f(a)}{b-a} - \frac{1}{b-a}\int_a^b f(t)dt \right| \leqslant \frac{b-a}{2}(\Gamma - S), \tag{3.2.23}$$

where $S = \frac{f(b)-f(a)}{b-a}$.

Proof. Let $p(x,t)$ be the mapping defined by (3.2.7). Integrating by parts, we have

$$\frac{1}{b-a}\int_a^b p(x,t)f'(t)dt = f(x) - \frac{1}{b-a}\int_a^b f(t)dt. \tag{3.2.24}$$

We also have

$$\frac{1}{b-a}\int_a^b p(x,t)dt = x - \frac{a+b}{2}, \tag{3.2.25}$$

and

$$\int_a^b f'(t)dt = f(b) - f(a). \tag{3.2.26}$$

From (3.2.24)–(3.2.26), it follows that

$$f(x) - \left(x - \frac{a+b}{2}\right) \frac{f(b)-f(a)}{b-a} - \frac{1}{b-a}\int_a^b f(t)dt$$

$$= \frac{1}{b-a}\int_a^b p(x,t)f'(t)dt - \frac{1}{(b-a)^2}\int_a^b f'(t)dt \int_a^b p(x,t)dt. \tag{3.2.27}$$

We denote

$$R_n(x) = \frac{1}{b-a}\int_a^b p(x,t)f'(t)dt - \frac{1}{(b-a)^2}\int_a^b f'(t)dt \int_a^b p(x,t)dt. \tag{3.2.28}$$

If $C \in \mathbb{R}$ is an arbitrary constant, then we have

$$R_n(x) = \frac{1}{b-a}\int_a^b (f'(t) - C)\left[p(x,t) - \frac{1}{b-a}\int_a^b p(x,s)ds\right] dt, \tag{3.2.29}$$

since

$$\int_a^b \left[p(x,t) - \frac{1}{b-a}\int_a^b p(x,s)ds\right] dt = 0. \tag{3.2.30}$$

First, we choose $C = \gamma$ in (3.2.29). Then we have

$$R_n(x) = \frac{1}{b-a}\int_a^b (f'(t) - \Gamma)\left[p(x,t) - \frac{1}{b-a}\int_a^b p(x,s)ds\right] dt,$$

and

$$|R_n(x)| \leqslant \frac{1}{b-a} \max_{t\in[a,b]} \left| p(x,t) - \left(x - \frac{a+b}{2} \right) \right| \int_a^b |f'(t) - \gamma| dt. \tag{3.2.31}$$

Since

$$\max_{t\in[a,b]} \left| p(x,t) - \left(x - \frac{a+b}{2} \right) \right| = \frac{b-a}{2}, \tag{3.2.32}$$

and

$$\int_a^b |f'(t) - \gamma| dt = f(b) - f(a) - \gamma(b-a) = (S-\gamma)(b-a),$$

from (3.2.31), we get

$$|R_n(x)| \leqslant \frac{b-a}{2}(S-\gamma). \tag{3.2.33}$$

From (3.2.27), (3.2.28), and (3.2.33), we eaisly get (3.2.22).

Second, we choose $C = \Gamma$ in (3.2.29). Then we have

$$R_n(x) = \frac{1}{b-a} \int_a^b (f'(t) - \Gamma) \left[p(x,t) - \frac{1}{b-a} \int_a^b p(x,s) ds \right] dt,$$

and

$$|R_n(x)| \leqslant \frac{1}{b-a} \max_{t\in[a,b]} \left| p(x,t) - \left(x - \frac{a+b}{2} \right) \right| \int_a^b |f'(t) - \Gamma| dt. \tag{3.2.34}$$

Since

$$\int_a^b |f'(t) - \Gamma| dt = \Gamma(b-a) - f(b) + f(a) = (\Gamma - S)(b-a). \tag{3.2.35}$$

From (3.2.34), (3.2.32), and (3.2.35), we get

$$|R_n(x)| \leqslant \frac{b-a}{2}(\Gamma - S). \tag{3.2.36}$$

From (3.2.27), (3.2.28), and (3.2.36), we eaisly get (3.2.23). The proof is complete.

Theorem 3.2.5. Let $I \subset \mathbb{R}$ be an open interval and $a,\ b \in I$, $a < b$. If $f : I \to \mathbb{R}$ is a differentiable function such that $\gamma \leqslant f'(t) \leqslant \Gamma$, for all $t \in [a,b]$ and some constants $\gamma,\ \Gamma \in \mathbb{R}$, then

$$\left| (b-a) \left[\frac{\lambda}{2}(f(a)+f(b)) + (1-\lambda) f(x) - \gamma(1-\lambda) \left(x - \frac{a+b}{2} \right) \right] - \int_a^b f(t) dt \right|$$
$$\leqslant (S-\gamma) \max \left\{ \lambda \frac{b-a}{2}, x - a - \lambda \frac{b-a}{2}, b - x - \lambda \frac{b-a}{2} \right\} (b-a), \tag{3.2.37}$$

$$\left| (b-a) \left[\frac{\lambda}{2}(f(a)+f(b)) + (1-\lambda) f(x) - \Gamma(1-\lambda) \left(x - \frac{a+b}{2} \right) \right] - \int_a^b f(t) dt \right|$$
$$\leqslant (\Gamma - S) \max \left\{ \lambda \frac{b-a}{2}, x - a - \lambda \frac{b-a}{2}, b - x - \lambda \frac{b-a}{2} \right\} (b-a), \tag{3.2.38}$$

where $S = \frac{f(b)-f(a)}{b-a}$ and $a + \lambda \frac{b-a}{2} \leqslant x \leqslant b - \lambda \frac{b-a}{2}$, for $\lambda \in [0,1]$.

Proof. Define the mapping

$$k(x,t)=\begin{cases} t-\left(a+\lambda\dfrac{b-a}{2}\right), & t\in[a,x],\\ t-\left(b-\lambda\dfrac{b-a}{2}\right), & t\in(x,b]. \end{cases} \tag{3.2.39}$$

Integrating by parts, we have

$$\begin{aligned}\int_a^b k(x,t)f'(t)dt &= \int_a^x \left[t-\left(a+\lambda\frac{b-a}{2}\right)\right]f'(t)dt+\int_x^b\left[t-\left(b-\lambda\frac{b-a}{2}\right)\right]f'(t)dt\\ &=(b-a)\left[\frac{\lambda}{2}(f(a)+f(b))+(1-\lambda)f(x)\right]-\int_a^b f(t)dt. \end{aligned}\tag{3.2.40}$$

We also have

$$\begin{aligned}\int_a^b k(x,t)dt &= \int_a^x\left[t-\left(a+\lambda\frac{b-a}{2}\right)\right]dt+\int_x^b\left[t-\left(b-\lambda\frac{b-a}{2}\right)\right]dt\\ &=\frac{1}{2}\left[(x-a)-\lambda\frac{b-a}{2}\right]^2-\frac{1}{2}\left[(x-b)+\lambda\frac{b-a}{2}\right]^2\\ &=(1-\lambda)(b-a)\left(x-\frac{a+b}{2}\right). \end{aligned}\tag{3.2.41}$$

Let $C\in\mathbb{R}$ be a constant. From (3.2.40) and (3.2.41) it follows that

$$\int_a^b k(x,t)[f'(t)-C]dt=\int_a^b k(x,t)f'(t)dt-C\int_a^b k(x,t)dt$$

$$=(b-a)\left[\frac{\lambda}{2}(f(a)+f(b))+(1-\lambda)f(x)\ -C(1-\lambda)\left(x-\frac{a+b}{2}\right)\right]-\int_a^b f(t)dt. \tag{3.2.42}$$

If we choose $C=\gamma$ in (3.2.42), then we get

$$(b-a)\left[\frac{\lambda}{2}(f(a)+f(b))+(1-\lambda)f(x)-\gamma(1-\lambda)\left(x-\frac{a+b}{2}\right)\right]-\int_a^b f(t)dt$$

$$=\int_a^b k(x,t)[f'(t)-\gamma]dt. \tag{3.2.43}$$

On the other hand, we have

$$\left|\int_a^b k(x,t)[f'(t)-\gamma]dt\right|\leqslant \max_{t\in[a,b]}|k(x,t)|\int_a^b|f'(t)-\Gamma|dt, \tag{3.2.44}$$

since

$$\max_{t\in[a,b]}|k(x,t)|=\max\left\{\lambda\frac{b-a}{2},x-a-\lambda\frac{b-a}{2},b-x-\lambda\frac{b-a}{2}\right\}, \tag{3.2.45}$$

and

$$\int_a^b|f'(t)-\Gamma|dt=f(b)-f(a)-\gamma(b-a)=(S-\gamma)(b-a). \tag{3.2.46}$$

From (3.2.43)–(3.2.46) it follows that (3.2.37) holds.

If we choose $C = \Gamma$ in (3.2.42), then we get

$$(b-a)\left[\frac{\lambda}{2}(f(a)+f(b))+(1-\lambda)f(x)-\Gamma(1-\lambda)\left(x-\frac{a+b}{2}\right)\right]-\int_a^b f(t)dt$$

$$=\int_a^b k(x,t)[f'(t)-\gamma]dt, \tag{3.2.47}$$

and

$$\int_a^b |f'(t)-\Gamma|dt = \Gamma(b-a)-(f(b)-f(a)) = (\Gamma-S)(b-a). \tag{3.2.48}$$

From (3.2.47), (3.2.45) and (3.2.48), we easily get (3.2.38). The proof is complete.

Corollary 3.2.1. Under the assumption of Theorem 3.2.5, we have

$$\left|f(x)(b-a)-\gamma(b-a)\left(x-\frac{a+b}{2}\right)-\int_a^b f(t)dt\right|$$

$$\leqslant (S-\gamma)\left[\frac{b-a}{2}+\left|x-\frac{a+b}{2}\right|\right](b-a), \tag{3.2.49}$$

$$\left|f(x)(b-a)-\Gamma(b-a)\left(x-\frac{a+b}{2}\right)-\int_a^b f(t)dt\right|$$

$$\leqslant (\Gamma-S)\left[\frac{b-a}{2}+\left|x-\frac{a+b}{2}\right|\right](b-a). \tag{3.2.50}$$

Proof. We set $\lambda = 0$ in (3.2.37) and (3.2.38). Then we have

$$\max\{x-a, b-x\} = \frac{1}{2}[b-a+|2x-a-b|] = \frac{b-a}{2}+\left|x-\frac{a+b}{2}\right|. \tag{3.2.51}$$

In the above proof, we used

$$\max\{A,B\} = \frac{1}{2}[A+B+|A-B|], \quad a,\, b \in \mathbb{R}.$$

Now, in view of (3.2.51), it is easy to see that (3.2.49) and (3.2.50) are valid.

Corollary 3.2.2. Under the assumptions of Theorem 3.2.5, we have

$$\left|\frac{b-a}{2}[f(a)+f(b)]-\int_a^b f(t)dt\right| \leqslant (S-\gamma)\frac{(b-a)^2}{2}, \tag{3.2.52}$$

$$\left|\frac{b-a}{2}[f(a)+f(b)]-\int_a^b f(t)dt\right| \leqslant (\Gamma-S)\frac{(b-a)^2}{2} \tag{3.2.53}$$

Proof. We set $\lambda = 1$ in (3.2.37) and (3.2.38). Then we have $x = \frac{a+b}{2}$ and

$$\max\left\{\lambda\frac{b-a}{2}, x-a-\lambda\frac{b-a}{2}, b-x-\lambda\frac{b-a}{2}\right\} = \frac{b-a}{2}.$$

The proof of (3.2.52) and (3.2.53) is now obvious.

Corollary 3.2.3. Under the assumptions of Theorem 3.2.5, we have

$$\left|(b-a)\left[\frac{f(a)+f(b)}{4}+\frac{1}{2}f(x)-\frac{\gamma}{2}\left(x-\frac{a+b}{2}\right)\right]-\int_a^b f(t)dt\right|$$
$$\leqslant (S-\gamma)\left[\frac{b-a}{4}+\left|x-\frac{a+b}{2}\right|\right](b-a), \tag{3.2.54}$$

$$\left|(b-a)\left[\frac{f(a)+f(b)}{4}+\frac{1}{2}f(x)-\frac{\Gamma}{2}\left(x-\frac{a+b}{2}\right)\right]-\int_a^b f(t)dt\right|$$
$$\leqslant (\Gamma-S)\left[\frac{b-a}{4}+\left|x-\frac{a+b}{2}\right|\right](b-a). \tag{3.2.55}$$

Proof. We set $\lambda = \frac{1}{2}$ in (3.2.37) and (3.2.38). Then we have

$$\max\left\{\frac{b-a}{4}, x-\frac{3a+b}{4}, \frac{a+3b}{4}-x\right\}$$
$$= \max\left\{\frac{1}{2}\left(x-a+\left|x-\frac{a+b}{2}\right|\right), \frac{1}{2}\left(b-x+\left|x-\frac{a+b}{2}\right|\right)\right\}$$
$$= \frac{1}{4}\left[b-a+2\left|x-\frac{a+b}{2}\right|+|2x-(a+b)|\right]$$
$$= \frac{b-a}{4}+\left|x-\frac{a+b}{2}\right|.$$

The proof of (3.2.54) and (3.2.55) is now obvious.

Corollary 3.2.4. Under the assumptions of Theorem 3.2.5, we have

$$\left|\frac{b-a}{6}[f(a)+4f(x)+f(b)]-\frac{2\gamma}{3}\left(x-\frac{a+b}{2}\right)-\int_a^b f(t)dt\right|$$
$$\leqslant (S-\gamma)\left[\frac{b-a}{3}+\left|x-\frac{a+b}{2}\right|\right](b-a), \tag{3.2.56}$$

$$\left|\frac{b-a}{6}[f(a)+4f(x)+f(b)]-\frac{2\Gamma}{3}\left(x-\frac{a+b}{2}\right)-\int_a^b f(t)dt\right|$$
$$\leqslant (\Gamma-S)\left[\frac{b-a}{3}+\left|x-\frac{a+b}{2}\right|\right](b-a). \tag{3.2.57}$$

Proof. We set $\lambda = \frac{1}{3}$ in (3.2.37) and (3.2.38). Then we have

$$\max\left\{\lambda\frac{b-a}{2}, x-a-\lambda\frac{b-a}{2}, b-x-\lambda\frac{b-a}{2}\right\}$$

$$= \max\left\{\frac{b-a}{6}, x-\frac{5a+b}{6}, \frac{a+5b}{6}-x\right\}$$

$$= \max\left\{\frac{1}{2}\left(x-a+\left|x-\frac{2a+b}{3}\right|\right), \frac{1}{2}\left(b-x+\left|x-\frac{a+2b}{3}\right|\right)\right\}$$

$$= \max\left\{\frac{b-a}{6}, \frac{b-a}{3}+\left|x-\frac{a+b}{2}\right|\right\}$$

$$= \frac{1}{2}\left[\frac{b-a}{2}+\left|x-\frac{a+b}{2}\right|+\left|\frac{b-a}{6}+\left(x-\frac{a+b}{2}\right)\right|\right]$$

$$= \frac{b-a}{3}+\left|x-\frac{a+b}{2}\right|.$$

The proof of (3.2.56) and (3.2.57) is now obvious.

Remark 3.2.3. If we set $x = \frac{a+b}{2}$ in (3.2.49) and (3.2.50); (3.2.54) and (3.2.55); (3.2.56) and (3.2.57), then we get corresponding inequalities which do not depend on x.

3.3 Ostrowski-type inequalities involving the product of two functions

In this section, we shall deal with some Ostrowski-type inequalities recently established by Pachpatte in [93,109,114,139], involving product of two functions.

We start with the following Theorem which contains the Ostrowski-type integral inequalities established in [114].

Theorem 3.3.1. Let $f,\ g \in C^1([a,b],\mathbb{R})$, $[a,b] \in \mathbb{R}$, $a < b$. Then

$$\left|f(x)g(x)-\frac{1}{2}[g(x)F+f(x)G]\right| \leqslant \frac{1}{4}\left[|g(x)|\int_a^b |f'(t)|dt+|f(x)|\int_a^b |g'(t)|dt\right], \tag{3.3.1}$$

and

$$|f(x)g(x)-[g(x)F+f(x)G]+FG| \leqslant \frac{1}{4}\left(\int_a^b |f'(t)|dt\right)\left(\int_a^b |g'(t)|dt\right), \tag{3.3.2}$$

for all $x \in [a,b]$, where

$$F = \frac{f(a)+f(b)}{2}, \quad G = \frac{g(a)+g(b)}{2}.$$

The constant $\frac{1}{4}$ in (3.3.1) and (3.3.2) is sharp.

Proof. From the hypotheses, we have the following identities (see [88], [108, p. 267]):

$$f(x)-F=\frac{1}{2}\left[\int_a^x f'(t)dt-\int_x^b f'(t)dt\right], \tag{3.3.3}$$

and

$$g(x)-G=\frac{1}{2}\left[\int_a^x g'(t)dt-\int_x^b g'(t)dt\right]. \tag{3.3.4}$$

Multiplying both sides of (3.3.3) and (3.3.4) by $g(x)$ and $f(x)$ respectively, adding the resulting identities and rewriting, we have

$$f(x)g(x)-\frac{1}{2}[g(x)F+f(x)G]$$
$$=\frac{1}{4}\left[g(x)\left[\int_a^x f'(t)dt-\int_x^b f(t)dt\right]+f(x)\left[\int_a^x g'(t)dt-\int_x^b g'(t)dt\right]\right]. \tag{3.3.5}$$

From (3.3.5) and using the properties of modulus and integrals, we have

$$\left|f(x)g(x)-\frac{1}{2}[g(x)F+f(x)G]\right|\leqslant\frac{1}{4}\left[|g(x)|\int_a^b|f'(t)|dt+|f(x)|\int_a^b|g'(t)|dt\right].$$

This is the required inequality in (3.3.1).

Multiplying the left hand sides and right hand sides of (3.3.3) and (3.3.4), we get

$$f(x)g(x)-[g(x)F+f(x)G]+FG$$
$$=\frac{1}{4}\left[\int_a^x f'(t)dt-\int_x^b f'(t)dt\right]\left[\int_a^x g'(t)dt-\int_x^b g'(t)dt\right]. \tag{3.3.6}$$

From (3.3.6) and using the properties of modulus and integrals, we have

$$|f(x)g(x)-[g(x)F+f(x)G]+FG|\leqslant\frac{1}{4}\left[\int_a^b|f'(t)|dt\right]\left[\int_a^b|g'(t)|dt\right].$$

This is the desired inequality in (3.3.2).

To prove the sharpness of the constant $\frac{1}{4}$ in (3.3.1) and (3.3.2), assume that the inequalities (3.3.1) and (3.3.2) hold with constants $c>0$ and $k>0$. That is,

$$\left|f(x)g(x)-\frac{1}{2}\left[|g(x)|F+|f(x)|G\right]\right|\leqslant c\left[|g(x)|\int_a^b|f'(t)|dt+|f(x)|\int_a^b|g'(t)|dt\right], \tag{3.3.7}$$

and

$$|f(x)g(x)-[|g(x)|F+|f(x)|G]+FG|\leqslant k\left(\int_a^b|f'(t)|dt\right)\left(\int_a^b|g'(t)|dt\right), \tag{3.3.8}$$

for $x\in[a,b]$. In (3.3.7) and (3.3.8), choose $f(x)=g(x)=x$ and hence $f'(x)=g'(x)=1$, $F=G=\frac{a+b}{2}$. Then by simple computation, we get

$$\left|x-\frac{1}{2}(a+b)\right|\leqslant 2c(b-a), \tag{3.3.9}$$

and

$$\left|x(x-(a+b))+\left(\frac{a+b}{2}\right)^2\right|\leqslant k(b-a)^2. \tag{3.3.10}$$

By taking $x=b$, from (3.3.9) we observe that $c\geqslant\frac{1}{4}$ and from (3.3.10) it is easy to observe that $k\geqslant\frac{1}{4}$, which proves the sharpness of the constants in (3.3.1) and (3.3.2). The proof is complete.

Remark3.3.1. Dividing both sides of (3.3.5) and (3.3.6) by $b-a$, then integrating both sides with respect to x over $[a,b]$ and closely looking at the proof of Theorem 3.3.1, we get

$$\left|\frac{1}{b-a}\int_a^b f(x)g(x)-\frac{1}{2(b-a)}\left[F\int_a^b g(x)dx+G\int_a^b f(x)dx\right]\right|$$

$$\leqslant \frac{1}{4(b-a)}\left[\left(\int_a^b |g(x)|dx\right)\left(\int_a^b |f'(x)|dx\right)+\left(\int_a^b |f(x)|dx\right)\left(\int_a^b |g'(x)|dx\right)\right], \tag{3.3.11}$$

and

$$\left|\frac{1}{b-a}\int_a^b f(x)g(x)-\frac{1}{b-a}\left[F\int_a^b g(x)dx+G\int_a^b f(x)dx-FG\right]\right|$$

$$\leqslant \frac{1}{4}\left(\int_a^b |f'(x)|dx\right)\left(\int_a^b |g'(x)|dx\right). \tag{3.3.12}$$

We note that the inequalities (3.3.11) and (3.3.12) are similar to those of the well-known inequalities due to Grüss-and Čebyšev, see [61,13].

The next Theorem deals with the Ostrowski-type inequalities proved in [109].

Theorem 3.3.2. Let $f,\ g:[a,b]\to\mathbb{R}$ be continuous functions on $[a,b]$ and differentiable on (a,b), with derivatives $f',\ g':(a,b)\to\mathbb{R}$ being bounded on (a,b). Then

$$\left|f(x)g(x)-\frac{1}{2(b-a)}\left[g(x)\int_a^b f(y)dy+f(x)\int_a^b g(y)dy\right]\right|$$

$$\leqslant \frac{1}{2}\left\{|g(x)|\|f'\|_\infty+|f(x)|\|g'\|_\infty\right\}\left[\frac{1}{4}+\frac{\left(x-\frac{a+b}{2}\right)^2}{(b-a)^2}\right](b-a), \tag{3.3.13}$$

and

$$\left|f(x)g(x)-\frac{1}{b-a}\left[g(x)\int_a^b f(y)dy+f(x)\int_a^b g(y)dy\right]+\frac{1}{b-a}\int_a^b f(y)g(y)dy\right|$$

$$\leqslant \frac{1}{b-a}\|f'\|_\infty\|g'\|_\infty\left[\frac{(x-a)^3+(b-x)^3}{3}\right], \tag{3.3.14}$$

for all $x\in[a,b]$.

Proof. For any $x,\ y\in[a,b]$ we have the following identities:

$$f(x)-f(y)=\int_y^x f'(t)dt, \tag{3.3.15}$$

and

$$g(x)-g(y)=\int_y^x g'(t)dt. \tag{3.3.16}$$

Multiplying both sides of (3.3.15) and (3.3.16) by $g(x)$ and $f(x)$ respectively and adding, we get

$$2f(x)g(x) - [g(x)f(y) + f(x)g(y)] = g(x)\int_y^x f'(t)dt + f(x)\int_y^x g'(t)dt. \tag{3.3.17}$$

Integrating both sides of (3.3.17) with respect to y over $[a,b]$ and rewriting, we have

$$\begin{aligned} &f(x)g(x) - \frac{1}{2(b-a)}\left[g(x)\int_a^b f(y)dy + f(x)\int_a^b g(y)dy\right] \\ &= \frac{1}{2(b-a)}\int_a^b \left\{g(x)\int_y^x f'(t)dt + f(x)\int_y^x g'(t)dt\right\}dy. \end{aligned} \tag{3.3.18}$$

From (3.3.18) and using the properties of modulus, we have

$$\begin{aligned} &\left|f(x)g(x) - \frac{1}{2(b-a)}\left[g(x)\int_a^b f(y)dy + f(x)\int_a^b g(y)dy\right]\right| \\ &\leqslant \frac{1}{2(b-a)}\int_a^b \{|g(x)|\|f'\|_\infty |x-y| + |f(x)|\|g'\|_\infty |x-y|\}dy \\ &= \frac{1}{2(b-a)}\{|g(x)|\|f'\|_\infty + |f(x)|\|g'\|_\infty\}\left[\frac{(x-a)^2+(b-x)^2}{2}\right] \\ &= \frac{1}{2}\{|g(x)|\|f'\|_\infty + |f(x)|\|g'\|_\infty\}\left[\frac{1}{4} + \frac{\left(x-\frac{a+b}{2}\right)^2}{(b-a)^2}\right](b-a) \end{aligned}$$

This is required inequality in (3.3.13).

Multiplying the left hand and right hand sides of (3.3.15) and (3.3.16), we get

$$f(x)g(x) - [g(x)f(y) + f(x)g(y)] + f(y)g(y) = \left\{\int_y^x f'(t)\right\}\left\{\int_y^x g'(t)\right\}. \tag{3.3.19}$$

Integrating both sides of (3.3.19) with respect to y over $[a,b]$ and rewriting, we have

$$\begin{aligned} &f(x)g(x) - \frac{1}{b-a}\left[g(x)\int_a^b f(y)dy + f(x)\int_a^b g(y)dy\right] + \frac{1}{b-a}\int_a^b f(y)g(y)dy \\ &= \frac{1}{b-a}\int_a^b \left\{\int_y^x f'(t)dt\right\}\left\{\int_y^x g'(t)dt\right\}dy. \end{aligned} \tag{3.3.20}$$

From (3.3.20) and using the properties of modulus, we obtain

$$\begin{aligned} &\left|f(x)g(x) - \frac{1}{b-a}\left[g(x)\int_a^b f(y)dy + f(x)\int_a^b g(y)dy\right] + \frac{1}{b-a}\int_a^b f(y)g(y)dy\right| \\ &\leqslant \frac{1}{b-a}\|f'\|_\infty\|g'\|_\infty\int_a^b |x-y|^2dy = \frac{1}{b-a}\|f'\|_\infty\|g'\|_\infty\left[\frac{(x-a)^3+(b-x)^3}{3}\right]. \end{aligned}$$

This is the desired inequality in (3.3.14). The proof is complete.

Remark 3.3.2. Dividing both sides of (3.3.18) and (3.3.20) by $b-a$, then integrating both sides with respect to x over $[a,b]$ and using the properties of modulus and by elementry calculations, we get

$$\left|\frac{1}{b-a}\int_a^b f(x)g(x)dx-\left(\frac{1}{b-a}\int_a^b f(x)dx\right)\left(\frac{1}{b-a}\int_a^b g(x)dx\right)\right|$$

$$\leqslant \frac{1}{2(b-a)^2}\int_a^b\left[\int_a^b\{|g(x)|\|f'\|_\infty+|f(x)|\|g'\|_\infty\}|x-y|dy\right]dx, \tag{3.3.21}$$

and

$$\left|\frac{1}{b-a}\int_a^b f(x)g(x)dx-\left(\frac{1}{b-a}\int_a^b f(x)dx\right)\left(\frac{1}{b-a}\int_a^b g(x)dx\right)\right|$$

$$\leqslant \frac{1}{12}(b-a)^2\|f'\|_\infty\|g'\|_\infty. \tag{3.3.22}$$

Here, it is to be noted that the inequality (3.3.21) is similar to the well-known Grüss inequality in (3) and the inequality (3.3.22) is the well-known Čebyšev inequality in (1). We also note that, by taking $g(x)=1$ and hence $g'(x)=0$ in (3.3.13), we recapture the celebrated Ostrowski's inequality (7).

In the following Theorem we present the results given in [139] where the derivatives of the functions belong to L_p spaces.

Theorem 3.3.3. Let $f,\ g:[a,b]\to\mathbb{R}$ be absolutely continous functions with derivatives $f',\ g'\in L_p[a,b]$, $p>1$. Then

$$\left|f(x)g(x)-\frac{1}{2(b-a)}\left[g(x)\int_a^b f(t)dt+f(x)\int_a^b g(t)dt\right]\right|$$

$$\leqslant \frac{1}{2(b-a)}\left[|g(x)|\|f'\|_p+|f(x)|\|g'\|_p\right](B(x))^{\frac{1}{q}}, \tag{3.3.23}$$

and

$$\left|f(x)g(x)-\frac{1}{b-a}\left[g(x)\int_a^b f(t)dt+f(x)\int_a^b g(t)dt\right]\right.$$

$$\left.+\left(\frac{1}{b-a}\int_a^b f(t)dt\right)\left(\frac{1}{b-a}\int_a^b g(t)dt\right)\right|$$

$$\leqslant \frac{1}{(b-a)^2}\|f'\|_p\|g'\|_p(B(x))^{\frac{2}{q}}, \tag{3.3.24}$$

for all $x\in[a,b]$, where

$$B(x)=\frac{1}{q+1}\left[(x-a)^{q+1}+(b-x)^{q+1}\right], \tag{3.3.25}$$

and $\frac{1}{p}+\frac{1}{q}=1$.

Proof. From the hypothesis, we have the following identities (see, Theorem 1.2.3):

$$f(x) - \frac{1}{b-a}\int_a^b f(t)dt = \frac{1}{b-a}\int_a^b p(x,t)f'(t)dt, \tag{3.3.26}$$

and

$$g(x) - \frac{1}{b-a}\int_a^b g(t)dt = \frac{1}{b-a}\int_a^b p(x,t)g'(t)dt, \tag{3.3.27}$$

for $x \in [a,b]$, where $p(x,t)$ is defined by (1.2.11). Multiplying both sides of (3.3.26) and (3.3.27) by $g(x)$ and $f(x)$ respectively, adding the resulting identities and rewriting, we have

$$f(x)g(x) - \frac{1}{2(b-a)}\left[g(x)\int_a^b f(t)dt + f(x)\int_a^b g(t)dt\right]$$

$$= \frac{1}{2(b-a)}\left[g(x)\int_a^b p(x,t)f'(t)dt + f(x)\int_a^b p(x,t)g'(t)dt\right]. \tag{3.3.28}$$

From (3.3.28), using the properties of modulus and Hölder's integral inequality, we obtain

$$\left|f(x)g(x) - \frac{1}{2(b-a)}\left[g(x)\int_a^b f(t)dt + f(x)\int_a^b g(t)dt\right]\right|$$

$$\leqslant \frac{1}{2(b-a)}\left[|g(x)|\int_a^b |p(x,t)||f'(t)|dt + |f(x)|\int_a^b |p(x,t)||g'(t)|dt\right]$$

$$\leqslant \frac{1}{2(b-a)}\left[|g(x)|\left(\int_a^b |p(x,t)|^q dt\right)^{\frac{1}{q}}\left(\int_a^b |f'(t)|^p dt\right)^{\frac{1}{p}}\right.$$

$$\left. + |f(x)|\left(\int_a^b |p(x,t)|^q dt\right)^{\frac{1}{q}}\left(\int_a^b |g'(t)|^p dt\right)^{\frac{1}{p}}\right]$$

$$= \frac{1}{2(b-a)}\left[|g(x)|\|f'\|_p + |f(x)|\|g'\|_p\right]\left(\int_a^b |p(x,t)|^q dt\right)^{\frac{1}{q}}. \tag{3.3.29}$$

By a simple calculation, we have

$$\begin{aligned}\int_a^b |p(x,t)|^q dt &= \int_a^x |t-a|^q dt + \int_x^b |t-b|^q dt \\ &= \int_a^x (t-a)^q dt + \int_x^b (b-t)^q dt \\ &= \frac{1}{q+1}\left[(x-a)^{q+1} + (b-x)^{q+1}\right] = B(x).\end{aligned} \tag{3.3.30}$$

Using (3.3.30) in (3.3.29), we get (3.3.23).

Multiplying the left hand sides and right hand sides of (3.3.26) and (3.3.27), we have

$$f(x)g(x) - \frac{1}{b-a}\left[g(x)\int_a^b f(t)dt + f(x)\int_a^b g(t)dt\right]$$

$$+\left(\frac{1}{b-a}\int_a^b f(t)dt\right)\left(\frac{1}{b-a}\int_a^b g(t)dt\right)$$

$$=\frac{1}{(b-a)^2}\left(\int_a^b p(x,t)f'(t)dt\right)\left(\int_a^b p(x,t)g'(t)dt\right). \tag{3.3.31}$$

From (3.3.31), using the properties of modulus, Hölder's integral inequality and (3.3.30), we have

$$\left|f(x)g(x)-\frac{1}{b-a}\left[g(x)\int_a^b f(t)dt+f(x)\int_a^b g(t)dt\right]\right.$$

$$\left.+\left(\frac{1}{b-a}\int_a^b f(t)dt\right)\left(\frac{1}{b-a}\int_a^b g(t)dt\right)\right|$$

$$\leqslant\frac{1}{(b-a)^2}\left(\int_a^b |p(x,t)||f'(t)|dt\right)\left(\int_a^b |p(x,t)||g'(t)|dt\right)$$

$$\leqslant\frac{1}{(b-a)^2}\left(\int_a^b |p(x,t)|^q dt\right)^{\frac{1}{q}}\left(\int_a^b |f'(t)|^p dt\right)^{\frac{1}{p}}\left(\int_a^b |p(x,t)|^q dt\right)^{\frac{1}{q}}\left(\int_a^b |g'(t)|^p dt\right)^{\frac{1}{p}}$$

$$=\frac{1}{(b-a)^2}\|f'\|_p\|g'\|_p\left(B(x)\right)^{\frac{2}{q}}.$$

This is the required inequality in (3.3.24). The proof is complete.

Remark 3.3.3. By taking $g(x)=1$ and hence $g'(x)=0$ in (3.3.23) and by simple calculation, we get

$$\left|f(x)-\frac{1}{b-a}\int_a^b f(t)dt\right|$$

$$\leqslant\frac{1}{(q+1)^{\frac{1}{q}}}\left[\left(\frac{x-a}{b-a}\right)^{q+1}+\left(\frac{b-x}{b-a}\right)^{q+1}\right](b-a)^{\frac{1}{q}}\|f'\|_p, \tag{3.3.32}$$

for all $x\in[a,b]$. We note that the inequality (3.3.32) is established by Dragomir and Wang in [27].

At the end of this section, we give the following Theorem which contains the inequalities proved in [93].

For continous function $z:[a,b]\to\mathbb{R}$ and $\lambda\in[0,1]$, we use the notation

$$L[z(x)]=(b-a)\left[\frac{\lambda}{2}(z(a)+z(b))+(1-\lambda)z(x)\right].$$

Theorem 3.3.4. Let $f,\ g : [a,b] \to \mathbb{R}$ be continous functions on $[a,b]$, differentiable on (a,b) with derivatives $f',\ g' : [a,b] \to \mathbb{R}$ being bounded on (a,b). Then

$$\left| g(x)L[f(x)] + f(x)L[g(x)] - g(x)\int_a^b f(t)dt - f(x)\int_a^b g(t)dt \right|$$

$$\leqslant \left[|g(x)| \|f'\|_\infty + |f(x)| \|g'\|_\infty \right] M(x), \tag{3.3.33}$$

and

$$\left| L[f(x)]L[g(x)] - L[g(x)]\int_a^b f(t)dt - L[f(x)]\int_a^b g(t)dt + \left(\int_a^b f(t)dt\right)\left(\int_a^b g(t)dt\right) \right|$$

$$\leqslant \|f'\|_\infty \|g'\|_\infty (M(x))^2, \tag{3.3.34}$$

for $a + \lambda\frac{b-a}{2} \leqslant x \leqslant b - \lambda\frac{b-a}{2}$, $\lambda \in [0,1]$, where

$$M(x) = \frac{1}{4}(b-a)^2\left[\lambda^2 + (\lambda-1)^2\right] + \left(x - \frac{a+b}{2}\right)^2. \tag{3.3.35}$$

Proof. From the hypotheses, we have the following identities (see, Theorem 1.4.1 and Theorem 3.2.5):

$$L[f(x)] - \int_a^b f(t)dt = \int_a^b K(x,t)f'(t)dt, \tag{3.3.36}$$

and

$$L[g(x)] - \int_a^b g(t)dt = \int_a^b k(x,t)g'(t)dt, \tag{3.3.37}$$

where $k(x,t)$ is defined by (3.2.39). Multiplying (3.3.36) and (3.3.37) by $g(x)$ and $f(x)$ respectively and adding the resulting identities, we have

$$g(x)L[f(x)] + f(x)L[g(x)] - g(x)\int_a^b f(t)dt - f(x)\int_a^b g(t)dt$$

$$= g(x)\int_a^b k(x,t)f'(t)dt + f(x)\int_a^b k(x,t)g'(t)dt. \tag{3.3.38}$$

From (3.2.38) and using the properties of modulus, we have

$$\left| g(x)L[f(x)] + f(x)L[g(x)] - g(x)\int_a^b f(t)dt - f(x)\int_a^b g(t)dt \right|$$

$$\leqslant \left[|g(x)| \int_a^b |k(x,t)||f'(t)|dt + |f(x)| \int_a^b |k(x,t)||g'(t)|dt \right]$$

$$\leqslant \left[|g(x)| \|f'\|_\infty + |f(x)| \|g'\|_\infty \right] \int_a^b |k(x,t)|dt$$

$$= \left[|g(x)| \|f'\|_\infty + |f(x)| \|g'\|_\infty \right]$$

$$\times\left[\int_a^x\left|t-\left(a+\lambda\frac{b-a}{2}\right)\right|dt+\int_x^b\left|t-\left(b-\lambda\frac{b-a}{2}\right)\right|dt\right]. \tag{3.3.39}$$

Now, by evaluating the integrals on the right side of (3.2.39) as in the proof of Theorem 1.4.1, we get the required inequality in (3.2.33).

Multiplying the left hand sides and right hand sides of (3.2.36) and (3.2.37), we get

$$L[f(x)]L[g(x)]-L[g(x)]\int_a^b f(t)dt-L[f(x)]\int_a^b g(t)dt+\left(\int_a^b f(t)dt\right)\left(\int_a^b g(t)dt\right)$$
$$=\left(\int_a^b k(x,t)f'(t)\right)\left(\int_a^b k(x,t)g'(t)\right). \tag{3.3.40}$$

From (3.2.40) and following the proof of inequality (3.2.33) with suitable changes, we get the desired inequality in (3.2.34). The proof is complete.

Remark 3.3.4. By taking $g(x)=1$ and hence $g'(x)=0$ in (3.3.29), we get the following inequality established by Dragomir, Cerone and Roumeliotics in [47]

$$\left|L[f(x)]-\int_a^b f(t)dt\right|\leqslant M(x)\|f'\|_\infty,$$

for all $\lambda\in[0,1]$ and $a+\lambda\frac{b-a}{2}\leqslant x\leqslant b-\lambda\frac{b-a}{2}$ and in addition if we choose (i) $\lambda=0$, we get the Ostrowski's inequality (7) and (ii) $\lambda=1, x=\frac{a+b}{2}$, we get the Trapezoid-type inequality.

3.4 Inequalities of the Ostrowski-and Grüss-type

In this section, we offer some inequalities of the Ostrowski-and Grüss-type, that have recently given by Pachpatte [100,131] and Cerone, Dragomir and Roumeliotis [15].

In the following Theorems we present the inequalities proved by Pachpatte in [131].

For suitable functions $f,\ g:[a,b]\to\mathbb{R}$, we use the following notation to simplify the details of presentation:

$$S(f,g)=f(x)g(x)-\frac{1}{2(b-a)}\left[g(x)\int_a^b f(t)dt+f(x)\int_a^b g(t)dt\right]$$
$$-\frac{1}{2}\left(x-\frac{a+b}{2}\right)[Fg(x)+Gf(x)],$$

$$H(f,g)=\frac{1}{b-a}\int_a^b f(x)g(x)dx-\left(\frac{1}{b-a}\int_a^b f(x)dx\right)\left(\frac{1}{b-a}\int_a^b g(x)dx\right)$$
$$-\frac{1}{2(b-a)}\int_a^b\left(x-\frac{a+b}{2}\right)[Fg(x)+Gf(x)]dx,$$

in which

$$F=\frac{f(b)-f(a)}{b-a},\quad G=\frac{g(b)-g(a)}{b-a}.$$

Theorem 3.4.1. Let $f,\ g:[a,b]\to\mathbb{R}$ be absolutely continuous functions with derivatives $f',g'\in L_2[a,b]$. Then we have

$$|S(f,g)|\leqslant\frac{b-a}{4\sqrt{3}}\left[|g(x)|\left(\frac{1}{b-a}\|f'\|_2^2-F^2\right)^{\frac{1}{2}}+|f(x)|\left(\frac{1}{b-a}\|g'\|_2^2-G^2\right)^{\frac{1}{2}}\right]dx,\tag{3.4.1}$$

for all $x\in[a,b]$ and

$$|H(f,g)|\leqslant\frac{1}{4\sqrt{3}}\int_a^b\left[|g(x)|\left(\frac{1}{b-a}\|f'\|_2^2-F^2\right)^{\frac{1}{2}}+f(x)|\left(\frac{1}{b-a}\|g'\|_2^2-G^2\right)^{\frac{1}{2}}\right]dx.\tag{3.4.2}$$

Theorem 3.4.2. Let the assumptions of Theorem 3.4.1 hold. If $\gamma\leqslant f'(x)\leqslant\Gamma$, $\phi\leqslant g'(x)\leqslant\Phi$ for $x\in[a,b]$; where $\gamma,\ \Gamma,\ \phi,\ \Phi$ are real constants. Then we have

$$|S(f,g)|\leqslant\frac{b-a}{8\sqrt{3}}\left[|g(x)|(\Gamma-\gamma)+|f(x)|(\Phi-\phi)\right],\tag{3.4.3}$$

for all $x\in[a,b]$ and

$$|H(f,g)|\leqslant\frac{1}{8\sqrt{3}}\int_a^b\left[|g(x)|(\Gamma-\gamma)+|f(x)|(\Phi-\phi)\right]dx.\tag{3.4.4}$$

Remark 3.4.1. If we take $g(x)=1$ and hence $g'(x)=0$ in (3.4.1) and (3.4.3), then by simple computation, we get the inequality established by Barnett, Dragomir and Sofo in [7, Theorem 2.1] and if we set $x=\frac{a+b}{2}$ in (3.4.1) and (3.4.3), then we get the corresponding midpoint inequalities.

Proofs of Theorems 3.4.1 and 3.4.2. Define the function

$$p(x,t)=\begin{cases}t-a & \text{if } t\in[a,x],\\ t-b & \text{if } t\in(x,b].\end{cases}$$

By using the well-known Korkine's identity (see [79]) for mappings $f,\ g:[a,b]\to\mathbb{R}$, which can be easily proved by direct computation:

$$T(f,g)=\frac{1}{2(b-a)^2}\int_a^b\int_a^b(f(t)-f(s))(g(t)-g(s))dtds,$$

where $T(f,g)$ is defined by (2), we obtain

$$\frac{1}{b-a}\int_a^b p(x,t)f'(t)dt-\left(\frac{1}{b-a}\int_a^b p(x,t)dt\right)\left(\frac{1}{b-a}\int_a^b f'(t)dt\right)$$

$$=\frac{1}{2(b-a)^2}\int_a^b\int_a^b(p(x,t)-p(x,s))(f'(t)-f'(s))dtds.\tag{3.4.5}$$

By simple calculation, we obtain

$$\frac{1}{b-a}\int_a^b p(x,t)f'(t)dt = f(x) - \frac{1}{b-a}\int_a^b f(t),$$

$$\frac{1}{b-a}\int_a^b p(x,t)dt = x - \frac{a+b}{2},$$

and

$$\frac{1}{b-a}\int_a^b f'(t)dt = \frac{f(b)-f(a)}{b-a} = F.$$

Using these facts in (3.4.5), we get the following identity (see [7]):

$$f(x) - \frac{1}{b-a}\int_a^b f(t)dt - F\left(x - \frac{a+b}{2}\right)$$

$$= \frac{1}{2(b-a)^2}\int_a^b\int_a^b (p(x,t)-p(x,s))(f'(t)-f'(s))dtds, \tag{3.4.6}$$

for all $x \in [a,b]$. Similarly, we get

$$g(x) - \frac{1}{b-a}\int_a^b g(t)dt - G\left(x - \frac{a+b}{2}\right)$$

$$= \frac{1}{2(b-a)^2}\int_a^b\int_a^b (p(x,t)-p(x,s))(g'(t)-g'(s))dtds. \tag{3.4.7}$$

Multiplying (3.4.6) and (3.4.7) by $g(x)$ and $f(x)$ respectively, adding the resulting identities and rewriting, we get

$$S(f,g) = \frac{1}{2}\left[g(x)\frac{1}{2(b-a)^2}\int_a^b\int_a^b (p(x,t)-p(x,s))(f'(t)-f'(s))dtds\right.$$

$$\left. + f(x)\frac{1}{2(b-a)^2}\int_a^b\int_a^b (p(x,t)-p(x,s))(g'(t)-g'(s))dtds\right]. \tag{3.4.8}$$

From (3.4.8) and using the properties of modulus, we get

$$|S(f,g)| \leqslant \frac{1}{2}\left[|g(x)|\frac{1}{2(b-a)^2}\int_a^b\int_a^b |p(x,t)-p(x,s)||f'(t)-f'(s)|dtds\right.$$

$$\left. + |f(x)|\frac{1}{2(b-a)^2}\int_a^b\int_a^b |p(x,t)-p(x,s)||g'(t)-g'(s)|dtds\right]. \tag{3.4.9}$$

By using the Cauchy-Schwarz inequality for double integrals, we observe that

$$\frac{1}{2(b-a)^2}\int_a^b\int_a^b |p(x,t)-p(x,s)||f'(t)-f'(s)|dtds$$

$$\leqslant \left(\frac{1}{2(b-a)^2}\int_a^b\int_a^b (p(x,t)-p(x,s))^2 dtds\right)^{\frac{1}{2}}$$

$$\times\left(\frac{1}{2(b-a)^2}\int_a^b\int_a^b (f'(t)-f'(s))^2 dtds\right)^{\frac{1}{2}}. \tag{3.4.10}$$

It is easy to observe that

$$\frac{1}{2(b-a)^2}\int_a^b\int_a^b (p(x,t)-p(x,s))^2 dtds = \frac{1}{b-a}\int_a^b p^2(x,t) - \left(\frac{1}{b-a}\int_a^b p(x,t)dt\right)^2$$

$$= \frac{1}{b-a}\left[\int_a^x (t-a)^2 dt + \int_x^b (b-t)^2 dt\right] - \left(x-\frac{a+b}{2}\right)^2$$

$$= \frac{1}{b-a}\left[\frac{(x-a)^3+(b-x)^3}{3}\right] - \left(x-\frac{a+b}{2}\right)^2 = \frac{1}{12}(b-a)^2, \tag{3.4.11}$$

and

$$\frac{1}{2(b-a)^2}\int_a^b\int_a^b (f'(t)-f'(s))^2 dtds = \frac{1}{b-a}\|f'\|_2^2 - \left(\frac{f(b)-f(a)}{b-a}\right)^2 = \frac{1}{b-a}\|f'\|_2^2 - F^2. \tag{3.4.12}$$

Using (3.4.11) and (3.4.12) in (3.4.10), we get

$$\frac{1}{2(b-a)^2}\int_a^b\int_a^b |p(x,t)-p(x,s)||f'(t)-f'(s)|dtds$$

$$\leqslant \frac{b-a}{2\sqrt{3}}\left(\frac{1}{b-a}\|f'\|_2^2 - F^2\right)^{\frac{1}{2}}. \tag{3.4.13}$$

Similarly, we get

$$\frac{1}{2(b-a)^2}\int_a^b\int_a^b |p(x,t)-p(x,s)||g'(t)-g'(s)|dtds$$

$$\leqslant \frac{b-a}{2\sqrt{3}}\left(\frac{1}{b-a}\|g'\|_2^2 - G^2\right)^{\frac{1}{2}}. \tag{3.4.14}$$

Using (3.4.13) and (3.4.14) in (3.4.9), we get the desired inequality in (3.4.1).

Integrating both sides of (3.4.8) with respect to x over $[a,b]$ and dividing throughout by $(b-a)$ we get

$$H(f,g) = \frac{1}{2(b-a)}\int_a^b\left[\frac{g(x)}{2(b-a)^2}\int_a^b\int_a^b (p(x,t)-p(x,s))(f'(t)-f'(s))dtds\right.$$

$$\left.+\frac{f(x)}{2(b-a)^2}\int_a^b\int_a^b (p(x,t)-p(x,s))(g'(t)-g'(s))dtds\right]dx. \tag{3.4.15}$$

From (3.4.15) and using the properties of modulus, we have

$$|H(f,g)| \leqslant \frac{1}{2(b-a)}\int_a^b\left[\frac{|g(x)|}{2(b-a)^2}\int_a^b\int_a^b |p(x,t)-p(x,s)||f'(t)-f'(s)|dtds\right.$$

$$+\frac{|f(x)|}{2(b-a)^2}\int_a^b\int_a^b |p(x,t)-p(x,s)||g'(t)-g'(s)|dtds\Bigg]\,dx. \tag{3.4.16}$$

Using (3.4.13) and (3.4.14) in (3.4.16), we get the required inequality in (3.4.2). The proof of Theorem 3.4.1 is complete.

By using the Grüss inequality (3), it is easy to observe that

$$0\leqslant \frac{1}{b-a}\int_a^b (f'(t))^2dt-\left(\frac{1}{b-a}\int_a^b f'(t)dt\right)^2\leqslant \frac{1}{4}(\Gamma-\gamma)^2,$$

i.e.,

$$0\leqslant \frac{1}{b-a}\|f'\|_2^2-F^2\leqslant \frac{1}{4}(\Gamma-\gamma)^2. \tag{3.4.17}$$

Similarly, we have

$$0\leqslant \frac{1}{b-a}\|g'\|_2^2-G^2\leqslant \frac{1}{4}(\Phi-\phi)^2. \tag{3.4.18}$$

Using (3.4.17), (3.4.18) in (3.4.1) and (3.4.2), we get the required inequalities in (3.4.3) and (3.4.4) and the proof of Theorem 3.4.2 is complete.

In [15], Cerone, Dragomir and Roumeliotis have obtained the following Ostrowski-Grüss-type inequality for twice differentiable mappings.

Theorem 3.4.3. Let $f:[a,b]\to\mathbb{R}$ be continous on $[a,b]$ and twice differentiable on (a,b), and assume that the second derivative $f'':(a,b)\to R$ satisfies the condition $\phi\leqslant f''(x)\leqslant\Phi$ for all $x\in(a,b)$. Then we have the inequality

$$\Bigg|f(x)-\left(x-\frac{a+b}{2}\right)f'(x)+\left[\frac{(b-a)^2}{24}+\frac{1}{2}\left(x-\frac{a+b}{2}\right)^2\right]\frac{f'(b)-f'(a)}{b-a}$$

$$-\frac{1}{b-a}\int_a^b f(t)dt\Bigg|\leqslant\frac{1}{8}(\Phi-\phi)\left[\frac{1}{2}(b-a)+\left|x-\frac{a+b}{2}\right|\right]^2, \tag{3.4.19}$$

for all $x\in[a,b]$.

Proof. From the hypotheses, we have the following identity (see Theorem 1.2.5):

$$\frac{1}{b-a}\int_a^b k(x,t)f''(t)dt=\frac{1}{b-a}\int_a^b f(t)dt+\left(x-\frac{a+b}{2}\right)f'(x)-f(x), \tag{3.4.20}$$

where the kernel $k:[a,b]^2\to\mathbb{R}$ is defined by

$$k(x,t)=\begin{cases}\dfrac{(t-a)^2}{2} & \text{if } t\in[a,x],\\[2mm] \dfrac{(t-b)^2}{2} & \text{if } t\in(x,b].\end{cases}$$

It is easy to observe that the kernel k satisfies the estimation

$$0 \leqslant k(x,t) \leqslant \begin{cases} \dfrac{(b-x)^2}{2}, & x \in \left[a, \dfrac{a+b}{2}\right), \\ \dfrac{(x-a)^2}{2}, & x \in \left[\dfrac{a+b}{2}, b\right], \end{cases} \tag{3.4.21}$$

for all $t \in [a,b]$. Applying Grüss integral inequality for the mappings $f''(\cdot)$ and $k(x,\cdot)$, we get

$$\left|\frac{1}{b-a}\int_a^b k(x,t)f''(t) - \frac{1}{b-a}\int_a^b k(x,t)dt \frac{1}{b-a}\int_a^b f''(t)dt\right|$$

$$\leqslant \frac{1}{4}(\Phi-\phi)\times \begin{cases} \dfrac{(b-x)^2}{2}, & x \in \left[a, \dfrac{a+b}{2}\right), \\ \dfrac{(x-a)^2}{2}, & x \in \left[\dfrac{a+b}{2}, b\right]. \end{cases} \tag{3.4.22}$$

We observe that

$$\int_a^b k(x,t)dt = \int_a^x \frac{(t-a)^2}{2}dt + \int_x^b \frac{(t-b)^2}{2}dt = \frac{1}{6}\left[(x-a)^3 + (b-x)^3\right].$$

Also, a simple computation shows that

$$(x-a)^3 + (b-x)^3 = (b-a)\left[(x-a)^2 + (b-x)^2 - (x-a)(b-x)\right]$$

$$= (b-a)\left[(b-a)^2 - 3(x-a)(b-x)\right]$$

$$= (b-a)\left[(b-a)^2 + 3\left[x^2 - (a+b)x + ab\right]\right]$$

$$= (b-a)\left[(b-a)^2 + 3\left[\left(x - \frac{a+b}{2}\right)^2 - \left(\frac{b-a}{2}\right)^2\right]\right]$$

$$= (b-a)\left[\frac{(b-a)^2}{4} + 3\left(x - \frac{a+b}{2}\right)^2\right].$$

Consequently,

$$\int_a^b k(x,t)dt = (b-a)\left[\frac{(b-a)^2}{24} + \frac{1}{2}\left(x - \frac{a+b}{2}\right)^2\right].$$

Using (3.4.22), we can state

$$\left|\frac{1}{b-a}\int_a^b k(x,t)f''(t)dt - \left[\frac{(b-a)^2}{24} + \frac{1}{2}\left(x - \frac{a+b}{2}\right)^2\right]\frac{f'(b)-f'(a)}{b-a}\right|$$

$$\leqslant \frac{1}{4}(\Phi-\phi)\times \begin{cases} \dfrac{(b-x)^2}{2} & \text{if } x \in \left[a, \dfrac{a+b}{2}\right), \\ \dfrac{(x-a)^2}{2} & \text{if } x \in \left[\dfrac{a+b}{2}, b\right]. \end{cases} \tag{3.4.23}$$

From (3.4.20) and (3.4.23), we get

$$\left|\frac{1}{b-a}\int_a^b f(t)dt - f(x) + \left(x-\frac{a+b}{2}\right)f'(x) - \left[\frac{(b-a)^2}{24}+\frac{1}{2}\left(x-\frac{a+b}{2}\right)^2\right]\frac{f'(b)-f'(a)}{b-a}\right|$$

$$\leqslant \frac{1}{4}(\Phi-\phi)\times\begin{cases}\dfrac{(b-x)^2}{2} & \text{if } x\in\left[a,\dfrac{a+b}{2}\right),\\[2mm] \dfrac{(x-a)^2}{2} & \text{if } x\in\left[\dfrac{a+b}{2},b\right].\end{cases}$$

Now, let us observe that

$$\max\left\{\frac{(b-x)^2}{2},\frac{(x-a)^2}{2}\right\} = \begin{cases}\dfrac{(b-x)^2}{2} & \text{if } x\in\left[a,\dfrac{a+b}{2}\right),\\[2mm] \dfrac{(x-a)^2}{2} & \text{if } x\in\left[\dfrac{a+b}{2},b\right].\end{cases}$$

On the other hand,

$$\begin{aligned}\max\left\{\frac{(b-x)^2}{2},\frac{(x-a)^2}{2}\right\} &= \frac{1}{2}\left[\frac{(b-x)^2+(x-a)^2}{2}+\frac{1}{2}\left|(b-x)^2-(x-a)^2\right|\right]\\ &= \frac{1}{2}\left[\frac{(b-a)^2}{4}+\left(x-\frac{a+b}{2}\right)^2+(b-a)\left|x-\frac{a+b}{2}\right|\right]\\ &= \frac{1}{2}\left[\frac{1}{2}(b-a)+\left|x-\frac{a+b}{2}\right|\right]^2,\end{aligned}$$

and the inequality (3.4.19) is proved.

The following corollaries hold.

Corollary 3.4.1. Let f be as in Theorem 3.4.3. Then we have the perturbed midpoint inequality:

$$\left|f\left(\frac{a+b}{2}\right)+\frac{1}{24}(b-a)(f'(b)-f'(a))-\frac{1}{b-a}\int_a^b f(t)dt\right| \leqslant \frac{1}{32}(\Phi-\phi)(b-a)^2. \tag{3.4.24}$$

By setting $x=\frac{a+b}{2}$ in (3.4.19), we get (3.4.24).

Remark 3.4.2. The classical midpoint inequality states that

$$\left|f\left(\frac{a+b}{2}\right)-\frac{1}{b-a}\int_a^b f(t)dt\right| \leqslant \frac{1}{24}(b-a)^2\|f''\|_\infty, \tag{3.4.25}$$

where $\|f''\|_\infty = \sup_{t\in(a,b)}|f''(t)| < \infty$. We note that if $\Phi-\phi\leqslant\frac{4}{3}\|f''\|_\infty$, then the estimation provided by (3.4.24) is better than the estimation given in (3.4.25). A sufficient condition for the assumption $\Phi-\phi\leqslant\frac{4}{3}\|f''\|_\infty$ to be true is $0\leqslant\phi\leqslant\Phi$.

Corollary 3.4.2. Let f be as in Theorem 3.4.3. Then we have the following perturbed trapezoid inequality:

$$\left|\frac{f(a)+f(b)}{2}-\frac{1}{12}(b-a)(f'(b)-f'(a))-\frac{1}{b-a}\int_a^b f(t)dt\right|$$

$$\leqslant \frac{1}{8}(\Phi-\phi)(b-a)^2. \tag{3.4.26}$$

Proof. Put in (3.4.19) $x=a$ and $x=b$, to get

$$\left|f(a)+\frac{b-a}{2}f'(a)+\frac{1}{6}(b-a)(f'(b)-f'(a))-\frac{1}{b-a}\int_a^b f(t)dt\right|$$

$$\leqslant \frac{1}{8}(\Phi-\phi)(b-a)^2, \tag{3.4.27}$$

and

$$\left|f(b)+\frac{b-a}{2}f'(b)+\frac{1}{6}(b-a)(f'(b)-f'(a))-\frac{1}{b-a}\int_a^b f(t)dt\right|$$

$$\leqslant \frac{1}{8}(\Phi-\phi)(b-a)^2, \tag{3.4.28}$$

respectively. Summing (3.4.27) and (3.4.28), using the triangle inequality and dividing by 2, we get the desired inequality in (3.4.26).

Remark 3.4.3. The classical trapezoid inequality states that

$$\left|\frac{f(a)+f(b)}{2}-\frac{1}{b-a}\int_a^b f(t)dt\right| \leqslant \frac{1}{12}(b-a)^2\|f''\|_\infty. \tag{3.4.29}$$

Now, if we assume that $(\Phi-\phi)\leqslant \frac{2}{3}\|f''\|_\infty$, and this condition holds if we assume that the infimum and supremum of the second derivative f'' are close enough, then the estimation provided by (3.4.26) is better than the estimation in the classical trapezoid inequality (3.4.29).

Next, we present the inequalities established by Pachpatte in [100], involving twice differentiable mappings.

Theorem 3.4.4. Let $f,\ g:[a,b]\to\mathbb{R}$ be twice differentiable mappings on (a,b) and $f'',\ g'':(a,b)\to\mathbb{R}$ are bounded. Then

$$\left|2\left(\frac{1}{b-a}\int_a^b f(t)dt\right)\left(\frac{1}{b-a}\int_a^b g(t)dt\right)-\left[f(x)-\left(x-\frac{a+b}{2}\right)f'(x)\right]\right.$$

$$\left.\times\left(\frac{1}{b-a}\int_a^b g(t)dt\right)-\left[g(x)-\left(x-\frac{a+b}{2}\right)g'(x)\right]\left(\frac{1}{b-a}\int_a^b f(t)dt\right)\right|$$

$$\leqslant E(x)\left[\|f''\|_\infty\left(\frac{1}{b-a}\int_a^b|g(t)|dt\right)+\|g''\|_\infty\left(\frac{1}{b-a}\int_a^b|f(t)|dt\right)\right], \tag{3.4.30}$$

and

$$\left|\left(\frac{1}{b-a}\int_a^b f(t)dt\right)g(x)+\left(\frac{1}{b-a}\int_a^b g(t)dt\right)f(x)+\left(x-\frac{a+b}{2}\right)(fg)'(x)-2f(x)g(x)\right|$$

$$\leqslant E(x)\left[\|f''\|_\infty|g(x)|+\|g''\|_\infty|f(x)|\right], \tag{3.4.31}$$

for $x\in[a,b]$, where

$$E(x)=\frac{1}{24}(b-a)^2+\frac{1}{2}\left(x-\frac{a+b}{2}\right)^2. \tag{3.4.32}$$

Theorem 3.4.5. Let f, g be as in Theorem 3.4.4. Then

$$\left|f(x)\left(\frac{1}{b-a}\int_a^b g(t)dt\right)+g(x)\left(\frac{1}{b-a}\int_a^b f(t)dt\right)-2\left(\frac{1}{b-a}\int_a^b f(t)dt\right)\right.$$

$$\times\left(\frac{1}{b-a}\int_a^b g(t)dt\right)-\left[\frac{f(b)-f(a)}{b-a}\left(x-\frac{a+b}{2}\right)\left(\frac{1}{b-a}\int_a^b g(t)dt\right)\right.$$

$$\left.\left.+\frac{g(b)-g(a)}{b-a}\left(x-\frac{a+b}{2}\right)\left(\frac{1}{b-a}\int_a^b f(t)dt\right)\right]\right|$$

$$\leqslant L(x)\left[\|f''\|_\infty\left(\frac{1}{b-a}\int_a^b|g(t)|dt\right)+\|g''\|_\infty\left(\frac{1}{b-a}\int_a^b|f(t)|dt\right)\right], \tag{3.4.33}$$

and

$$\left|2f(x)g(x)-\left\{\left[\frac{1}{b-a}\int_a^b f(t)dt+\frac{f(b)-f(a)}{b-a}\left(x-\frac{a+b}{2}\right)\right]g(x)\right.\right.$$

$$\left.\left.+\left[\frac{1}{b-a}\int_a^b g(t)dt+\frac{g(b)-g(a)}{b-a}\left(x-\frac{a+b}{2}\right)\right]f(x)\right\}\right|$$

$$\leqslant L(x)\left[\|f''\|_\infty|g(x)|+\|g''\|_\infty|f(x)|\right], \tag{3.4.34}$$

for $x\in[a,b]$, where

$$L(x)=\frac{1}{2}\left\{\left[\frac{\left(x-\frac{a+b}{2}\right)^2}{(b-a)^2}+\frac{1}{4}\right]^2+\frac{1}{12}\right\}(b-a)^2. \tag{3.4.35}$$

Remark 3.4.4. It is easy to observe that, by taking $g(x)=1$ and hence $g'(x)=0$, $g''(x)=0$ in Theorems 3.4.4 and 3.4.5, we recapture respectively the main inequalities established by Cerone, Dragomir and Roumeliotis in [14, Theorem 2.1] and Dragomir and Barnett in [28, Theorem 2.1].

Proofs of Theorems 3.4.4 and 3.4.5. From the hypotheses, we have the following identities (See, Theorem 1.2.5):

$$\frac{1}{b-a}\int_a^b f(t)dt = \left[f(x) - \left(x - \frac{a+b}{2}\right)f'(x)\right] + \frac{1}{b-a}\int_a^b k(x,t)f''(t)dt, \qquad (3.4.36)$$

and

$$\frac{1}{b-a}\int_a^b g(t)dt = \left[g(x) - \left(x - \frac{a+b}{2}\right)g'(x)\right] + \frac{1}{b-a}\int_a^b k(x,t)g''(t)dt, \qquad (3.4.37)$$

for $x \in [a,b]$, where $k(x,t)$ is given by (1.2.28). Multiplying both sides of (3.4.36) and (3.4.37) by $\frac{1}{b-a}\int_a^b g(t)dt$ and $\frac{1}{b-a}\int_a^b f(t)dt$ respectively and adding the resulting identities, we get

$$\begin{aligned}
&2\left(\frac{1}{b-a}\int_a^b f(t)dt\right)\left(\frac{1}{b-a}\int_a^b g(t)dt\right) \\
&= \left[f(x) - \left(x - \frac{a+b}{2}\right)f'(x)\right]\left(\frac{1}{b-a}\int_a^b g(t)dt\right) \\
&+ \left[g(x) - \left(x - \frac{a+b}{2}\right)g'(x)\right]\left(\frac{1}{b-a}\int_a^b f(t)dt\right) \\
&+ \left(\frac{1}{b-a}\int_a^b k(x,t)f''(t)dt\right)\left(\frac{1}{b-a}\int_a^b g(t)dt\right) \\
&+ \left(\frac{1}{b-a}\int_a^b k(x,t)g''(t)dt\right)\left(\frac{1}{b-a}\int_a^b f(t)dt\right),
\end{aligned} \qquad (3.4.38)$$

for $x \in [a,b]$. From (3.4.38) and using the properties of modulus, we have

$$\begin{aligned}
&\Bigg|2\left(\frac{1}{b-a}\int_a^b f(t)dt\right)\left(\frac{1}{b-a}\int_a^b g(t)dt\right) \\
&- \left[f(x) - \left(x - \frac{a+b}{2}\right)f'(x)\right]\left(\frac{1}{b-a}\int_a^b g(t)dt\right) \\
&- \left[g(x) - \left(x - \frac{a+b}{2}\right)g'(x)\right]\left(\frac{1}{b-a}\int_a^b f(t)dt\right)\Bigg| \\
&\leqslant \left[\|f''\|_\infty\left(\frac{1}{b-a}\int_a^b |g(t)|dt\right) + \|g''\|_\infty\left(\frac{1}{b-a}\int_a^b |f(t)|dt\right)\right] \\
&\times \left(\frac{1}{b-a}\int_a^b |k(x,t)|dt\right).
\end{aligned} \qquad (3.4.39)$$

By using the elementary calculations (see, the proof of Theorem 3.4.3), we obtain

$$\frac{1}{b-a}\int_a^b |k(x,t)|dt = \frac{(b-a)^2}{24} + \frac{1}{2}\left(x - \frac{a+b}{2}\right)^2$$

$$= E(x), \tag{3.4.40}$$

for $x \in [a,b]$. Using (3.4.40) in (3.4.39), we get the required inequality in (3.4.30).

Rewriting (3.4.36) and (3.4.37) as

$$f(x) = \frac{1}{b-a}\int_a^b f(t)dt + \left(x - \frac{a+b}{2}\right) f'(x) - \frac{1}{b-a}\int_a^b k(x,t) f''(t)dt, \tag{3.4.36$'$}$$

and

$$g(x) = \frac{1}{b-a}\int_a^b g(t)dt + \left(x - \frac{a+b}{2}\right) g'(x) - \frac{1}{b-a}\int_a^b k(x,t) g''(t)dt, \tag{3.4.37$'$}$$

for $x \in [a,b]$.

Multiplying both sides of (3.4.36′) and (3.4.37′) by $g(x)$ and $f(x)$ respectively and adding the resulting identities, we get

$$2f(x)g(x) = \left(\frac{1}{b-a}\int_a^b f(t)dt\right) g(x) + \left(\frac{1}{b-a}\int_a^b g(t)dt\right) f(x)$$

$$+\left(x - \frac{a+b}{2}\right)(fg)'(x) - \left(\frac{1}{b-a}\int_a^b k(x,t) f''(t)dt\right) g(x)$$

$$-\left(\frac{1}{b-a}\int_a^b k(x,t) g''(t)dt\right) f(x). \tag{3.4.41}$$

Rewriting (3.4.41) and using the properties of modulus and (3.4.40), we get the desired inequality in (3.4.31). The proof of Theorem 3.4.4 is complete.

From the hypotheses, we have the following identities (See, Lemma 1.5.2):

$$f(x) = \frac{1}{b-a}\int_a^b f(t)dt + \frac{f(b)-f(a)}{b-a}\left(x - \frac{a+b}{2}\right)$$

$$+\frac{1}{(b-a)^2}\int_a^b\int_a^b p(x,t)p(t,s) f''(s)dsdt, \tag{3.4.42}$$

and

$$g(x) = \frac{1}{b-a}\int_a^b g(t)dt + \frac{g(b)-g(a)}{b-a}\left(x - \frac{a+b}{2}\right)$$

$$+\frac{1}{(b-a)^2}\int_a^b\int_a^b p(x,t)p(t,s) g''(s)dsdt, \tag{3.4.43}$$

for $x \in [a,b]$, where $p(x,t)$ is given by (1.2.11).

Multiplying both sides of (3.4.42) and (3.4.43) by $\frac{1}{b-a}\int_a^b g(t)dt$ and $\frac{1}{b-a}\int_a^b f(t)dt$ respectively and adding the resulting identities, we get

$$f(x)\left(\frac{1}{b-a}\int_a^b g(t)dt\right) + g(x)\left(\frac{1}{b-a}\int_a^b f(t)dt\right)$$

$$
\begin{aligned}
&= 2\left(\frac{1}{b-a}\int_a^b f(t)dt\right)\left(\frac{1}{b-a}\int_a^b g(t)dt\right) \\
&+\frac{f(b)-f(a)}{b-a}\left(x-\frac{a+b}{2}\right)\left(\frac{1}{b-a}\int_a^b g(t)dt\right) \\
&+\frac{g(b)-g(a)}{b-a}\left(x-\frac{a+b}{2}\right)\left(\frac{1}{b-a}\int_a^b f(t)dt\right) \\
&+\left(\frac{1}{(b-a)^2}\int_a^b\int_a^b p(x,t)p(t,s)f''(s)dsdt\right)\left(\frac{1}{b-a}\int_a^b g(t)dt\right) \\
&+\left(\frac{1}{(b-a)^2}\int_a^b\int_a^b p(x,t)p(t,s)g''(s)dsdt\right)\left(\frac{1}{b-a}\int_a^b f(t)dt\right),
\end{aligned}
\tag{3.4.44}
$$

for $x \in [a,b]$. From (3.4.44) and using the properties of modulus, we have

$$
\begin{aligned}
&\left| f(x)\left(\frac{1}{b-a}\int_a^b g(t)dt\right)+g(x)\left(\frac{1}{b-a}\int_a^b f(t)dt\right)\right. \\
&-2\left(\frac{1}{b-a}\int_a^b f(t)dt\right)\left(\frac{1}{b-a}\int_a^b g(t)dt\right) \\
&-\frac{f(b)-f(a)}{b-a}\left(x-\frac{a+b}{2}\right)\left(\frac{1}{b-a}\int_a^b g(t)dt\right) \\
&\left.-\frac{g(b)-g(a)}{b-a}\left(x-\frac{a+b}{2}\right)\left(\frac{1}{b-a}\int_a^b f(t)dt\right)\right| \\
&\leqslant \left[\|f''\|_\infty\left(\frac{1}{b-a}\int_a^b |g(t)|dt\right)+\|g''\|_\infty\left(\frac{1}{b-a}\int_a^b |f(t)|dt\right)\right] \\
&\times\left(\frac{1}{(b-a)^2}\int_a^b\int_a^b |p(x,t)|\,|p(t,s)|dsdt\right).
\end{aligned}
\tag{3.4.45}
$$

By using (1.2.11) and simple algebraic manipulations (see [28]), we obtain

$$
\frac{1}{(b-a)^2}\int_a^b\int_a^b |p(x,t)|\,|p(t,s)|dsdt = L(x),
\tag{3.4.46}
$$

for $x \in [a,b]$, where $L(x)$ is given by (3.4.35). Using (3.4.46) in (3.4.45), we get the inequality (3.4.33).

To prove the inequality (3.4.34), we multiply both sides of (3.4.42) and (3.4.43) by $g(x)$ and $f(x)$ respectively and adding the resulting identities, we get

$$
\begin{aligned}
2f(x)g(x) &= \left[\frac{1}{b-a}\int_a^b f(t)dt+\frac{f(b)-f(a)}{b-a}\left(x-\frac{a+b}{2}\right)\right]g(x) \\
&+\left[\frac{1}{b-a}\int_a^b g(t)dt+\frac{g(b)-g(a)}{b-a}\left(x-\frac{a+b}{2}\right)\right]f(x) \\
&+\left(\frac{1}{(b-a)^2}\int_a^b\int_a^b p(x,t)p(t,s)f''(s)dsdt\right)g(x) \\
&+\left(\frac{1}{(b-a)^2}\int_a^b\int_a^b p(x,t)p(t,s)g''(s)dsdt\right)f(x).
\end{aligned}
\tag{3.4.47}
$$

Rewriting (3.4.47) and using the properties of modulus and (3.4.46), we get the required inequality in (3.4.34). The proof of Theorem 3.4.5 is complete.

3.5 Further inequalities of the Ostrowski-type

In this section, we will give some Ostrowski-type inequalities recently established by various investigators in [16,73,101,116] involving n-time differentiable mappings.

In [16], Cerone, Dragomir and Roumeliotis proved the following inequality.

Theorem 3.5.1. Let $f : [a,b] \to \mathbb{R}$ be a mapping such that $f^{(n-1)}$ is absolutely continuous on $[a,b]$ and $f^{(n)} \in L_\infty[a,b]$. Then for all $x \in [a,b]$, we have the inequality

$$\left| \int_a^b f(t)dt - \sum_{k=0}^{n-1} \left[\frac{(b-x)^{k+1} + (-1)^k (x-a)^{k+1}}{(k+1)!} \right] f^{(k)}(x) \right|$$

$$\leqslant \frac{\|f^{(n)}\|_\infty}{(n+1)!} \left[(x-a)^{n+1} + (b-x)^{n+1} \right] \leqslant \frac{\|f^{(n)}\|_\infty}{(n+1)!} (b-a)^{n+1} \tag{3.5.1}$$

where $\|f^{(n)}\|_\infty = \sup_{t \in [a,b]} \left| f^{(n)}(t) \right| < \infty$.

Proof. From the hypotheses, we have the following identity (see Lemma 1.5.3):

$$\int_a^b f(t)dt = \sum_{k=0}^{n-1} \left[\frac{(b-x)^{k+1} + (-1)^k (x-a)^{k+1}}{(k+1)!} \right] f^{(k)}(x)$$

$$+ (-1)^n \int_a^b E_n(x,t) f^{(n)}(t)dt, \tag{3.5.2}$$

for all $x \in [a,b]$, where $E_n(x,t)$ is given by (1.5.23). From (3.5.2) and (1.5.23), we have

$$\left| \int_a^b f(t)dt - \sum_{k=0}^{n-1} \left[\frac{(b-x)^{k+1} + (-1)^k (x-a)^{k+1}}{(k+1)!} \right] f^{(k)}(x) \right|$$

$$= \left| \int_a^b E_n(x,t) f^{(n)}(t)dt \right| \leqslant \|f^{(n)}\|_\infty \int_a^b |E_n(x,t)|dt$$

$$= \|f^{(n)}\|_\infty \left[\int_a^x \frac{(t-a)^n}{n!} dt + \int_x^b \frac{(b-t)^n}{n!} dt \right] = \frac{\|f^{(n)}\|_\infty}{(n+1)!} \left[(x-a)^{n+1} + (b-x)^{n+1} \right],$$

and the inequality (3.5.1) is proved. To prove the second inequality in (3.5.1), we observe that

$$(x-a)^{n+1} + (b-x)^{n+1} \leqslant (b-a)^{n+1},$$

for $x \in [a,b]$.

Remark 3.5.1 By taking $x = \frac{a+b}{2}$ in (3.5.1), we have the inequality

$$\left| \int_a^b f(t)dt - \sum_{k=0}^{n-1} \left[\frac{1+(-1)^k}{(k+1)!} \right] \frac{(b-a)^{k+1}}{2^{k+1}} f^{(k)}\left(\frac{a+b}{2}\right) \right| \leqslant \frac{\|f^{(n)}\|_\infty}{2^n(n+1)!}(b-a)^{n+1}. \quad (3.5.3)$$

If we choose $n = 1$ in the inequality (3.5.1), then by simple calculation, we get the well-known Ostrowski's inequality in (7).

Another result analogous to Theorem 3.5.1 obtained by Matić, Pečarić and Ujević [73], is embodied in the following theorem.

Theorem 3.5.2. Let $f : I \to \mathbb{R}$, where $I \subseteq \mathbb{R}$ is an interval. Suppose that f is n-time differentiable in the $\mathring{I}$ (the interior of I), and let $a,\ b \in \mathring{I}$, $a < b$. Let $f^{(n)}$ is integrable on $[a,b]$ and suppose that $\gamma \leqslant f^{(n)} \leqslant \Gamma$ for all $x \in [a,b]$, where γ and Γ are real constants. For $x \in [a,b]$, define

$$R_n(x) = f(x) + \frac{1}{b-a} \sum_{k=1}^{n-1} \frac{(b-x)^{k+1} + (-1)^k (x-a)^{k+1}}{(k+1)!} f^{(k)}(x)$$

$$+ \frac{(b-x)^{n+1} + (-1)^n (x-a)^{n+1}}{(n+1)!(b-a)^2} \left[f^{(n-1)}(b) - f^{(n-1)}(a) \right] - \frac{1}{b-a} \int_a^b f(t)dt.$$

Then for all $x \in [a,b]$,

$$|R_n(x)| \leqslant \frac{\Gamma - \gamma}{2(n!)} \left[\frac{(x-a)^{2n+1} - (x-b)^{2n+1}}{(b-a)(2n+1)} - \left(\frac{(x-a)^{n+1} - (x-b)^{n+1}}{(b-a)(n+1)} \right)^2 \right]^{\frac{1}{2}}. \quad (3.5.4)$$

Proof. From the hypotheses, the identity (3.5.2) holds (see, Lemma 1.5.3). We can rewrite (3.5.2) as

$$\int_a^b f(t)dt = (b-a)f(x) + \sum_{k=1}^{n-1} \frac{(b-x)^{k+1} + (-1)^k (x-a)^{k+1}}{(k+1)!} f^{(k)}(x)$$

$$+ (-1)^n \int_a^b E_n(x,t) f^{(n)}(t)dt,$$

or

$$\frac{(-1)^{n+1}}{b-a} \int_a^b E_n(x,t) f^{(n)}(t)dt$$

$$= f(x) + \frac{1}{b-a} \sum_{k=1}^{n-1} \frac{(b-x)^{k+1} + (-1)^k (x-a)^{k+1}}{(k+1)!} f^{(k)}(x) - \frac{1}{b-a} \int_a^b f(t)dt. \quad (3.5.5)$$

Also,

$$\int_a^b E_n(x,t)dt = \int_a^x \frac{(t-a)^n}{n!}dt + \int_x^b \frac{(t-b)^n}{n!}dt = \frac{(x-a)^{n+1} - (x-b)^{n+1}}{(n+1)!}$$

$$= (-1)^{n+1}\frac{(b-x)^{n+1}+(-1)^n(x-a)^{n+1}}{(n+1)!},$$

and

$$\int_a^b f^{(n)}(t)dt = f^{(n-1)}(b) - f^{(n-1)}(a).$$

So, we have

$$-\frac{(-1)^{n+1}}{(b-a)^2}\int_a^b E_n(x,t)dt\int_a^b f^{(n)}(t)dt$$

$$= \frac{(b-x)^{n+1}+(-1)^n(x-a)^{n+1}}{(n+1)!(b-a)^2}\left[f^{(n-1)}(b)-f^{(n-1)}(a)\right]. \tag{3.5.6}$$

Using (3.5.5) and (3.5.6), we see that

$$(-1)^{n+1}\left[\frac{1}{b-a}\int_a^b E_n(x,t)f^{(n)}(t)dt - \frac{1}{(b-a)^2}\int_a^b E_n(x,t)dt\int_a^b f^{(n)}(t)dt\right],$$

is equal to $R_n(x)$. We now apply Theorem 1.2.1 with $E_n(x,\cdot)$ and $f^{(n)}(\cdot)$ in place of f and g, respectively, to obtain

$$|R_n(x)| \leqslant \frac{1}{2}(\Gamma-\gamma)\sqrt{T(E_n(x,\cdot),E_n(x,\cdot))}, \tag{3.5.7}$$

where $T(\cdot,\cdot)$ is given by (2). We have already calculated

$$\int_a^b E_n(x,t)dt = \frac{(x-a)^{n+1}-(x-b)^{n+1}}{(n+1)!}.$$

Similar calculation gives

$$\int_a^b E_n^2(x,t)dt = \frac{(x-a)^{2n+1}-(x-b)^{2n+1}}{(n!)^2(2n+1)},$$

so that

$$T(E_n(x,\cdot),E_n(x,\cdot)) = \frac{1}{b-a}\int_a^b E_n^2(x,t)dt - \frac{1}{(b-a)^2}\left(\int_a^b E_n(x,t)dt\right)^2$$

$$= \frac{1}{(n!)^2}\left[\frac{(x-a)^{2n+1}-(x-b)^{2n+1}}{(b-a)(2n+1)} - \left(\frac{(x-a)^{n+1}-(x-b)^{n+1}}{(b-a)(n+1)}\right)^2\right]. \tag{3.5.8}$$

Combining (3.5.7) and (3.5.8), we get (3.5.4). The proof is complete.

In a recent paper [101], Pachpatte has established the following new generalization of Milovanović, Pečarić inequality [77], involving a pair of n-time differentiable mappings.

Theorem 3.5.3. Let $f,\ g:[a,b]\to\mathbb{R}$ be continous functions on $[a,b]$ and n-times differentiable on (a,b), and with derivatives $f^{(n)},\ g^{(n)}:(a,b)\to\mathbb{R}$ being bounded on (a,b), i.e. $\|f^{(n)}\|_\infty=\sup_{t\in(a,b)}\left|f^{(n)}(t)\right|<\infty$, $\|g^{(n)}\|_\infty=\sup_{t\in(a,b)}\left|g^{(n)}(t)\right|<\infty$. Then

$$\left|f(x)g(x)-\frac{1}{2(b-a)}[g(x)I_0+f(x)J_0]-\frac{1}{2(b-a)}\left[g(x)\sum_{k=1}^{n-1}I_k+f(x)\sum_{k=1}^{n-1}J_k\right]\right|$$

$$\leqslant\frac{1}{2(n+1)!}\left[|g(x)|\|f^{(n)}\|_\infty+|f(x)|\|g^{(n)}\|_\infty\right]\left[\frac{(x-a)^{n+1}+(b-x)^{n+1}}{b-a}\right],\tag{3.5.9}$$

for all $x\in[a,b]$, where $I_k,\ I_0$ and $J_k,\ J_0$ are respectively given by (1.5.9) and (1.5.10) in Section 1.5.

Proof. Let $x\in[a,b]$, $y\in(a,b)$. From the hypotheses, by using Taylor's formula with the Lagrange form of the remainder (see[77]), we have

$$f(x)=f(y)+\sum_{k=1}^{n-1}\frac{f^k(y)}{k!}(x-y)^k+\frac{1}{n!}f^{(n)}(\xi)(x-y)^n,\tag{3.5.10}$$

and

$$g(x)=g(y)+\sum_{k=1}^{n-1}\frac{g^k(y)}{k!}(x-y)^k+\frac{1}{n!}g^{(n)}(\sigma)(x-y)^n,\tag{3.5.11}$$

where $\xi=y+\alpha(x-y)\ (0<\alpha<1)$ and $\sigma=y+\beta(x-y)\ (0<\beta<1)$. Let $F_k(x)$ and $G_k(x)$ be respectively given by (1.5.6) and (1.5.7) in Section 1.5. From the definitions of I_k, J_k and integration by parts (see [77]), we have the relations

$$I_0+\sum_{k=1}^{n-1}I_k=nI_0-(b-a)\sum_{k=1}^{n-1}F_k(x),\tag{3.5.12}$$

$$J_0+\sum_{k=1}^{n-1}J_k=nJ_0-(b-a)\sum_{k=1}^{n-1}G_k(x).\tag{3.5.13}$$

Multiplying (3.5.10) and (3.5.11) by $g(x)$ and $f(x)$ respectively, adding the resulting identities and rewriting, we have

$$f(x)g(x)=\frac{1}{2}g(x)f(y)+\frac{1}{2}f(x)g(y)$$

$$+\frac{1}{2}g(x)\sum_{k=1}^{n-1}\frac{f^{(k)}(y)}{k!}(x-y)^k+\frac{1}{2}f(x)\sum_{k=1}^{n-1}\frac{g^{(k)}(y)}{k!}(x-y)^k$$

$$+\frac{1}{2}\frac{1}{n!}g(x)f^{(n)}(\xi)(x-y)^n+\frac{1}{2}\frac{1}{n!}f(x)g^{(n)}(\sigma)(x-y)^n.\tag{3.5.14}$$

Integrating (3.5.14) with respect to y over (a,b) and rewriting,we obtain

$$f(x)g(x) = \frac{1}{2(b-a)}[g(x)I_0 + f(x)J_0] + \frac{1}{2(b-a)}\left[g(x)\sum_{k=1}^{n-1} I_k + f(x)\sum_{k=1}^{n-1} J_k\right]$$

$$+\frac{1}{2(b-a)}\frac{1}{n!}\left[g(x)\int_a^b f^{(n)}(\xi)(x-y)^n dy + f(x)\int_a^b g^{(n)}(\sigma)(x-y)^n dy\right]. \tag{3.5.15}$$

From (3.5.15) and using the properties of modulus, we have

$$\left| f(x)g(x) - \frac{1}{2(b-a)}[g(x)I_0 + f(x)J_0] - \frac{1}{2(b-a)}\left[g(x)\sum_{k=1}^{n-1} I_k + f(x)\sum_{k=1}^{n-1} J_k\right]\right|$$

$$\leqslant \frac{1}{2(b-a)}\frac{1}{n!}\left[|g(x)|\int_a^b \left|f^{(n)}(\xi)\right| |x-y|^n dy + |f(x)|\int_a^b \left|g^{(n)}(\sigma)\right| |x-y|^n dy\right]$$

$$\leqslant \frac{1}{2(b-a)}\frac{1}{n!}\left[|g(x)|\|f^{(n)}\|_\infty + |f(x)|\|g^{(n)}\|_\infty\right]\int_a^b |x-y|^n dy$$

$$= \frac{1}{2}\frac{1}{(n+1)!}\left[|g(x)|\|f^{(n)}\|_\infty + |f(x)|\|g^{(n)}\|_\infty\right]\left[\frac{(x-a)^{n+1}+(b-x)^{n+1}}{b-a}\right],$$

which is the required inequality in (3.5.9). The proof is complete.

The following Corollary holds.

Corollary 3.5.1. Let $f,\ g : [a,b] \to \mathbb{R}$ be continuous functions on $[a,b]$ and differentiable on (a,b) and with derivatives $f',\ g' : [a,b] \to \mathbb{R}$ being bounded on (a,b), i.e., $\|f'\|_\infty = \sup_{t\in(a,b)} |f'(t)| < \infty$, $\|g'\|_\infty = \sup_{t\in(a,b)} |g'(t)| < \infty$. Then

$$\left| f(x)g(x) - \frac{1}{2(b-a)}[g(x)I_0 + f(x)J_0]\right|$$

$$\leqslant \frac{1}{2}\left[|g(x)|\|f'\|_\infty + |f(x)|\|g'\|_\infty\right]\left[\frac{1}{4} + \frac{\left(x-\frac{a+b}{2}\right)^2}{(b-a)^2}\right](b-a), \tag{3.5.16}$$

for all $x \in [a,b]$, where I_0 and J_0 are as in Theorem 3.5.3.

Remark 3.5.2. We note that in the special cases, if we take (i) $g(x) = 1$ and hence $g^{(n)}(x) = 0$ in (3.5.9) and (ii) $g(x) = 1$ and hence $g'(x) = 0$ in (3.5.16), then by simple calculations we get the inequalities given by Milovanović and Pečarić [77] and Ostrowski [81] respectively.

The following Theorem contains the Ostrowski type inequalities, recently established by Pachpatte in [116], involving a harmonic sequence of polynomials and a pair of n-time differentiable functions.

Theorem 3.5.4. Let (P_n) be a harmonic sequence of polynomials, that is $P_n' = P_{n-1}, n \geqslant 1$, $P_0 = 1$. Let $f, g : [a,b] \to \mathbb{R}$ be such that $f^{(n-1)}$, $g^{(n-1)}$ are absolutely continuous for $n \geqslant 1$ and $f^{(n)}, g^{(n)} \in L_p[a,b]$, $1 \leqslant p \leqslant \infty$. Then

$$\left| g(x)B[f(x)] + f(x)B[g(x)] - \frac{1}{b-a}\left[g(x)\int_a^b f(t)dt + f(x)\int_a^b g(t)dt \right] \right|$$

$$\leqslant D(n,p,x)\left[|g(x)| \left\| f^{(n)} \right\|_p + |f(x)| \left\| g^{(n)} \right\|_p \right]. \tag{3.5.17}$$

and

$$\left| B[f(x)]B[g(x)] - \frac{1}{b-a}\left[B[g(x)]\int_a^b f(t)dt + B[f(x)]\int_a^b g(t)dt \right] \right.$$

$$\left. + \left(\frac{1}{b-a}\int_a^b f(t)dt \right)\left(\frac{1}{b-a}\int_a^b g(t)dt \right) \right| \leqslant \{D(n,p,x)\}^2 \left\| f^{(n)} \right\|_p \left\| g^{(n)} \right\|_p. \tag{3.5.18}$$

for all $x \in [a,b]$, where $B[\cdot]$ is given by (1.5.4),

$$D(n,p,x) = \frac{1}{n(b-a)} \left\| P_{n-1}e(\cdot,x) \right\|_{p'}, \tag{3.5.19}$$

$$e(t,x) = \begin{cases} t-a & \text{if } t \in [a,x], \\ t-b & \text{if } t \in (x,b], \end{cases}$$

and as usual $\frac{1}{p} + \frac{1}{p'} = 1$ with $p' = 1$ for $p = \infty$, $p' = \infty$ for $p = 1$ and $\| \cdot \|_p$ is the norm in $L_p[a,b]$.

Proof. From the hypotheses, we have the following identities (see. Lemma 1.5.4):

$$B[f(x)] - \frac{1}{b-a}\int_a^b f(t)dt = \frac{(-1)^{n-1}}{n(b-a)}\int_a^b P_{n-1}(t)e(t,x)f^{(n)}(t)dt, \tag{3.5.20}$$

and

$$B[g(x)] - \frac{1}{b-a}\int_a^b g(t)dt = \frac{(-1)^{n-1}}{n(b-a)}\int_a^b P_{n-1}(t)e(t,x)g^{(n)}(t)dt, \tag{3.5.21}$$

for $x \in [a,b]$. Multiplying (3.5.20) and (3.5.21) by $g(x)$ and $f(x)$ respectively and adding the resulting identities, we have

$$g(x)B[f(x)] + f(x)B[g(x)] - \frac{1}{b-a}\left[g(x)\int_a^b f(t)dt + f(x)\int_a^b g(t)dt \right]$$

$$= \frac{(-1)^{n-1}}{n(b-a)}\left[g(x)\int_a^b P_{n-1}(t)e(t,x)f^{(n)}(t)dt + f(x)\int_a^b P_{n-1}(t)e(t,x)g^{(n)}(t)dt \right]. \tag{3.5.22}$$

From (3.5.22) and using the properties of modulus and Hölder's integral inequality, we have

$$\left| g(x)B[f(x)] + f(x)B[g(x)] - \frac{1}{b-a}\left[g(x)\int_a^b f(t)dt + f(x)\int_a^b g(t)dt \right] \right|$$

$$\leqslant \frac{1}{n(b-a)}\left[|g(x)|\int_a^b \left|P_{n-1}(t)e(t,x)f^{(n)}(t)\right|dt+|f(x)|\int_a^b \left|P_{n-1}(t)e(t,x)g^{(n)}(t)\right|dt\right]$$

$$\leqslant \frac{1}{n(b-a)}\left[|g(x)|\left\{\int_a^b |P_{n-1}(t)e(t,x)|^{p'}dt\right\}^{\frac{1}{p'}}\left\{\int_a^b \left|f^{(n)}(t)\right|^{p}dt\right\}^{\frac{1}{p}}\right.$$

$$\left.+|f(x)|\left\{\int_a^b |P_{n-1}(t)e(t,x)|^{p'}dt\right\}^{\frac{1}{p'}}\left\{\int_a^b \left|g^{(n)}(t)\right|^{p}dt\right\}^{\frac{1}{p}}\right]$$

$$= D(n,p,x)\left[|g(x)|\left\|f^{(n)}\right\|_p+|f(x)|\left\|g^{(n)}\right\|_p\right].$$

This is the required inequality in (3.5.17).

Multiplying the left hand sides and right hand sides of (3.5.20) and (3.5.21), we get

$$B[f(x)]B[g(x)]-\frac{1}{b-a}\left[B[g(x)]\int_a^b f(t)dt+B[f(x)]\int_a^b g(t)dt\right]$$

$$+\left(\frac{1}{b-a}\int_a^b f(t)dt\right)\left(\frac{1}{b-a}\int_a^b g(t)dt\right)$$

$$=\frac{(-1)^{2n-2}}{n^2(b-a)^2}\left\{\int_a^b P_{n-1}(t)e(t,x)f^{(n)}(t)dt\right\}\left\{\int_a^b P_{n-1}(t)e(t,x)g^{(n)}(t)dt\right\}. \tag{3.5.23}$$

From (3.5.23) and following the proof of inequality (3.5.17) given above with suitable modifications, we get the required inequality in (3.5.18). The proof is complete.

Remark 3.5.3. If we take $g(t)=1$ and hence $g^{(n-1)}(t)=0$ for $n\geqslant 2$ in (3.5.17), then we get a variant of the Ostrowski-type inequality given by Dedić, Pečarić and Ujević in [22]. For many additional interesting results see [21].

3.6 Discrete Ostrowski-type inequalities

This section is devoted to the discrete Ostrowski-type inequalities that have recently investigated by Pachpatte in [88,105,114,133].

The first Theorem considers the Ostrowski-type inequalities proved in [88].

Theorem 3.6.1. Let $\{x_i\}$ for $i=0,1,\ldots,n$ $(n\in\mathbb{N})$ be a sequence of real numbers. Then the following inequalities hold

$$\left|\sum_{i=0}^{n-1} x_i - n\left(\frac{x_0+x_n}{2}\right)\right| \leqslant \frac{1}{2}n\sum_{i=0}^{n-1}|\Delta x_i|, \tag{3.6.1}$$

and

$$\left|\sum_{i=1}^{n-1} x_i^2 - n\left(\frac{x_0^2+x_n^2}{2}\right)\right| \leqslant n\sum_{i=0}^{n-1}|(x_{i+1}+x_i)\Delta x_i|, \tag{3.6.2}$$

where $\Delta x_i = x_{i+1}-x_i$.

Proof. It is easy to observe that the following identities hold (see [88]):

$$x_i = x_0 + \sum_{j=0}^{i-1} \Delta x_j, \tag{3.6.3}$$

$$x_i = x_n - \sum_{j=i}^{n-1} \Delta x_j, \tag{3.6.4}$$

$$x_i^2 = x_0^2 + \sum_{j=0}^{i-1} (x_{j+1} + x_j)(\Delta x_j), \tag{3.6.5}$$

$$x_i^2 = x_n^2 - \sum_{j=i}^{n-1} (x_{j+1} + x_j)(\Delta x_j). \tag{3.6.6}$$

From (3.6.3), (3.6.4) and (3.6.5), (3.6.6), we have

$$x_i = \frac{x_0 + x_n}{2} + \frac{1}{2}\sum_{j=0}^{i-1} \Delta x_j - \frac{1}{2}\sum_{j=i}^{n-1} \Delta x_j, \tag{3.6.7}$$

and

$$x_i^2 = \frac{x_0^2 + x_n^2}{2} + \frac{1}{2}\sum_{j=0}^{i-1} (x_{j+1} + x_j)(\Delta x_j) - \frac{1}{2}\sum_{j=i}^{n-1} (x_{j+1} + x_j)(\Delta x_j). \tag{3.6.8}$$

Summing both sides of (3.6.7) and (3.6.8) from $i = 0$ to $n-1$ and by making elementary calculations, we get the required inequalities in (3.6.1) and (3.6.2). The proof is complete.

The next Theorem contains the discrete Ostrowski-type inequalities given in [114], involving two sequences.

Theorem 3.6.2. Let $\{u_i\}$, $\{v_i\}$ for $i = 0, 1, \ldots, n$ $(n \in \mathbb{N})$ be sequences of real numbers. Then the following inequalities hold

$$\left| u_i v_i - \frac{1}{2}[v_i U + u_i V] \right| \leqslant \frac{1}{4}\left[|v_i| \sum_{j=0}^{n-1} |\Delta u_j| + |u_i| \sum_{j=0}^{n-1} |\Delta v_j| \right], \tag{3.6.9}$$

and

$$|u_i v_i - [v_i U + u_i V] + UV| \leqslant \frac{1}{4}\left(\sum_{j=0}^{n-1} |\Delta u_j| \right)\left(\sum_{j=0}^{n-1} |\Delta v_j| \right), \tag{3.6.10}$$

for $i = 0, 1, \ldots, n$, where

$$U = \frac{u_1 + u_n}{2}, \qquad V = \frac{v_1 + v_n}{2}, \tag{3.6.11}$$

and Δ is the forward difference operator.

Proof. From the hypotheses, we have the following identities (see [108, p. 352]):

$$u_i - U = \frac{1}{2}\left[\sum_{j=0}^{i-1} \Delta u_j - \sum_{j=i}^{n-1} \Delta u_j\right], \tag{3.6.12}$$

and

$$v_i - V = \frac{1}{2}\left[\sum_{j=0}^{i-1} \Delta v_j - \sum_{j=i}^{n-1} \Delta v_j\right]. \tag{3.6.13}$$

Multiplying both sides of (3.6.12) and (3.6.13) by v_i and u_i $(i = 0, 1, \dots, n)$ respectively, adding the resulting identities and rewriting, we get

$$u_i v_i - \frac{1}{2}[v_i U + u_i V] = \frac{1}{4}\left[v_i\left[\sum_{j=0}^{i-1} \Delta u_j - \sum_{j=i}^{n-1} \Delta u_j\right] + u_i\left[\sum_{j=0}^{i-1} \Delta v_j - \sum_{j=i}^{n-1} \Delta v_j\right]\right]. \tag{3.6.14}$$

Multiplying the left hand sides and right hand sides of (3.6.12) and (3.6.13), we have

$$u_i v_i - [v_i U + u_i V] + UV = \frac{1}{4}\left[\sum_{j=0}^{i-1} \Delta u_j - \sum_{j=i}^{n-1} \Delta u_j\right]\left[\sum_{j=0}^{i-1} \Delta v_j - \sum_{j=i}^{n-1} \Delta v_j\right]. \tag{3.6.15}$$

From (3.6.14), (3.6.15), using the properties of modulus and sums, we get the desired inequalities in (3.6.9) and (3.6.10). The proof is complete.

In the following theorem, we present the inequalities established in [133], similar to those of given in the above theorem, by using somewhat different representation.

Theorem 3.6.3. Let $\{u_k\}$, $\{v_k\}$ for $k = 1, \dots, n$ be two finite sequences of real numbers such that $\max_{1 \leqslant k \leqslant n-1}\{|\Delta u_k|\} = A$, $\max_{1 \leqslant k \leqslant n-1}\{|\Delta v_k|\} = B$, where A, B are nonnegative constants. Then the following inequalities hold

$$\left| u_k v_k - \frac{1}{2n}\left[v_k \sum_{i=1}^{n} u_i + u_k \sum_{i=1}^{n} v_i\right]\right| \leqslant \frac{1}{2}\left[|v_k| A + |u_k| B\right] H_n(K), \tag{3.6.16}$$

and

$$\left| u_k v_k - \frac{1}{n}\left[v_k \sum_{i=1}^{n} u_i + u_k \sum_{i=1}^{n} v_i\right] + \frac{1}{n^2}\left(\sum_{i=1}^{n} u_i\right)\left(\sum_{i=1}^{n} v_i\right)\right| \leqslant AB\{H_n(k)\}^2, \tag{3.6.17}$$

for $k = 1, \dots, n$, where $H_n(k)$ is given by (1.6.35).

Proof. Following the proof of Theorem 1.6.5, we have (1.6.38) and (1.6.40). From (1.6.38) and (1.6.40) and using the properties of modulus, we get the desired inequalities in (3.6.16) and (3.6.17).

Remark 3.6.1. By taking $v_k = 1$ and hence $\Delta v_k = 0$ for $k = 1,\dots,n$ in (3.6.16) and by simple computation, we get

$$\left| u_k - \frac{1}{n}\sum_{i=1}^{n} u_i \right| \leqslant H_n(k) \max_{1\leqslant k\leqslant n-1} \{|\Delta u_k|\}, \tag{3.6.18}$$

for $k = 1,\dots,n$. By simple computation (see [49]), we have

$$H_n(k) = \sum_{i=1}^{n-1} |D_n(k,i)| = \frac{1}{n}\left[\frac{n^2-1}{4} + \left(k - \frac{n+1}{2}\right)^2\right]. \tag{3.6.19}$$

In fact, the inequality (3.6.18) is established by Dragomir [49, Theorem 3.1] in a normed linear space.

In concluding this section we give the discrete Ostrowski-type inequality recently proved in [105].

Theorem 3.6.4. Let $N_{a,b} = \{a, a+1, \dots, a+n = b\}$ for $a \in \mathbb{R}$, $n \in \mathbb{N}$. Let $f(t)$, $g(t)$, $h(t)$ be real-valued functions defined on $N_{a,b}$ and are zero when $t \notin N_{a,b}$ and $|\Delta f(t)| \leqslant M_1$, $|\Delta g(t)| \leqslant M_2$, $|\Delta h(t)| \leqslant M_3$, on $N_{a,b}$, where M_1, M_2, M_3 are nonnegative constants. Then

$$\left| f(t)g(t)h(t) - \frac{1}{3(b-a)}\left[g(t)h(t)\sum_{s=a}^{b-1} f(s) + h(t)f(t)\sum_{s=a}^{b-1} g(s) + f(t)g(t)\sum_{s=a}^{b-1} h(s) \right]\right|$$

$$\leqslant \frac{1}{3}\left[|g(t)||h(t)|M_1 + |h(t)||f(t)|M_2 + |f(t)||g(t)|M_3\right] B(t), \tag{3.6.20}$$

for all $t \in \mathbb{N}_{a,b}$, where

$$B(t) = \left[\frac{1}{2} + \left| t - \frac{a+b}{2}\right|\right]. \tag{3.6.21}$$

Proof. For any t, $s \in \mathbb{N}_{a,b}$, it is easy to observe that the following identities hold:

$$f(t) - f(s) = \sum_{m=s}^{t-1} \Delta f(m), \tag{3.6.22}$$

$$g(t) - g(s) = \sum_{m=s}^{t-1} \Delta g(m), \tag{3.6.23}$$

$$h(t) - h(s) = \sum_{m=s}^{t-1} \Delta h(m). \tag{3.6.24}$$

Multiplying both sides of (3.6.22), (3.6.23) and (3.6.24) by $g(t)h(t)$, $h(t)f(t)$ and $f(t)g(t)$ respectively and adding the resulting identities, we get

$$3f(t)g(t)h(t) - [g(t)h(t)f(s) + h(t)f(t)g(s) + f(t)g(t)h(s)]$$

$$= g(t)h(t)\sum_{m=s}^{t-1} \Delta f(m) + h(t)f(t)\sum_{m=s}^{t-1} \Delta g(m) + f(t)g(t)\sum_{m=s}^{t-1} \Delta h(m). \tag{3.6.25}$$

Summing both sides of (3.6.25) over s from a to $b-1$ and rewriting, we have

$$f(t)g(t)h(t)-\frac{1}{3(b-a)}\left[g(t)h(t)\sum_{s=a}^{b-1}f(s)+h(t)f(t)\sum_{s=a}^{b-1}g(s)+f(t)g(t)\sum_{s=a}^{b-1}h(s)\right]$$

$$=\frac{1}{3(b-a)}\sum_{s=a}^{b-1}\left[g(t)h(t)\sum_{m=s}^{t-1}\Delta f(m)+h(t)f(t)\sum_{m=s}^{t-1}\Delta g(m)\right.$$

$$\left.+f(t)g(t)\sum_{m=s}^{t-1}\Delta h(m)\right]. \tag{3.6.26}$$

From (3.6.26) and using the properties of modulus, we have

$$\left|f(t)g(t)h(t)-\frac{1}{3(b-a)}\left[g(t)h(t)\sum_{s=a}^{b-1}f(s)+h(t)f(t)\sum_{s=a}^{b-1}g(s)+f(t)g(t)\sum_{s=a}^{b-1}h(s)\right]\right|$$

$$\leqslant\frac{1}{3(b-a)}\left[|g(t)||h(t)|M_1+|h(t)||f(t)|M_2+|f(t)||g(t)|M_3\right]\left|\sum_{s=a}^{b-1}(t-s)\right|. \tag{3.6.27}$$

By using the summation formula for Arithmetic Progression, it is easy to observe that

$$\left|\sum_{s=a}^{b-1}(t-s)\right|=\left|t(b-a)-\frac{b-a}{2}[2a+b-a-1]\right|$$

$$=\left[\frac{1}{2}+\left|t-\frac{a+b}{2}\right|\right](b-a)=B(t)(b-a). \tag{3.6.28}$$

Using (3.6.28) in (3.6.27), we get (3.6.20). The proof is complete.

Remark 3.6.2. By taking $h(t)=1$ and hence $\Delta h(t)=0$ in Theorem 3.6.4 and by simple computations, it is easy to see that the inequality (3.6.20) reduces to

$$\left|f(t)g(t)-\frac{1}{2(b-a)}\left[g(t)\sum_{s=a}^{b-1}f(s)+f(t)\sum_{s=a}^{b-1}g(s)\right]\right|$$

$$\leqslant\frac{1}{2}\left[|g(t)|M_1+|f(t)|M_2\right]B(t), \tag{3.6.29}$$

for all $t\in\mathbb{N}_{a,b}$. Further by taking $g(t)=1$ and hence $\Delta g(t)=0$ in (3.6.29), we get by simple computation

$$\left|f(t)-\frac{1}{b-a}\sum_{s=a}^{b-1}f(s)\right|\leqslant M_1B(t), \tag{3.6.30}$$

for all $t\in\mathbb{N}_{a,b}$.

3.7 Applications

The literature on the applications of inequalities related to the celebrated Ostrowski's inequality is vast and rapidly growing vaster. In this section, we present applications of certain inequalities given in earlier sections, which have been investigated during the past few years.

3.7.1 *Applications for some special means*

We present below, applications of Theorem 3.2.2 given by Dragomir and Wang in [26], to the estimation of error bounds for some special means. In [26], some important relationships between the following means are given.

(a) The arithmetic mean:

$$A = A(a,b) = \frac{a+b}{2}, \quad a,\ b \geqslant 0.$$

(b) The geometric mean:

$$G = G(a,b) = \sqrt{ab}, \quad a,\ b \geqslant 0.$$

(c) The harmonic mean:

$$H = H(a,b) = \frac{2}{\frac{1}{a}+\frac{1}{b}}, \quad a,\ b > 0.$$

(d) The logarithmic mean:

$$L = L(a,b) = \begin{cases} \dfrac{b-a}{\log b - \log a} & \text{if } a \neq b, \\ a & \text{if } a = b, \end{cases} \qquad a,\ b > 0.$$

(e) The identric mean:

$$I = I(a,b) = \begin{cases} \dfrac{1}{e}\left(\dfrac{b^b}{a^a}\right)^{\frac{1}{b-a}} & \text{if } a \neq b, \\ a & \text{if } a = b, \end{cases} \qquad a,\ b > 0.$$

(f) The p-logarithmic mean:

$$L_p = L_p(a,b) = \begin{cases} \left[\dfrac{b^{p+1}-a^{p+1}}{(p+1)(b-a)}\right]^{\frac{1}{p}} & \text{if } a \neq b, \\ a & \text{if } a = b, \end{cases} \qquad p \in \mathbb{R} - \{-1,0\}; \quad a,\ b > 0.$$

The following simple relationships between the means are well known in the literature:

$$H \leqslant G \leqslant L \leqslant I \leqslant A,$$

and L_p is monotonically increasing in $p \in \mathbb{R}$ with $L_0 = I$ and $L_{-1} = L$.

1. Applying the inequality (3.2.6) to the mapping $f(x) = x^p$ $(p > 1)$, $x \in [a,b] \subset (0,\infty)$, we get

$$\left|x^p - L_p^p - pL_{p-1}^{p-1}(x-a)\right| \leqslant \frac{1}{4}(b-a)^2(p-1)L_{p-2}^{p-2}. \tag{3.7.1}$$

If we choose $x = A$ and $x = I$ in (3.7.1), then we get, respectively,

$$\left|A^p - L_p^p\right| \leqslant \frac{1}{4}(b-a)^2(p-1)L_{p-2}^{p-2},$$

and

$$\left|I^p - L_p^p - pL_{p-1}^{p-1}(I-A)\right| \leqslant \frac{1}{4}(b-a)^2(p-1)L_{p-2}^{p-2}.$$

2. Choosing $f(x) = \frac{1}{x}$, $x \in [a,b] \subset (0,\infty)$ in (3.2.6) we obtain

$$\left|\frac{1}{x} - \frac{1}{L} - \frac{x-A}{G^2}\right| \leqslant \frac{A(b-a)^2}{2G^4}, \tag{3.7.2}$$

for all $x \in [a,b]$. Replacing x in (3.7.2) by A and L, we get

$$0 \leqslant A - L \leqslant \frac{A^2L(b-a)^2}{2G^4}, \tag{3.7.3}$$

and

$$0 \leqslant A - L \leqslant \frac{A(b-a)^2}{2G^2}, \tag{3.7.4}$$

respectively. Note that $\frac{A^2L}{G^4} \geqslant \frac{A}{G^2}$, since $AL \geqslant G^2$. Then the last term in (3.7.4) is a sharper bound for $A-L$ than that in (3.7.3).

3. We now apply (3.2.6), to the mapping $f(x) = -\log x$, $x \in [a,b] \subset (0,\infty)$, to get

$$\left|\log\left[\frac{I\left(\frac{b}{a}\right)^{\left(\frac{x-A}{b-a}\right)}}{x}\right]\right| \leqslant \log\left(\frac{b}{a}\right)^{\frac{b-a}{4}}, \tag{3.7.5}$$

for all $x \in [a,b]$. Putting $x = A$ and $x = I$ in (3.7.5), we obtain respectively, the following inequalities

$$1 \leqslant \frac{A}{I} \leqslant \left(\frac{b}{a}\right)^{\frac{b-a}{4}},$$

and

$$0 \leqslant A - I \leqslant \frac{1}{4}(b-a)^2.$$

We remark that one can also choose $x = L$, $x = G$, and $x = H$ in inequalities (3.7.1), (3.7.2) and (3.7.5). The resulting inequalities in L, G and H will be similar to those obtained above. We omit the details.

3.7.2 *Applications in numerical integration*

In [150], Ujević used the various special versions of Ostrowski-type inequality in Theorem 3.2.5 to study the numerical integration. We present below the results given in [150], which deals with the approximations of the integral $\int_a^b f(t)dt$.

Theorem 3.7.1. Let all assumptions of Theorem 3.2.5 hold. If $I_n = \{a = x_0 < x_1 < \cdots < x_n = b\}$ is a given subdivision of the interval $[a,b]$ and $h_i = x_{i+1} - x_i$, $i = 0,1,\ldots,n-1$, then

$$\int_a^b f(t)dt = A(I_n,\xi,f) + R_\gamma(I_n,\xi,f), \tag{3.7.6}$$

where

$$A(I_n,\xi,f) = \sum_{i=0}^{n-1}\left[f(\xi_i) - \gamma\left(\xi_i - \frac{x_i + x_{i+1}}{2}\right)\right]h_i, \tag{3.7.7}$$

for $x_i \leqslant \xi_i \leqslant x_{i+1}$, $i = 0,1,\ldots,n-1$. The remainder term satisfies

$$|R_\gamma(I_n,\xi,f)| \leqslant \sum_{i=0}^{n-1}(S_i - \gamma)\left[\frac{h_i}{2} + \left|\xi_i - \frac{x_i + x_{i+1}}{2}\right|\right]h_i, \tag{3.7.8}$$

where $S_i = \frac{1}{h_i}(f(x_{i+1}) - f(x_i))$, $i = 0,1,\ldots,n-1$. Also,

$$\int_a^b f(t)dt = A(I_n,\xi,f) + R_\Gamma(I_n,\xi,f), \tag{3.7.9}$$

where

$$|R_\Gamma(I_n,\xi,f)| \leqslant \sum_{i=0}^{n-1}(\Gamma - S_i)\left[\frac{h_i}{2} + \left|\xi_i - \frac{x_i + x_{i+1}}{2}\right|\right]h_i. \tag{3.7.10}$$

Proof. We apply inequality (3.2.49) in Corollary 3.2.1 to the interval $[x_i,x_{i+1}]$, then

$$\left|f(\xi_i)h_i - \gamma\left(\xi_i - \frac{x_i + x_{i+1}}{2}\right)h_i - \int_{x_i}^{x_{i+1}} f(t)dt\right|$$
$$\leqslant (S_i - \gamma)\left[\frac{h_i}{2} + \left|\xi_i - \frac{x_i + x_{i+1}}{2}\right|\right]h_i, \tag{3.7.11}$$

for $i = 0,1,\ldots,n-1$. We also have

$$f(\xi_i)h_i - \gamma\left(\xi_i - \frac{x_i + x_{i+1}}{2}\right)h_i - \int_{x_i}^{x_{i+1}} f(t)dt = \int_{x_i}^{x_{i+1}} \hat{k}(\xi_i,t)[f'(t) - \gamma]dt, \tag{3.7.12}$$

where

$$\hat{k}(\xi_i,t) = \begin{cases} t - x_i, & t \in [x_i,\xi_i], \\ t - x_{i+1}, & t \in (\xi_i,x_{i+1}], \end{cases} \tag{3.7.13}$$

for $i = 0,1,\ldots,n-1$. If we now sum (3.7.12) over i from 0 to $n-1$ and apply the triangle inequality and (3.7.11), then we get (3.7.6), (3.7.7) and (3.7.8). In a similar way, we can prove that (3.7.9) and (3.7.10) hold.

Remark 3.7.1. If we set $\xi_i = \frac{x_i+x_{i+1}}{2}$ in Theorem 3.7.1, then we get the composite midpoint rule.

Theorem 3.7.2. Let all assumptions of Theorem 3.2.5 hold. If $I_n = \{a = x_0 < x_1 < \cdots < x_n = b\}$ is a given subdivision of the interval $[a,b]$ and $h_i = x_{i+1} - x_i$, $i = 0,1,\ldots,n-1$, then

$$\int_a^b f(t)dt = A_{\mathrm{T}}(I_n,f) + R_{\mathrm{T}\gamma}(I_n,f), \tag{3.7.14}$$

where

$$A_{\mathrm{T}}(I_n,f) = \sum_{i=0}^{n-1} \frac{1}{2}\left[f(x_i) + f(x_{i+1})\right] h_i, \tag{3.7.15}$$

$$\left|R_{\mathrm{T}\gamma}(I_n,f)\right| \leqslant \frac{1}{2}\sum_{i=0}^{n-1}(S_i-\gamma)h_i^2, \tag{3.7.16}$$

and S_i is as given in Theorem 3.7.1. Also,

$$\int_a^b f(t)dt = A_{\mathrm{T}}(I_n,f) + R_{\mathrm{T}\Gamma}(I_n,f), \tag{3.7.17}$$

where

$$|R_{\mathrm{T}\Gamma}(I_n,f)| \leqslant \frac{1}{2}\sum_{i=0}^{n-1}(\Gamma - S_i)h_i^2. \tag{3.7.18}$$

Proof. We apply inequality (3.2.52) in Corollary 3.2.2 to the interval $[x_i,x_{i+1}]$, then

$$\left|\frac{f(x_i)+f(x_{i+1})}{2}h_i - \int_{x_i}^{x_{i+1}} f(t)dt\right| \leqslant \frac{1}{2}(S_i-\gamma)h_i^2, \tag{3.7.19}$$

for $i = 0,1,\ldots,n-1$. We also have

$$\frac{f(x_i)+f(x_{i+1})}{2}h_i - \int_{x_i}^{x_{i+1}} f(t)dt = \int_{x_i}^{x_{i+1}} \hat{k}_i(t)[f'(t)-\gamma]dt, \tag{3.7.20}$$

where

$$\hat{k}_i(t) = t - \frac{x_i + x_{i+1}}{2}, \tag{3.7.21}$$

for $i = 0,1,\ldots,n-1$. If we now sum (3.7.20) over i from 0 to $n-1$ and apply the triangle inequality and (3.7.19), then we get (3.7.14), (3.7.15) and (3.7.16). In a similar way, we can prove that (3.7.17) and (3.7.18) hold.

Theorem 3.7.3. Let the assumptions of Theorem 3.7.1 hold. Then

$$\int_a^b f(t)dt = A_C(I_n,\xi,f) + R_{C\gamma}(I_n,\xi,f), \tag{3.7.22}$$

where

$$A_C(I_n,\xi,f) = \frac{1}{2}\sum_{i=0}^{n-1}\frac{f(x_i)+f(x_{i+1})}{2}h_i$$

$$+\frac{1}{2}\sum_{i=0}^{n-1} f(\xi_i)h_i - C\sum_{i=0}^{n-1}\left(\xi_i - \frac{x_i+x_{i+1}}{2}\right)h_i, \tag{3.7.23}$$

and

$$|R_{C\gamma}(I_n,\xi,f)| \leqslant \sum_{i=0}^{n-1}(S_i-\gamma)\left[\frac{h_i}{4}+\left|\xi_i - \frac{x_i+x_{i+1}}{2}\right|\right]h_i, \tag{3.7.24}$$

for $C = \frac{\gamma}{2}$, where S_i is as given in Theorem 3.7.1. Also,

$$\int_a^b f(t)dt = A_C(I_n,\xi,f) + R_{C\Gamma}(I_n,\xi,f), \tag{3.7.25}$$

where

$$|R_{C\Gamma}(I_n,\xi,f)| \leqslant \sum_{i=0}^{n-1}(\Gamma - S_i)\left[\frac{h_i}{4}+\left|\xi_i - \frac{x_i+x_{i+1}}{2}\right|\right]h_i, \tag{3.7.26}$$

for $C = \frac{\Gamma}{2}$.

Proof. We apply inequality (3.2.54) in Corollary 3.2.3 to the interval $[x_i, x_{i+1}]$, then

$$\left|\frac{f(x_i)+f(x_{i+1})}{4}h_i + \frac{1}{2}f(\xi_i)h_i - \frac{\gamma}{2}\left(\xi_i - \frac{x_i+x_{i+1}}{2}\right)h_i - \int_{x_i}^{x_{i+1}} f(t)dt\right|$$

$$\leqslant (S_i-\gamma)\left[\frac{h_i}{4}+\left|\xi_i - \frac{x_i+x_{i+1}}{2}\right|\right]h_i, \tag{3.7.27}$$

for $i = 0,1,\ldots,n-1$. We also have

$$\frac{f(x_i)+f(x_{i+1})}{4}h_i + \frac{1}{2}f(\xi_i)h_i - \frac{\gamma}{2}\left(\xi_i - \frac{x_i+x_{i+1}}{2}\right)h_i - \int_{x_i}^{x_{i+1}} f(t)dt$$

$$= \int_{x_i}^{x_{i+1}} k(\xi_i,t)[f'(t)-\gamma]dt, \tag{3.7.28}$$

where

$$k(\xi_i,t) = \begin{cases} t - \dfrac{3x_i+x_{i+1}}{4}, & t \in [x_i,\xi_i], \\ t - \dfrac{x_i+3x_{i+1}}{4}, & t \in (\xi_i, x_{i+1}], \end{cases} \tag{3.7.29}$$

for $i = 0,1,\ldots,n-1$. If we now sum (3.7.28) over i from 0 to $n-1$ and apply the triangle inequality and (3.7.27), then we get (3.7.22),(3.7.23) and (3.7.24). In a similar way we can prove that (3.7.25) and (3.7.26) hold.

Remark 3.7.2. If we set $\xi_i = \frac{x_i+x_{i+1}}{2}$ in Theorem 3.7.3, then we get corresponding composite rules which do not depend on ξ.

Theorem 3.7.4. Let the assumptions of Theorem 3.7.1 hold. Then

$$\int_a^b f(t)dt = A_S(I_n,\xi,f) + R_{S\gamma}(I_n,\xi,f), \tag{3.7.30}$$

where

$$A_S(I_n,\xi,f) = \frac{1}{6}\sum_{i=0}^{n-1}\left[f(x_i)+4f(\xi_i)+f(x_{i+1})\right]h_i - S\sum_{i=0}^{n-1}\left(\xi_i - \frac{x_i+x_{i+1}}{2}\right)h_i, \tag{3.7.31}$$

and

$$\left|R_{S\gamma}(I_n,\xi,f)\right| \leqslant \sum_{i=0}^{n-1}(S_i-\gamma)\left[\frac{h_i}{3} + \left|\xi_i = \frac{x_i+x_{i+1}}{2}\right|\right]h_i, \tag{3.7.32}$$

for $S = \frac{2\gamma}{3}$, where S_i is as given in Theorem 3.7.1. Also,

$$\int_a^b f(t)dt = A_S(I_n,\xi,f) + R_{S\Gamma}(I_n,\xi,f), \tag{3.7.33}$$

where

$$\left|R_{S\Gamma}(I_n,\xi,f)\right| \leqslant \sum_{i=0}^{n-1}(\Gamma-S_i)\left[\frac{h_i}{3} + \left|\xi_i = \frac{x_i+x_{i+1}}{2}\right|\right]h_i, \tag{3.7.34}$$

and $S = \frac{2\Gamma}{3}$.

Proof. We apply inequality (3.2.56) in Corollary 3.2.4 to the interval $[x_i, x_{i+1}]$, then

$$\left|\frac{1}{6}\left[f(x_i)+4f(\xi_i)+f(x_{i+1})\right]h_i - \frac{2\gamma}{3}\left(\xi_i - \frac{x_i+x_{i+1}}{2}\right)h_i - \int_{x_i}^{x_{i+1}} f(t)dt\right|$$
$$\leqslant (S_i-\gamma)\left[\frac{h_i}{3} + \left|\xi_i - \frac{x_i+x_{i+1}}{2}\right|\right]h_i, \tag{3.7.35}$$

for $i = 0,1,\ldots,n-1$. We also have

$$\frac{1}{6}\left[f(x_i)+4f(\xi_i)+f(x_{i+1})\right]h_i - \frac{2\gamma}{3}\left(\xi_i = \frac{x_i+x_{i+1}}{2}\right)h_i - \int_{x_i}^{x_{i+1}} f(t)dt$$
$$= \int_{x_i}^{x_{i+1}} k(\xi_i,t)\left[f'(t)-\gamma\right]dt, \tag{3.7.36}$$

where

$$k(\xi_i,t) = \begin{cases} t - \dfrac{5x_i+x_{i+1}}{6}, & t \in [x_i,\xi_i], \\ t - \dfrac{x_i+5x_{i+1}}{6}, & t \in (\xi_i,x_{i+1}], \end{cases} \tag{3.7.37}$$

for $i = 0,1,\ldots,n-1$. If we now sum (3.7.36) over i from 0 to $n-1$ and apply the triangle inequality and (3.7.35), then we get (3.7.30), (3.7.31) and (3.7.32). In a similar way we can prove that (3.7.33) and (3.7.34) hold.

Remark 3.7.3. If we set $\xi_i = \frac{x_i+x_{i+1}}{2}$ in Theorem 3.7.4, then we get the composite Simpson's rule.

3.7.3 *More applications in numerical integration*

Consider the partition $I_m : a = x_0 < x_1 < \cdots < x_m = b$ of the interval $[a,b]$ and the intermediate points $\xi = (\xi_0,\ldots,\xi_{m-1})$ where $\xi_j \in [x_j,x_{j+1}]$, $j = 0,1,\ldots,m-1$. Define the formula

$$F_{m,k}(f,I_m,\xi) = \sum_{j=0}^{m-1}\sum_{k=0}^{n-1} \frac{\left[(x_{j+1}-\xi_j)^{k+1}+(-1)^k(\xi_j-x_j)^{k+1}\right]}{(k+1)!} f^{(k)}(\xi_i),$$

which can be regarded as a perturbation of Riemann's sum

$$\Gamma(f,I_m,\xi) = \sum_{j=0}^{m-1} f(\xi_j)h_j,$$

where $h_j = x_{j+1}-x_j$, $j = 0,1,\ldots,m-1$.
The following Theorem holds (see [16]).

Theorem 3.7.5. Let $f : [a,b] \to \mathbb{R}$ be a mapping such that $f^{(n-1)}$ is absolutely continuous on $[a,b]$ and I_m a partitioning of $[a,b]$ as above. Then we have the quadrature formula

$$\int_a^b f(x)dx = F_{m,k}(f,I_m,\xi) + R_{m,k}(f,I_m,\xi), \tag{3.7.38}$$

where $F_{m,k}$ is as defined above and the remainder $R_{m,k}$ satisfies the estimation

$$\begin{aligned} \left|R_{m,k}(f,I_m,\xi)\right| &\leqslant \frac{\left\|f^{(n)}\right\|_\infty}{(n+1)!} \sum_{j=0}^{m-1} \left[(\xi_j-x_j)^{n+1}+(x_{j+1}-\xi_j)^{n+1}\right] \\ &\leqslant \frac{\left\|f^{(n)}\right\|_\infty}{(n+1)!} \sum_{j=0}^{m-1} h_j^{n+1},. \end{aligned} \tag{3.7.39}$$

for all ξ as above.

Proof. Apply Theorem 3.5.1 on the interval $[x_j,x_{j+1}]$, to get

$$\left| \int_{x_j}^{x_{j+1}} f(t)dt - \sum_{k=0}^{m-1} \frac{\left[(x_{j+1}-\xi_j)^{k+1}+(-1)^k(\xi_j-x_j)^{k+1}\right]}{(k+1)!} f^{(k)}(\xi_i) \right|$$

$$\leqslant \frac{\left\|f^{(n)}\right\|_\infty}{(n+1)!} \left[(\xi_j-x_j)^{n+1}+(x_{j+1}-\xi_j)^{n+1}\right] \leqslant \frac{\left\|f^{(n)}\right\|_\infty}{(n+1)!} h_j^{n+1},$$

for all $j = 0,1,\ldots,m-1$.
Summing over j from 0 to $m-1$ and using the generalized triangle inequality, we deduce the desired estimation (3.7.39).

As an interesting particular case, we can consider the following perturbed midpoint formula

$$M_{m,k}(f,I_m)=\sum_{j=0}^{m-1}\sum_{k=0}^{n-1}\left[\frac{1+(-1)^k}{(k+1)!}\right]\frac{h_j^{k+1}}{2^{k+1}}f^{(k)}\left(\frac{x_j+x_{j+1}}{2}\right),$$

which in effect involves only even k.

We state the following result concerning the estimation of the remainder term.

Corollary 3.7.1. Let f and I_m be as in Theorem 3.7.5. Then we have

$$\int_a^b f(x)dx=M_{m,k}(f,I_m)+R_{m,k}(f,I_m), \tag{3.7.40}$$

where $M_{m,k}$ is as defined above and the remainder term $R_{m,k}$ satisfies the estimation

$$\left|R_{m,k}(f,I_m)\right|\leqslant\frac{\left\|f^{(n)}\right\|_\infty}{2^n(n+1)!}\sum_{j=0}^{m-1}h_j^{n+1}. \tag{3.7.41}$$

3.8 Miscellaneous inequalities

3.8.1 *Dragomir, Barnett and Wang [33]*

Let X be a random variable with the probability density function $f:[a,b]\to\mathbb{R}_+$ and with cumulative distribution function $F(x)=\Pr(X\leqslant x)$. If $f\in L_p[a,b]$, $p>1$, then we have the inequality

$$\begin{aligned}\left|\Pr(X\leqslant x)-\frac{b-E(x)}{b-a}\right|&\leqslant\frac{q}{q+1}\|f\|_p(b-a)^{\frac{1}{q}}\left[\left(\frac{x-a}{b-a}\right)^{\frac{1+q}{q}}+\left(\frac{b-x}{b-a}\right)^{\frac{1+q}{q}}\right]\\&\leqslant\frac{q}{q+1}\|f\|_p(b-a)^{\frac{1}{q}},\end{aligned}$$

for all $x\in[a,b]$, where $\frac{1}{p}+\frac{1}{q}=1$.

3.8.2 *Cheng [19]*

Let $f:I\subset\mathbb{R}\to\mathbb{R}$ be a mapping differentiable in $\overset{\circ}{I}$ (interior of I), and let $a,\ b\in\overset{\circ}{I}$, $a<b$. If f' is integrable and $\gamma\leqslant f'(t)\leqslant\Gamma$ for all $t\in[a,b]$ and some constants $\gamma,\ \Gamma\in\mathbb{R}$, then we have

$$\left|f(x)-\left(x-\frac{a+b}{2}\right)\frac{f(b)-f(a)}{b-a}-\frac{1}{b-a}\int_a^b f(t)dt\right|\leqslant\frac{b-a}{8}(\Gamma-\gamma),$$

for all $t\in[a,b]$.

3.8.3 *Dragomir, Cerone and Roumeliotis [47]*

Let $f:[a,b]\to\mathbb{R}$ be continuous on $[a,b]$, differentiable on (a,b) with derivative $f':(a,b)\to\mathbb{R}$ being bounded on (a,b). Then

$$\left|\int_a^b f(t)dt-\left[f(x)(1-\lambda)+\frac{f(a)+f(b)}{2}\lambda\right](b-a)\right|$$

$$\leqslant\left[\frac{1}{4}(b-a)^2\left(\lambda^2+(1-\lambda)^2\right)+\left(x-\frac{a+b}{2}\right)^2\right]\|f'\|_\infty,$$

for all $\lambda\in[0,1]$ and $a+\lambda\frac{b-a}{2}\leqslant x\leqslant b-\lambda\frac{b-a}{2}$.

3.8.4 *Ujević [155]*

Let $I\subset\mathbb{R}$ be an open interval and $a,b\in I$, $a<b$. If $f:I\to\mathbb{R}$ is a differentiable function such that $\gamma\leqslant f'(t)\leqslant\Gamma$, for all $t\in[a,b]$ and some constants γ, $\Gamma\in\mathbb{R}$, then we have

$$\left|(b-a)\left[\frac{\lambda}{2}(f(a)+f(b))+(1-\lambda)f(x)-\frac{\Gamma+\gamma}{2}(1-\lambda)\left(x-\frac{a+b}{2}\right)\right]-\int_a^b f(t)dt\right|$$

$$\leqslant\frac{\Gamma-\gamma}{2}\left[\frac{1}{4}(b-a)^2\left(\lambda^2+(1-\lambda)^2\right)+\left(x-\frac{a+b}{2}\right)^2\right],$$

where $a+\lambda\frac{b-a}{2}\leqslant x\leqslant b-\lambda\frac{b-a}{2}$ and $\lambda\in[0,1]$.

3.8.5 *Pachpatte [123]*

Let $f,\ g,\ h:[a,b]\to\mathbb{R}$ be continuous on $[a,b]$, $a<b$ and differentiable on (a,b) and $w:[a,b]\to[0,\infty)$ be integrable function such that $\int_a^b w(y)dy>0$. If $h'(t)\neq 0$ for each $t\in(a,b)$. Then

$$\left|f(x)g(x)-\frac{1}{2\int_a^b w(y)dy}\left[f(x)\int_a^b w(y)g(y)dy+g(x)\int_a^b w(y)f(y)dy\right]\right|$$

$$\leqslant\frac{1}{2}\left[\left\|\frac{f'}{h'}\right\|_\infty|g(x)|+\left\|\frac{g'}{h'}\right\|_\infty|f(x)|\right]\left|h(x)-\frac{1}{\int_a^b w(y)dy}\int_a^b w(y)h(y)dy\right|,$$

for all $x\in[a,b]$, where

$$\left\|\frac{f'}{h'}\right\|_\infty=\sup_{t\in(a,b)}\left|\frac{f'(t)}{h'(t)}\right|<\infty,\qquad\left\|\frac{g'}{h'}\right\|_\infty=\sup_{t\in(a,b)}\left|\frac{g'(t)}{h'(t)}\right|<\infty.$$

3.8.6 *Pachpatte [105]*

Let $f,\ g,\ h : [a,b] \to \mathbb{R}$ be continuous functions on $[a,b]$, $a<b$ and differentiable on (a,b) with derivatives $f', g', h' : (a,b) \to \mathbb{R}$ being bounded on (a,b). Then

$$\left| f(x)g(x)h(x) - \frac{1}{3(b-a)}\left[g(x)h(x)\int_a^b f(y)dy \right.\right.$$

$$\left.\left. + h(x)f(x)\int_a^b g(y)dy + f(x)g(x)\int_a^b h(y)dy \right]\right|$$

$$\leqslant \frac{1}{3}\left[|g(x)||h(x)|\|f'\|_\infty + |h(x)||f(x)|\|g'\|_\infty + |f(x)||g(x)|\|h'\|_\infty \right] A(x),$$

for all $x \in [a,b]$, where

$$A(x) = \left[\frac{1}{4} + \frac{\left(x - \frac{a+b}{2}\right)^2}{(b-a)^2} \right](b-a).$$

3.8.7 *Cerone,Dragomir and Roumeliotis [14]*

Let $f : [a,b] \to \mathbb{R}$ be a twice differentiable mapping on (a,b) and $f'' : (a,b) \to \mathbb{R}$ is bounded on (a,b). Then we have the inequality

$$\left| f(x) - \frac{1}{b-a}\int_a^b f(t)dt - \left(x - \frac{a+b}{2}\right) f'(x) \right| \leqslant \left[\frac{1}{24}(b-a)^2 + \frac{1}{2}\left(x - \frac{a+b}{2}\right)^2 \right] \|f''\|_\infty$$

$$\leqslant \frac{(b-a)^2}{6}\|f''\|_\infty,$$

for all $x \in [a,b]$.

3.8.8 *Dragomir and Barnett [28]*

Let $f : [a,b] \to \mathbb{R}$ be a continuous function on $[a,b]$ and twice differentiable on (a,b), with second derivative $f'' : (a,b) \to \mathbb{R}$ being bounded on (a,b). Then we have the inequality

$$\left| f(x) - \frac{1}{b-a}\int_a^b f(t)dt - \frac{f(b)-f(a)}{b-a}\left(x - \frac{a+b}{2}\right) \right|$$

$$\leqslant \frac{1}{2}\left\{ \left[\frac{\left(x - \frac{a+b}{2}\right)^2}{(b-a)^2} + \frac{1}{4} \right]^2 + \frac{1}{12} \right\}(b-a)^2\|f''\|_\infty \leqslant \frac{(b-a)^2}{6}\|f''\|_\infty,$$

for all $x \in [a,b]$.

3.8.9 *Ujević [151]*

Let $f : I \subset \mathbb{R} \to \mathbb{R}$ be a twice continuously differentiable mapping in $\overset{\circ}{I}$ (the interior of I) with $f'' \in L_2(a,b)$ and let $a, b \in \overset{\circ}{I}$, $a < b$. Then we have

$$\left| f(x) - \left(x - \frac{a+b}{2}\right)\frac{f(b)-f(a)}{b-a} - \frac{1}{b-a}\int_a^b f(t)dt \right| \leqslant \frac{(b-a)^{\frac{3}{2}}}{2\pi\sqrt{3}}\|f''\|_2,$$

for $x \in [a,b]$.

3.8.10 *Pachpatte [100]*

Let $f,\ g : [a,b] \to \mathbb{R}$ be mappings with first derivatives being absolutely continuous on $[a,b]$ and assume that the second derivatives $f'',\ g'' \in L_\infty[a,b]$. Then

$$\left| 2\left(\frac{1}{b-a}\int_a^b f(t)dt\right)\left(\frac{1}{b-a}\int_a^b g(t)dt\right) \right.$$

$$-\left\{\frac{1}{2}\left[f(x) + \frac{f(a)+f(b)}{2}\right] - \left(x - \frac{a+b}{2}\right)f'(x)\right\}\left(\frac{1}{b-a}\int_a^b g(t)dt\right)$$

$$\left. -\left\{\frac{1}{2}\left[g(x) + \frac{g(a)+g(b)}{2}\right] - \left(x - \frac{a+b}{2}\right)g'(x)\right\}\left(\frac{1}{b-a}\int_a^b f(t)dt\right)\right|$$

$$\leqslant M(x)\left[\|f''\|_\infty\left(\frac{1}{b-a}\int_a^b g(t)dt\right) + \|g''\|_\infty\left(\frac{1}{b-a}\int_a^b f(t)dt\right)\right],$$

$$\left|\left(\frac{1}{b-a}\int_a^b f(t)dt\right)g(x) + \left(\frac{1}{b-a}\int_a^b g(t)dt\right)f(x) - f(x)g(x)\right.$$

$$\left. -\frac{1}{2}\left[\frac{f(a)+f(b)}{2}g(x) + \frac{g(a)+g(b)}{2}f(x)\right] - \left(x - \frac{a+b}{2}\right)(fg)'(x)\right|$$

$$\leqslant M(x)\left[\|f''\|_\infty|g(x)| + \|g''\|_\infty|f(x)|\right],$$

for $x \in [a,b]$, where

$$M(x) = \frac{1}{2(b-a)}\int_a^b |p(x,t)|\left|t - \frac{a+b}{2}\right|dt,$$

in which $p(x,t)$ is given by (3.2.7).

3.8.11 *Pachpatte [118]*

Let $f,\ g:[a,b]\to\mathbb{R}$ be twice differentiable functions on (a,b) and $f'',\ g'':(a,b)\to\mathbb{R}$ are bounded on (a,b). Then

$$\left|[g(x)L[f(x)]+f(x)L[g(x)]]-\frac{1}{b-a}\left[g(x)\int_a^b f(t)dt+f(x)\int_a^b g(t)dt\right]\right|$$
$$\leqslant\frac{1}{b-a}\left[|g(x)|\|f''\|_\infty+|f(x)|\|g''\|_\infty\right]E(x),$$

and

$$\left|L[f(x)]L[g(x)]-\frac{1}{b-a}\left[L[g(x)]\int_a^b f(t)dt+L[f(x)]\int_a^b g(t)dt\right]\right.$$
$$\left.+\frac{1}{(b-a)^2}\left(\int_a^b f(t)dt\right)\left(\int_a^b g(t)dt\right)\right|$$
$$\leqslant\frac{1}{(b-a)^2}\|f''\|_\infty\|g''\|_\infty(E(x))^2,$$

for $x\in[a,b]$, where

$$E(x)=\int_a^b|k(x,t)|dt, \tag{3.8.1}$$

in which $k(x,t)$ is defined as in the proof of Theorem 3.4.3 and for a suitable function $h:[a,b]\to\mathbb{R}$, the notation

$$L[h(x)]=h(x)-\left(x-\frac{a+b}{2}\right)h'(x),$$

is set to simplify the presentation.

3.8.12 *Pachpatte [124]*

Let $f,\ g,\ h:[a,b]\to\mathbb{R}$ be twice differentiable functions on (a,b) and $f'',\ g'',\ h'':(a,b)\to\mathbb{R}$ are bounded on (a,b). Then

$$\left|f(x)g(x)h(x)-\frac{1}{3(b-a)}A[f,g,h](x)-\frac{1}{3}\left(x-\frac{a+b}{2}\right)(f(x)g(x)h(x))'\right|$$
$$\leqslant\frac{1}{3(b-a)}B[f,g,h](x)E(x),$$

for all $x\in[a,b]$, where

$$A[f,g,h](x)=g(x)h(x)\int_a^b f(t)dt$$
$$+h(x)f(x)\int_a^b g(t)dt+f(x)g(x)\int_a^b h(t)dt,$$
$$B[f,g,h](x)=|g(x)||h(x)|\|f''\|_\infty$$
$$+|h(x)||f(x)|\|g''\|_\infty+|f(x)||g(x)|\|h''\|_\infty,$$

and $E(x)$ is given by (3.8.1).

3.8.13 *Fink [59]*

Let $f^{(n-1)}(t)$ be absolutely continuous on $[a,b]$ with $f^{(n)} \in L_p[a,b]$. Then the inequality

$$\left|\frac{1}{n}\left(f(x)+\sum_{k=1}^{n-1} F_k(x)\right)-\frac{1}{b-a}\int_a^b f(y)dy\right|$$

$$\leqslant K(n,p,x)\left\|f^{(n)}\right\|_p, \tag{3.8.2}$$

holds for all $x \in [a,b]$, with $F_k(x)$ is given by (1.5.6),

$$K(n,p,x) = \frac{\left[(x-a)^{np'+1}+(b-x)^{np'+1}\right]^{\frac{1}{p'}}}{n!(b-a)} B\left((n-1)p'+1,p'+1\right)^{\frac{1}{p'}},$$

where $1 < p \leqslant \infty$, B is the beta function, and

$$K(n,1,x) = \frac{(n-1)^{n-1}}{n^n n!(b-a)} \max\left\{(x-a)^n,(b-x)^n\right\}.$$

Moreover, for $p > 1$ the inequality (3.8.2) is the best possible in the strong sense that for any $x \in (a,b)$ there is an f for which equality holds at x.

3.8.14 *Pachpatte [102]*

Let f, $g : [a,b] \to \mathbb{R}$ be mappings such that $f^{(n-1)}, g^{(n-1)}$ are absolutely continuous on $[a,b]$ and $f^{(n)}$, $g^{(n)} \in L_\infty[a,b], n \geqslant 1$ is a natural number. Then

$$\left|2\left(\int_a^b f(t)dt\right)\left(\int_a^b g(t)dt\right)-\left[F(x)\int_a^b g(t)dt+G(x)\int_a^b f(t)dt\right]\right|$$

$$\leqslant \left[\left\|f^{(n)}\right\|_\infty \int_a^b |g(t)|dt+\left\|g^{(n)}\right\|_\infty \int_a^b |f(t)|dt\right]$$

$$\times \frac{1}{(n+1)!}\left[(x-a)^{n+1}+(b-x)^{n+1}\right],$$

for all $x \in [a,b]$, where

$$F(x) = \sum_{k=0}^{n-1}\left[\frac{(b-x)^{k+1}+(-1)^k(x-a)^{k+1}}{(k+1)!}\right] f^{(k)}(x),$$

$$G(x) = \sum_{k=0}^{n-1}\left[\frac{(b-x)^{k+1}+(-1)^k(x-a)^{k+1}}{(k+1)!}\right] g^{(k)}(x).$$

3.8.15 *Pachpatte [135]*

Let $f,\ g \in C^{n+1}([a,b],\mathbb{R})$, $n \in \mathbb{N}$ and $x \in [a,b]$ be fixed, such that $f^{(k)}(x) = 0$, $g^{(k)}(x) = 0$, $k = 1,\ldots,n$. Then

$$\left|2\int_a^b f(y)g(y)dy - \left[f(x)\int_a^b g(y)dy + g(x)\int_a^b f(y)dy\right]\right|$$

$$\leqslant \frac{1}{(n+1)!}\int_a^b \left[|g(y)|\left\|f^{(n+1)}\right\|_\infty + |f(y)|\left\|g^{(n+1)}\right\|_\infty\right]|y-x|^{n+1}dy,$$

and

$$\left|\int_a^b f(y)g(y)dy - \left[f(x)\int_a^b g(y)dy + g(x)\int_a^b f(y)dy\right] + (b-a)f(x)g(x)\right|$$

$$\leqslant \left\{\frac{1}{(n+1)!}\right\}^2 \left\|f^{(n+1)}\right\|_\infty \left\|g^{(n+1)}\right\|_\infty \left(\frac{(x-a)^{2n+3} + (b-x)^{2n+3}}{2n+3}\right).$$

3.8.16 *Pachpatte [121]*

Let $f,\ g,\ h : [a,b] \to \mathbb{R}$ be continuous functions on $[a,b]$ and n-times differentiable on (a,b) with derivatives $f^{(n)}, g^{(n)}, h^{(n)} : (a,b) \to \mathbb{R}$ being bounded on (a,b). Then

$$\left|f(x)g(x)h(x) - \frac{1}{3(b-a)}\left[g(x)h(x)\left\{I_0 + \sum_{k=1}^{n-1} I_k\right\}\right.\right.$$

$$\left.\left. + h(x)f(x)\left\{J_0 + \sum_{k=1}^{n-1} J_k\right\} + f(x)g(x)\left\{L_0 + \sum_{k=1}^{n-1} L_k\right\}\right]\right|$$

$$\leqslant \frac{1}{3(b-a)}\left[|g(x)||h(x)|\|f^{(n)}\|_\infty\right.$$

$$\left. + |h(x)||f(x)|\|g^{(n)}\|_\infty + |f(x)||g(x)|\|h^{(n)}\|_\infty\right] M_n(x),$$

for all $x \in [a,b]$, where $I_0, I_k, J_0, J_k, L_0, L_k$ and $M_n(x)$ are as given in Theorem 1.5.5.

3.8.17 *Pachpatte [120]*

Let $f,\ g : [a,b] \to \mathbb{R}$ be functions such that $f^{(n-1)}, g^{(n-1)}$ are absolutely continuous on $[a,b]$ and $f^{(n)},\ g^{(n)} \in L_\infty[a,b], n \geqslant 1$ is a natural number. Then

$$\left|g(x)A[f(x)] + f(x)A[g(x)] - \frac{1}{b-a}\left[g(x)\int_a^b f(t)dt + f(x)\int_a^b g(t)dt\right]\right|$$

$$\leqslant \frac{1}{b-a}\left[|g(x)|\|f^{(n)}\|_\infty + |f(x)|\|g^{(n)}\|_\infty\right] H_n(x),$$

and

$$\left|A[f(x)]A[g(x)] - \frac{1}{b-a}\left[A[g(x)]\int_a^b f(t)dt + A[f(x)]\int_a^b g(t)dt\right]\right.$$
$$\left. + \frac{1}{(b-a)^2}\left(\int_a^b f(t)dt\right)\left(\int_a^b g(t)dt\right)\right|$$
$$\leqslant \frac{1}{(b-a)^2}\|f^{(n)}\|_\infty \|g^{(n)}\|_\infty (H_n(x))^2,$$

for all $x \in [a,b]$, where

$$H_n(x) = \int_a^b |E_n(x,t)|dt,$$

in which $E_n(x,t)$ is defined by (1.5.23) and for a suitable function $h : [a,b] \to \mathbb{R}$, the notation $A[h(x)]$ is given by (1.5.3).

3.8.18 *Pachpatte [120]*

Let (P_n) be a harmonic sequence of polynomials and $f,\ g : [a,b] \to \mathbb{R}$, be functions such that $f^{(n-1)},\ g^{(n-1)}$ are respectively L-Lipschitz and M-Lipschitz functions (for some $n \geqslant 1$ is a natural number), i.e.,

$$\left|f^{(n-1)}(x) - f^{(n-1)}(y)\right| \leqslant L|x-y|,$$

$$\left|g^{(n-1)}(x) - g^{(n-1)}(y)\right| \leqslant M|x-y|,$$

for all $x,\ y \in [a,b]$, where $L,\ M$ are nonnegative constants. Then

$$\left|g(x)B[f(x)] + f(x)B[g(x)] - \frac{1}{b-a}\left[g(x)\int_a^b f(t)dt + f(x)\int_a^b g(t)dt\right]\right|$$
$$\leqslant \frac{1}{b-a}[L|g(x)| + M|f(x)|]D_n(x),$$

and

$$\left|B[f(x)]B[g(x)] - \frac{1}{b-a}\left[B[g(x)]\int_a^b f(t)dt + B[f(x)]\int_a^b g(t)dt\right]\right.$$
$$\left. + \frac{1}{(b-a)^2}\left(\int_a^b f(t)dt\right)\left(\int_a^b g(t)dt\right)\right|$$
$$\leqslant \frac{1}{(b-a)^2}LM(D_n(x))^2,$$

for all $x \in [a,b]$, where

$$D_n(x) = \frac{1}{n}\int_a^b |P_{n-1}(t)p(x,t)|\,dt,$$

in which $p(x,t)$ is defined by (1.2.11) and for a suitable function $h : [a,b] \to \mathbb{R}$, the notation $B[h(x)]$ is given by (1.5.4).

3.8.19 *Dragomir [49]*

Let $(X,\|\cdot\|)$ be a normed linear space and x_i $(i=1,\ldots,n)$ be vectors in X. Then we have the inequality

$$\left\|x_i-\frac{1}{n}\sum_{k=1}^{n}x_k\right\|\leqslant\frac{1}{n}\left[\left(i-\frac{n+1}{2}\right)^2+\frac{n^2-1}{4}\right]\max_{k=1,\ldots,n-1}|\Delta x_k|,$$

for all $i\in\{1,\ldots,n\}$. The constant $\frac{1}{4}$ in the right hand side is the best possible.

3.8.20 *Aglic Aljinović and Pečarić [1]*

Let $(X,\|\cdot\|)$ be a normed linear space, $\{x_1,\ldots,x_n\}$ a finite sequence of vectors in X, $\{w_1,\ldots,w_n\}$ a finite sequence of positive real numbers. Let (p,q) be a pair of conjugate exponents, that is $1<p,q<\infty$, $\frac{1}{p}+\frac{1}{q}=1$, $m\in\{2,3,\ldots,n-1\}$ $k\in\{1,\ldots,n\}$. Then the following inequality holds

$$\left\|x_k-\frac{1}{W_n}\sum_{i=1}^{n}w_ix_i-\sum_{r=1}^{m-1}\frac{1}{n-r}\left(\sum_{i=1}^{n-r}\Delta^r x_i\right)\right.$$
$$\left.\times\left(\sum_{i_1=1}^{n-1}\sum_{i_2=1}^{n-2}\cdots\sum_{i_r=1}^{n-r}D_w(k,i_1)D_{n-1}(i_1,i_2)\cdots D_{n-r+1}(i_{r-1},i_r)\right)\right\|$$
$$\leqslant\left\|\sum_{i_1=1}^{n-1}\sum_{i_2=1}^{n-2}\cdots\sum_{i_{m-1}=1}^{n-m+1}D_w(k,i_1)D_{n-1}(i_1,i_2)\cdots D_{n-m+1}(i_{m-1},\cdot)\right\|_q\|\Delta^m x\|_p,$$

where

$$\|\Delta^m x\|_p=\begin{cases}\left(\sum_{i=1}^{n-m}\|\Delta^m x_i\|^p\right)^{\frac{1}{p}} & \text{if } 1\leqslant p<\infty,\\ \max_{1\leqslant i\leqslant n-m}\|\Delta^m x_i\| & \text{if } p=\infty,\end{cases}$$

and W_n, $D_w(k,i)$, $D_n(k,i)$ are as given in Lemma 1.6.1.

3.9 Notes

A number of authors have written about extensions, generalizations and variants of the Ostrowski's inequality. Theorem 3.2.1 deals with the generalization of the Ostrowski's inequality for Lipschitzian mappings and is taken from Dragomir [52]. Theorem 3.2.2 contains the Ostrowski-type inequality and is due to Dragomir and Wang [26] and Theorem 3.2.3 is taken from Ujević [154]. The inequalities in Theorems 3.2.4 and 3.2.5 are adapted from Ujević [151] and [150]. Section 3.3 contains some Ostrowski-type inequalities involving two functions and their derivatives established by Pachpatte in [93,109, 114,139].

Section 3.4 is devoted to the inequalities of Ostrowski- and Grüss-type involving functions and their derivatives. The results in Theorems 3.4.1 and 3.4.2 and Theorems 3.4.4 and 3.4.5 are due to Pachpatte and taken from [131] and [100]. Theorem 3.4.3 is adapted from Cerone, Dragomir and Roumeliotis [15]. Section 3.5 contains further inequalities of the Ostrowski-type involving functions and their higher order derivatives. Theorem 3.5.1 is taken from Cerone, Dragomir and Roumeliotis [16] and Theorem 3.5.2 is adapted from Matić, Pečarić and Ujević [73]. The results in Theorems 3.5.3 and 3.5.4 are taken from Pachpatte [101,116]. The discrete Ostrowski-type inequalities in Theorems 3.6.1–3.6.4 are due to Pachpatte and taken from [88,105,114,133]. Section 3.7 contains applications of some of the inequalities given in earlier sections and taken from Dragomir and Wang [26], Ujević [150] and Cerone, Dragomir and Roumeliotis [16]. Section 3.8 deals with some useful miscellaneous inequalities established by various investigators.

Chapter 4

Multidimensional Ostrowski-type inequalities

4.1 Introduction

The Ostrowski inequality (7) has been generalized over the last years in a number of ways. The first multidimensional version of the Ostrowski's inequality was given by G.V. Milovanović in [76] (see also [80, p. 468]). Recently a number of authors have written about multidimensional generalizations, extensions and variants of the Ostrowski's inequality, see [8,29,37,44,64,65,83,86,87,91,94,115]. In this way, some new multidimensional Ostrowski-type inequalities have been found in the literature. Inspired and motivated by the recent work going on in this direction, in this chapter, we present some new multidimensional Ostrowski-type inequalities, recently investigated in order to achieve various goals. We also present some immediate applications of certain inequalities. In our subsequent discussion, we make use of some of the notation and definitions given in Chapter 2 without further mention.

4.2 Ostrowski-type inequalities in two variables

In this section we shall give some fundamental Ostrowski-type inequalities involving functions of two independent variables recently investigated in [8,37,64,115,125].
We start with the Ostrowski-type inequality established by Barnett and Dragomir [8] for mappings of two variables.

Theorem 4.2.1. Let $\Delta = [a,b] \times [c,d]$ and $f : \Delta \to \mathbb{R}$ be continuous on Δ, $D_2D_1f(x,y)$ exists on $(a,b) \times (c,d)$ and is bounded, then we have the inequality

$$\left| \int_a^b \int_c^d f(s,t)dtds - \left[(b-a) \int_c^d f(x,t)dt + (d-c) \int_a^b f(s,y)ds - (d-c)(b-a)f(x,y) \right] \right|$$

$$\leqslant \left[\frac{1}{4}(b-a)^2+\left(x-\frac{a+b}{2}\right)^2\right]\left[\frac{1}{4}(d-c)^2+\left(y-\frac{c+d}{2}\right)^2\right]\|D_2D_1f\|_\infty, \tag{4.2.1}$$

for all $(x,y)\in\Delta$.

Proof. From the hypotheses, we have the following identity (see Lemma 2.3.2):

$$(d-c)(b-a)f(x,y)-(b-a)\int_c^d f(x,t)dt-(d-c)\int_a^b f(s,y)ds+\int_a^b\int_c^d f(s,t)dtds$$

$$=\int_a^b\int_c^d p(x,s)q(y,t)D_2D_1f(s,t)dtds, \tag{4.2.2}$$

for all $(x,y)\in\Delta$. From (4.2.2) we get

$$\left|\int_a^b\int_c^d f(s,t)dtds-\left[(b-a)\int_c^d f(x,t)dt+(d-c)\int_a^b f(s,y)ds-(d-c)(b-a)f(x,y)\right]\right|$$

$$\leqslant \int_a^b\int_c^d |p(x,s)||Q(y,t)||D_2D_1f(s,t)|dtds$$

$$\leqslant \|D_2D_1f\|_\infty\int_a^b\int_c^d |p(x,s)||Q(y,t)|dtds. \tag{4.2.3}$$

Now, observe that

$$\int_a^b |p(x,s)|ds=\int_a^x (s-a)ds+\int_x^b (b-s)ds$$

$$=\frac{(x-a)^2+(b-x)^2}{2}=\frac{1}{4}(b-a)^2+\left(x-\frac{a+b}{2}\right)^2, \tag{4.2.4}$$

and, similarly,

$$\int_c^d |Q(y,t)|dt=\frac{1}{4}(d-c)^2+\left(y-\frac{c+d}{2}\right)^2. \tag{4.2.5}$$

Using (4.2.4), (4.2.5) in (4.2.3), we get the required inequality in (4.2.1).

Remark 4.2.1. The constants $\frac{1}{4}$ from the first and the second bracket on the right hand side in (4.2.1) are optimal in the sense that not both of them can be less than $\frac{1}{4}$. Indeed, if we had assumed that there exists $c_1,c_2\in\left(0,\frac{1}{4}\right)$ so that

$$\left|\int_a^b\int_c^d f(s,t)dtds-\left[(b-a)\int_c^d f(x,t)dt+(d-c)\int_a^b f(s,y)ds-(d-c)(b-a)f(x,y)\right]\right|$$

$$\leqslant\left[c_1(b-a)^2+\left(x-\frac{a+b}{2}\right)^2\right]$$

$$\times\left[c_2(d-c)^2+\left(y-\frac{c+d}{2}\right)^2\right]\|D_2D_1f\|_\infty, \tag{4.2.6}$$

for all f as in Theorem 4.2.1 and $(x,y)\in\Delta$, then we would have had for $f(s,t)=st$ and $x=a$, $y=c$ that

$$\int_a^b\int_c^d f(s,t)dtds=\frac{(b^2-a^2)(d^2-c^2)}{4},\qquad \int_c^d f(x,t)dt=a\frac{(d^2-c^2)}{2},$$

$$\int_a^b f(s,y)ds=c\frac{(b^2-a^2)}{2},\qquad \|D_2D_1f\|=1,$$

and by (4.2.6), the inequality

$$\left|\frac{(b^2-a^2)(d^2-c^2)}{4}-(b-a)a\frac{(d^2-c^2)}{2}-(d-c)c\frac{(b^2-a^2)}{2}+(d-c)(b-a)ac\right|$$

$$\leqslant(b-a)^2\left(c_1+\frac{1}{4}\right)(d-c)^2\left(c_2+\frac{1}{4}\right),$$

i.e.,

$$\frac{(b^2-a^2)(d^2-c^2)}{4}\leqslant(b-a)^2\left(c_1+\frac{1}{4}\right)(d-c)^2\left(c_2+\frac{1}{4}\right),$$

i.e.,

$$\frac{1}{4}\leqslant\left(c_1+\frac{1}{4}\right)\left(c_2+\frac{1}{4}\right). \tag{4.2.7}$$

Now, as we have assumed that $c_1,\ c_2\in\left(0,\frac{1}{4}\right)$, we get

$$c_1+\frac{1}{4}<\frac{1}{2},\quad c_2+\frac{1}{4}<\frac{1}{2},$$

and then $\left(c_1+\frac{1}{4}\right)\left(c_2+\frac{1}{4}\right)<\frac{1}{4}$ which contradicts the inequality (4.2.7), and the statement in Remark 4.2.1 is proved.

A particular case which is of interest is embodied in the following corollary.

Corollary 4.2.1. Assume that the hypotheses of Theorem 4.2.1 hold. Then we have the inequality

$$\left|\int_a^b\int_c^d f(s,t)dtds-\left[(b-a)\int_c^d f\left(\frac{a+b}{2},t\right)dt\right.\right.$$

$$\left.\left.+(d-c)\int_a^b f\left(s,\frac{c+d}{2}\right)ds-(d-c)(b-a)f\left(\frac{a+b}{2},\frac{c+d}{2}\right)\right]\right|$$

$$\leqslant\frac{1}{16}(b-a)^2(d-c)^2\|D_2D_1f\|_\infty. \tag{4.2.8}$$

Remark 4.2.2. If we assume that $f(s,t) = h(s)h(t)$, $h : [a,b] \to \mathbb{R}$, h is continuous and suppose that $\|h'\|_\infty < \infty$, then from (4.2.1) we get (for $x = y$)

$$\left| \int_a^b h(s)ds \int_a^b h(s)ds - h(x)(b-a) \int_a^b h(s)ds - h(x)(b-a) \int_a^b h(s)ds + (b-a)^2 h^2(x) \right|$$
$$\leqslant \left[\frac{1}{4}(b-a)^2 + \left(x - \frac{a+b}{2} \right)^2 \right]^2 \|h'\|_\infty^2,$$

i.e.,

$$\left[\int_a^b h(s)ds - h(x)(b-a) \right]^2 \leqslant \left[\frac{1}{4}(b-a)^2 + \left(x - \frac{a+b}{2} \right)^2 \right]^2 \|h'\|_\infty^2,$$

which is clearly equivalent to Ostrowski's inequality. Consequently (4.2.1) can be also regarded as a generalization for double integrals of the classical result due to Ostrowski given in (7).

The following inequality of the Ostrowski-type which holds for mappings of two independent variables is given by Dragomir, Cerone, Barnett and Roumeliotis in [37].

Theorem 4.2.2. Let $f : \Delta \to \mathbb{R}$ be such that the partial derivatives $D_1 f(x,y)$ $D_2 f(x,y)$, $D_2 D_1 f(x,y)$ exist and are continuous on Δ. Then we have the inequality

$$\left| f(x,y) - \frac{1}{(b-a)(d-c)} \int_a^b \int_c^d f(t,s)dsdt \right| \leqslant M_1(x) + M_2(y) + M_3(x,y), \tag{4.2.9}$$

where

$$M_1(x) = \begin{cases} \dfrac{1}{b-a} \left[\dfrac{1}{4}(b-a)^2 + \left(x - \dfrac{a+b}{2} \right)^2 \right] \|D_1 f\|_\infty, & \text{if } D_1 f(x,y) \in L_\infty(\Delta); \\[2ex] \dfrac{\left[\dfrac{1}{q_1+1} \left[(b-x)^{q_1+1} + (x-a)^{q_1+1} \right] \right]^{\frac{1}{q_1}}}{[(b-a)(d-c)]^{\frac{1}{p_1}}} \|D_1 f\|_{p_1}, & \text{if } D_1 f(x,y) \in L_{p_1}(\Delta), \\ & \dfrac{1}{p_1} + \dfrac{1}{q_1} = 1,\ p_1 > 1; \\[2ex] \dfrac{1}{(b-a)(d-c)} \left[\dfrac{1}{2}(b-a) + \left| x - \dfrac{a+b}{2} \right| \right] \|D_1 f\|_1, & \text{if } D_1 f(x,y) \in L_1(\Delta), \end{cases}$$

$$M_2(y) = \begin{cases} \dfrac{1}{d-c} \left[\dfrac{1}{4}(d-c)^2 + \left(y - \dfrac{c+d}{2} \right)^2 \right] \|D_2 f\|_\infty, & \text{if } D_2 f(x,y) \in L_\infty(\Delta); \\[2ex] \dfrac{\left[\dfrac{1}{q_2+1} \left[(d-y)^{q_2+1} + (y-c)^{q_2+1} \right] \right]^{\frac{1}{q_2}}}{[(d-c)(b-a)]^{\frac{1}{p_2}}} \|D_2 f\|_{p_2}, & \text{if } D_2 f(x,y) \in L_{p_2}(\Delta), \\ & \dfrac{1}{p_2} + \dfrac{1}{q_2} = 1,\ p_2 > 1; \\[2ex] \frac{1}{(b-a)(d-c)} \left[\dfrac{1}{2}(d-c) + \left| y - \dfrac{c+d}{2} \right| \right] \|D_2 f\|_1, & \text{if } D_2 f(x,y) \in L_1(\Delta), \end{cases}$$

and

$$M_3(x,y) =$$

$$\begin{cases} \frac{1}{(b-a)(d-c)}\left[\frac{1}{4}(b-a)^2+\left(x-\frac{a+b}{2}\right)^2\right]\left[\frac{1}{4}(d-c)^2+\left(y-\frac{c+d}{2}\right)^2\right]\|D_2D_1f\|_\infty, \\ \qquad \text{if } D_2D_1f(x,y)\in L_\infty(\Delta); \\ \frac{\left[\frac{\left[(b-x)^{q_3+1}+(x-a)^{q_3+1}\right]}{q_3+1}\right]^{\frac{1}{q_3}}\left[\frac{\left[(d-y)^{q_3+1}+(y-c)^{q_3+1}\right]}{q_3+1}\right]^{\frac{1}{q_3}}}{(b-a)(d-c)}\|D_2D_1f\|_{p_3}, \\ \qquad \text{if } D_2D_1f(x,y)\in L_{p_3}(\Delta),\ \frac{1}{p_3}+\frac{1}{q_3}=1,\ p_3>1; \\ \frac{1}{(b-a)(d-c)}\left[\frac{1}{2}(b-a)+\left|x-\frac{a+b}{2}\right|\right]\left[\frac{1}{2}(d-c)+\left|y-\frac{c+d}{2}\right|\right]\|D_2D_1f\|_1, \\ \qquad \text{if } D_2D_1f(x,y)\in L_1(\Delta), \end{cases}$$

for all $(x,y)\in\Delta$, where $\|\cdot\|_p$ $(1\leqslant p<\infty)$ are the usual p-norms on Δ.

Proof. From the hypotheses we have the following identity (see, Lemma 2.3.1):

$$f(x,y)-\frac{1}{(b-a)(d-c)}\int_a^b\int_c^d f(t,s)dsdt = \frac{1}{(b-a)(d-c)}\left[\int_a^b\int_c^d p(x,t)D_1f(t,s)dsdt\right.$$
$$\left.+\int_a^b\int_c^d q(y,s)D_2f(t,s)dsdt+\int_a^b\int_c^d p(x,t)q(y,s)D_2D_1f(t,s)dsdt\right], \tag{4.2.10}$$

for all $(x,y)\in\Delta$. From (4.2.10),we have

$$\left|f(x,y)-\frac{1}{(b-a)(d-c)}\int_a^b\int_c^d f(t,s)dsdt\right|$$
$$\leqslant \frac{1}{(b-a)(d-c)}\left[\int_a^b\int_c^d |p(x,t)||D_1f(t,s)|dsdt\right.$$
$$\left.+\int_a^b\int_c^d |q(y,s)||D_2f(t,s)|dsdt+\int_a^b\int_c^d |p(x,t)||q(y,s)||D_2D_1f(t,s)|dsdt\right]. \tag{4.2.11}$$

We have that

$$\int_a^b\int_c^d |p(x,t)||D_1f(t,s)|dsdt \leqslant$$

$$\begin{cases} \|D_1f\|_\infty\int_a^b\int_c^d |p(x,t)|dsdt, & \text{if } D_1f(x,y)\in L_\infty(\Delta); \\ \|D_1f\|_{p_1}\left(\int_a^b\int_c^d |p(x,t)|^{q_1}dsdt\right)^{\frac{1}{q_1}}, & \text{if } D_1f(x,y)\in L_{p_1}(\Delta), \\ & \frac{1}{p_1}+\frac{1}{q_1}=1,\ p_1>1; \\ \|D_1f\|_1\sup_{t\in[a,b]}|p(x,t)|, & \text{if } D_1f(x,y)\in L_1(\Delta); \end{cases} \tag{4.2.12}$$

and as

$$\begin{aligned}
\int_a^b \int_c^d |p(x,t)|dsdt &= \int_c^d \left(\int_a^b |p(x,t)|dt \right) ds \\
&= (d-c)\left[\int_a^x |p(x,t)|dt + \int_x^b |p(x,t)|dt \right] \\
&= (d-c)\left[\int_a^x (t-a)dt + \int_x^b (b-t)dt \right] \\
&= (d-c)\left[\frac{(x-a)^2+(b-x)^2}{2} \right] \\
&= (d-c)\left[\frac{1}{4}(b-a)^2 + \left(x - \frac{a+b}{2} \right)^2 \right],
\end{aligned}$$

$$\begin{aligned}
\left[\int_a^b \int_c^d |p(x,t)|^{q_1} dsdt \right]^{\frac{1}{q_1}} &= \left[\int_c^d \left(\int_a^b |p(x,t)|^{q_1} dt \right) ds \right]^{\frac{1}{q_1}} \\
&= (d-c)^{\frac{1}{q_1}} \left[\int_a^x |p(x,t)|^{q_1} dt + \int_x^b |p(x,t)|^{q_1} dt \right]^{\frac{1}{q_1}} \\
&= (d-c)^{\frac{1}{q_1}} \left[\int_a^x (t-a)^{q_1} dt + \int_x^b (b-t)^{q_1} dt \right]^{\frac{1}{q_1}} \\
&= (d-c)^{\frac{1}{q_1}} \left[\frac{(b-x)^{q_1+1} + (x-a)^{q_1+1}}{q_1+1} \right]^{\frac{1}{q_1}},
\end{aligned}$$

and

$$\sup_{t\in[a,b]} |p(x,t)| = \max\{x-a, b-x\} = \frac{b-a}{2} + \left| x - \frac{a+b}{2} \right|,$$

then, by (4.2.12), we obtain

$$\int_a^b \int_c^d |p(x,t)||D_1 f(t,s)|dsdt \leqslant$$

$$\begin{cases}
(d-c)\left[\frac{1}{4}(b-a)^2 + \left(x - \frac{a+b}{2} \right)^2 \right] \|D_1 f\|_\infty, & \text{if } D_1 f(x,y) \in L_\infty(\Delta); \\
(d-c)^{\frac{1}{q_1}} \left[\frac{(b-x)^{q_1+1} + (x-a)^{q_1+1}}{q_1+1} \right]^{\frac{1}{q_1}} \|D_1 f\|_{p_1}, & \text{if } D_1 f(x,y) \in L_{p_1}(\Delta),\ \frac{1}{p_1} + \frac{1}{q_1} = 1,\ p_1 \\
\left[\frac{b-a}{2} + \left| x - \frac{a+b}{2} \right| \right] \|D_1 f\|_1, & \text{if } D_1 f(x,y) \in L_1(\Delta).
\end{cases} \tag{4.2.13}$$

In a similar fashion, we state that the following inequality holds

$$\int_a^b \int_c^d |q(y,s)||D_2 f(t,s)|dsdt \leqslant$$

$$\begin{cases} (b-a)\left[\frac{1}{4}(d-c)^2+\left(y-\frac{c+d}{2}\right)^2\right]\|D_2f\|_\infty, & \text{if } D_2f(x,y)\in L_\infty(\Delta);\\ (b-a)^{\frac{1}{q_2}}\left[\frac{(d-y)^{q_2+1}+(y-c)^{q_2+1}}{q_2+1}\right]^{\frac{1}{q_2}}\|D_2f\|_{p_2}, & \text{if } D_2f(x,y)\in L_{p_2}(\Delta),\\ & \frac{1}{p_2}+\frac{1}{q_2}=1,\ p_2>1;\\ \left[\frac{d-c}{2}+\left|y-\frac{c+d}{2}\right|\right]\|D_2f\|_1, & \text{if } D_2f(x,y)\in L_1(\Delta). \end{cases} \tag{4.2.14}$$

In addition, we have

$$\int_a^b\int_c^d |p(x,t)||q(y,s)||D_2D_1f(t,s)|dsdt$$

$$\leqslant \begin{cases} \|D_2D_1f\|_\infty \int_a^b |p(x,t)|dt\int_c^d |q(y,s)|ds,\\ \qquad \text{if } D_2D_1f(x,y)\in L_\infty(\Delta);\\ \|D_2D_1f\|_{p_3}\left(\int_a^b |p(x,t)|^{q_3}dt\right)^{\frac{1}{q_3}}\left(\int_c^d |q(y,s)|^{q_3}ds\right)^{\frac{1}{q_3}},\\ \qquad \text{if } D_2D_1f(x,y)\in L_{p_3}(\Delta);\ \frac{1}{p_3}+\frac{1}{q_3}=1,\ p_3>1;\\ \|D_2D_1f\|_1 \sup_{t\in[a,b]}|p(x,t)|\sup_{s\in[c,d]}|q(y,s)|,\\ \qquad \text{if } D_2D_1f(x,y)\in L_1(\Delta) \end{cases}$$

$$= \begin{cases} \left[\frac{1}{4}(b-a)^2+\left(x-\frac{a+b}{2}\right)^2\right]\left[\frac{1}{4}(d-c)^2+\left(y-\frac{c+d}{2}\right)^2\right]\|D_2D_1f\|_\infty,\\ \qquad \text{if } D_2D_1f(x,y)\in L_\infty(\Delta);\\ \left[\frac{(b-x)^{q_3+1}+(x-a)^{q_3+1}}{q_3+1}\right]^{\frac{1}{q_3}}\left[\frac{(d-y)^{q_3+1}+(y-c)^{q_3+1}}{q_3+1}\right]^{\frac{1}{q_3}}\|D_2D_1f\|_{p_3},\\ \qquad \text{if } D_2D_1f(x,y)\in L_{p_3}(\Delta);\ \frac{1}{p_3}+\frac{1}{q_3}=1,\ p_3>1;\\ \left[\frac{1}{2}(b-a)+\left|x-\frac{a+b}{2}\right|\right]\left[\frac{1}{2}(d-c)+\left|y-\frac{c+d}{2}\right|\right]\|D_2D_1f\|_1,\\ \qquad \text{if } D_2D_1f(x,y)\in L_1(\Delta). \end{cases} \tag{4.2.15}$$

The required inequality in (4.2.9) follows from (4.2.11), (4.2.13)–(4.2.15).

The following integral identity proved in [64] is useful in the proof of the next theorem.

Lemma 4.2.1. Let $f:\Delta\to\mathbb{R}$ be a continuous mapping such that the partial derivatives $\frac{\partial^{l+k}f(\cdot,\cdot)}{\partial x^k\partial y^l}$, $k=0,1,\ldots,n-1$; $l=0,1,\ldots,m-1$ exist and are continuous on $\Delta=[a,b]\times[c,d]$

and $K_n : [a,b]^2 \to \mathbb{R}$, $S_m : [c,d]^2 \to \mathbb{R}$ are given by

$$K_n(x,t) = \begin{cases} \dfrac{(t-a)^n}{n!}, & t \in [a,x], \\ \dfrac{(t-b)^n}{n!}, & t \in (x,b], \end{cases} \tag{4.2.16}$$

$$S_m(y,s) = \begin{cases} \dfrac{(s-c)^m}{m!}, & s \in [c,y], \\ \dfrac{(s-d)^m}{m!}, & s \in (y,d]. \end{cases} \tag{4.2.17}$$

Then for all $(x,y) \in \Delta$, we have the identity

$$\int_a^b \int_c^d f(t,s)dsdt = \sum_{k=0}^{n-1}\sum_{l=0}^{m-1} X_k(x)Y_l(y)\frac{\partial^{l+k} f(x,y)}{\partial x^k \partial y^l}$$

$$+(-1)^m \sum_{k=0}^{n-1} X_k(x) \int_c^d S_m(y,s)\frac{\partial^{k+m} f(x,s)}{\partial x^k \partial s^m}ds + (-1)^n \sum_{l=0}^{m-1} Y_l(y) \int_a^b K_n(x,t)\frac{\partial^{n+l} f(t,y)}{\partial t^n \partial y^l}dt$$

$$+(-1)^{m+n} \int_a^b \int_c^d K_n(x,t)S_m(y,s)\frac{\partial^{n+m} f(t,s)}{\partial t^n \partial s^m}dsdt, \tag{4.2.18}$$

where

$$X_k(x) = \frac{(b-x)^{k+1} + (-1)^k (x-a)^{k+1}}{(k+1)!}, \tag{4.2.19}$$

$$Y_l(y) = \frac{(d-y)^{l+1} + (-1)^l (y-c)^{l+1}}{(l+1)!}. \tag{4.2.20}$$

Proof. Applying the identity (see, Lemma 1.5.3)

$$\int_a^b g(t)dt = \sum_{k=0}^{n-1}\left[\frac{(b-x)^{k+1} + (-1)^k (x-a)^{k+1}}{(k+1)!}\right] g^{(k)}(x)$$

$$+(-1)^n \int_a^b P_n(x,t) g^{(n)}(t)dt, \tag{4.2.21}$$

where

$$P_n(x,t) = \begin{cases} \dfrac{(t-a)^n}{n!}, & t \in [a,x], \\ \dfrac{(t-b)^n}{n!}, & t \in (x,b], \end{cases}$$

for the partial mapping $f(\cdot,s)$, $s \in [c,d]$, we can write

$$\int_a^b f(t,s)dt = \sum_{k=0}^{n-1}\left[\frac{(b-x)^{k+1} + (-1)^k (x-a)^{k+1}}{(k+1)!}\right] \frac{\partial^k f(x,s)}{\partial x^k}$$

$$+(-1)^n \int_a^b K_n(x,t)\frac{\partial^n f(t,s)}{\partial t^n}dt, \tag{4.2.22}$$

for every $x \in [a,b]$ and $s \in [c,d]$.

Integrating (4.2.22) over s on $[c,d]$, we deduce

$$\int_a^b \int_c^d f(t,s)dsdt = \sum_{k=0}^{n-1} \left[\frac{(b-x)^{k+1} + (-1)^k (x-a)^{k+1}}{(k+1)!} \right] \int_c^d \frac{\partial^k f(x,s)}{\partial x^k} ds$$

$$+(-1)^n \int_a^b K_n(x,t) \left(\int_c^d \frac{\partial^n f(t,s)}{\partial t^n} ds \right) dt, \tag{4.2.23}$$

for all $x \in [a,b]$.

Applying the identity (4.2.21) again for the partial mapping $\frac{\partial^k f(x,\cdot)}{\partial x^k}$ on $[c,d]$, we obtain

$$\int_c^d \frac{\partial^k f(x,s)}{\partial x^k} ds = \sum_{l=0}^{m-1} \left[\frac{(d-y)^{l+1} + (-1)^l (y-c)^{l+1}}{(l+1)!} \right] \frac{\partial^l}{\partial y^l} \left(\frac{\partial^k f(x,y)}{\partial x^k} \right)$$

$$+(-1)^m \int_c^d S_m(y,s) \frac{\partial^m}{\partial s^m} \left(\frac{\partial^k f(x,s)}{\partial x^k} \right) ds$$

$$= \sum_{l=0}^{m-1} \left[\frac{(d-y)^{l+1} + (-1)^l (y-c)^{l+1}}{(l+1)!} \right] \frac{\partial^{l+k} f(x,y)}{\partial x^k \partial y^l}$$

$$+(-1)^m \int_c^d S_m(y,s) \frac{\partial^{k+m} f(x,s)}{\partial x^k \partial s^m} ds. \tag{4.2.24}$$

In addition, the identity (4.2.22) applied for the partial derivative $\frac{\partial^n f(t,\cdot)}{\partial t^n}$ also gives

$$\int_c^d \frac{\partial^n f(t,s)}{\partial t^n} ds = \sum_{l=0}^{m-1} \left[\frac{(d-y)^{l+1} + (-1)^l (y-c)^{l+1}}{(l+1)!} \right] \frac{\partial^{n+l} f(t,y)}{\partial t^n \partial y^l}$$

$$+(-1)^m \int_c^d S_m(y,s) \frac{\partial^{n+m} f(t,s)}{\partial t^n \partial s^m} ds. \tag{4.2.25}$$

Substituting (4.2.24) and (4.2.25) in (4.2.23) and rewriting will produce the desired identity in (4.2.18).

The inequality of the Ostrowski-type given by Hanna, Dragomir and Cerone [64] is embodied in the following theorem.

Theorem 4.2.3. Let $f : \Delta \to \mathbb{R}$ be continuous on Δ and assume that $\frac{\partial^{n+m} f(t,s)}{\partial t^n \partial s^m}$ exist on $(a,b) \times (c,d)$. Then we have the inequality

$$\left| \int_a^b \int_c^d f(t,s)dsdt - \sum_{k=0}^{n-1} \sum_{l=0}^{m-1} X_k(x) Y_l(y) \frac{\partial^{l+k} f(x,y)}{\partial x^k \partial y^l} \right.$$

$$\left. -(-1)^m \sum_{k=0}^{n-1} X_k(x) \int_c^d S_m(y,s) \frac{\partial^{k+m} f(x,s)}{\partial x^k \partial s^m} ds - (-1)^n \sum_{l=0}^{m-1} Y_l(y) \int_a^b K_n(x,t) \frac{\partial^{n+l} f(t,y)}{\partial t^n \partial y^l} dt \right|$$

$$\leqslant \begin{cases} \dfrac{1}{(n+1)!(m+1)!}\left[(x-a)^{n+1}+(b-x)^{n+1}\right] \\ \qquad \times\left[(y-c)^{m+1}+(d-y)^{m+1}\right]\left\|\dfrac{\partial^{n+m} f}{\partial t^n \partial s^m}\right\|_\infty, \\ \qquad\qquad \text{if } \dfrac{\partial^{n+m} f(t,s)}{\partial t^n \partial s^m}\in L_\infty(\Delta); \\ \dfrac{1}{n!m!}\left[\dfrac{(x-a)^{nq+1}+(b-x)^{nq+1}}{nq+1}\right]^{\frac{1}{q}} \\ \qquad \times\left[\dfrac{(y-c)^{mq+1}+(d-y)^{mq+1}}{mq+1}\right]^{\frac{1}{q}}\left\|\dfrac{\partial^{n+m} f}{\partial t^n \partial s^m}\right\|_p \\ \qquad\qquad \text{if } \dfrac{\partial^{n+m} f(t,s)}{\partial t^n \partial s^m}\in L_p(\Delta),\ \dfrac{1}{p}+\dfrac{1}{q}=1,\ p>1; \\ \dfrac{1}{4n!m!}\left[(x-a)^n+(b-x)^n+|(x-a)^n-(b-x)^n|\right] \\ \qquad \times\left[(y-c)^m+(d-y)^m+|(y-c)^m-(d-y)^m|\right]\left\|\dfrac{\partial^{n+m} f}{\partial t^n \partial s^m}\right\|_1, \\ \qquad\qquad \text{if } \dfrac{\partial^{n+m} f(t,s)}{\partial t^n \partial s^m}\in L_1(\Delta); \end{cases} \tag{4.2.26}$$

for all $(x,y)\in\Delta$, where

$$\left\|\frac{\partial^{n+m} f}{\partial t^n \partial s^m}\right\|_\infty = \sup_{(t,s)\in\Delta}\left|\frac{\partial^{n+m} f(t,s)}{\partial t^n \partial s^m}\right| < \infty,$$

$$\left\|\frac{\partial^{n+m} f}{\partial t^n \partial s^m}\right\|_p = \left(\int_a^b\int_c^d \left|\frac{\partial^{n+m} f(t,s)}{\partial t^n \partial s^m}\right|^p dsdt\right)^{\frac{1}{p}} < \infty.$$

Proof. From the hypotheses, the identity (4.2.18) holds. From (4.2.18), we get

$$\Bigg|\int_a^b\int_c^d f(t,s)dsdt-\sum_{k=0}^{n-1}\sum_{l=0}^{m-1} X_k(x)Y_l(y)\frac{\partial^{l+k} f(x,y)}{\partial x^k \partial y^l}$$

$$-(-1)^m\sum_{k=0}^{n-1} X_k(x)\int_c^d S_m(y,s)\frac{\partial^{k+m} f(x,s)}{\partial x^k \partial s^m}ds-(-1)^n\sum_{l=0}^{m-1} Y_l(y)\int_a^b K_n(x,t)\frac{\partial^{n+l} f(t,y)}{\partial t^n \partial y^l}dt\Bigg|$$

$$\leqslant \int_a^b\int_c^d |K_n(x,t)||S_m(y,s)|\left|\frac{\partial^{n+m} f(t,s)}{\partial t^n \partial s^m}\right| dsdt. \tag{4.2.27}$$

We observe that

$$\int_a^b\int_c^d |K_n(x,t)||S_m(y,s)|\left|\frac{\partial^{n+m} f(t,s)}{\partial t^n \partial s^m}\right| dsdt$$

$$\leqslant \begin{cases} \left\|\dfrac{\partial^{n+m} f}{\partial t^n \partial s^m}\right\|_\infty \displaystyle\int_a^b\int_c^d |K_n(x,t)||S_m(y,s)|dsdt, \\ \left\|\dfrac{\partial^{n+m} f}{\partial t^n \partial s^m}\right\|_p \left(\displaystyle\int_a^b\int_c^d |K_n(x,t)|^q|S_m(y,s)|^q dsdt\right)^{\frac{1}{q}}, \quad \dfrac{1}{p}+\dfrac{1}{q}=1,\ p>1; \\ \left\|\dfrac{\partial^{n+m} f}{\partial t^n \partial s^m}\right\|_1 \sup\limits_{(t,s)\in\Delta}|K_n(x,t)||S_m(y,s)|. \end{cases} \tag{4.2.28}$$

Now, using (4.2.16), (4.2.17), we have

$$\int_a^b \int_c^d |K_n(x,t)S_m(y,s)|dtds = \int_a^b |K_n(x,t)|dt \int_c^d |S_m(y,s)|ds$$

$$= \left[\int_a^x \frac{(t-a)^n}{n!}dt + \int_x^b \frac{(b-t)^n}{n!}dt\right]\left[\int_c^y \frac{(s-c)^m}{m!}ds + \int_y^d \frac{(d-s)^m}{m!}ds\right]$$

$$= \frac{\left[(x-a)^{n+1} + (b-x)^{n+1}\right]\left[(y-c)^{m+1} + (d-y)^{m+1}\right]}{(n+1)!(m+1)!}. \tag{4.2.29}$$

Using (4.2.29) in the first inequality in (4.2.28), we get the first inequality in (4.2.26). Further, using (4.2.16), (4.2.17) we have

$$\left(\int_a^b \int_c^d |K_n(x,t)S_m(y,s)|^q dtds\right)^{\frac{1}{q}} = \left(\int_a^b |K_n(x,t)|^q dt\right)^{\frac{1}{q}} \left(\int_c^d |S_m(y,s)|^q ds\right)^{\frac{1}{q}}$$

$$= \frac{1}{n!m!}\left[\int_a^x (t-a)^{nq}dt + \int_x^b (b-t)^{nq}dt\right]^{\frac{1}{q}} \left[\int_c^y (s-c)^{mq}ds + \int_y^d (d-s)^{mq}ds\right]^{\frac{1}{q}}$$

$$= \frac{1}{n!m!}\left[\frac{(x-a)^{nq+1} + (b-x)^{nq+1}}{nq+1}\right]^{\frac{1}{q}} \left[\frac{(y-c)^{mq+1} + (d-y)^{mq+1}}{mq+1}\right]^{\frac{1}{q}}. \tag{4.2.30}$$

Using (4.2.30) in the second inequality in (4.2.28), we get the second inequality in (4.2.26). Finally, using (4.2.16), (4.2.17), we have

$$\sup_{(t,s)\in\Delta} |K_n(x,t)S_m(y,s)| = \sup_{t\in[a,b]} |K_n(x,t)| \sup_{s\in[c,d]} |S_m(y,s)|$$

$$= \max\left\{\frac{(x-a)^n}{n!}, \frac{(b-x)^n}{n!}\right\}\left\{\frac{(y-c)^m}{m!}, \frac{(d-y)^m}{m!}\right\}$$

$$= \frac{1}{n!m!}\left[\frac{(x-a)^n + (b-x)^n}{2} + \left|\frac{(x-a)^n - (b-x)^n}{2}\right|\right]$$

$$\times \left[\frac{(y-c)^m + (d-y)^m}{2} + \left|\frac{(y-c)^m - (d-y)^m}{2}\right|\right], \tag{4.2.31}$$

where, we have used the fact that

$$\max\{X,Y\} = \frac{X+Y}{2} + \left|\frac{X-Y}{2}\right|.$$

Using (4.2.31) in the third inequality in (4.2.28), we get the third inequality in (4.2.26). The proof is complete.

In a recent paper [115], Pachpatte investigated the following Ostrowski-type inequality for double integrals.

Theorem 4.2.4. Let $f,\ g:\Delta\to\mathbb{R}$ be continuous mappings such that $D_2D_1f(x,y)$, $D_2D_1g(x,y)$ exist and are continuous on Δ. Then

$$\begin{aligned}&\Bigg|f(x,y)g(x,y)-\frac{1}{2}g(x,y)\left[\frac{1}{b-a}\int_a^b f(s,y)ds+\frac{1}{d-c}\int_a^b f(x,t)dt\right.\\&\left.\qquad-\frac{1}{(b-a)(d-c)}\int_a^b\int_c^d f(s,t)dtds\right]\\&-\frac{1}{2}f(x,y)\left[\frac{1}{b-a}\int_a^b g(s,y)ds+\frac{1}{d-c}\int_c^d g(x,t)dt-\frac{1}{(b-a)(d-c)}\int_a^b\int_c^d g(s,t)dtds\right]\Bigg|\\&\leqslant\frac{1}{2(b-a)(d-c)}\int_a^b\int_c^d\left[|g(x,y)|\left|\int_s^x\int_t^y D_2D_1f(\sigma,\tau)d\tau d\sigma\right|\right.\\&\left.\qquad+|f(x,y)|\left|\int_s^x\int_t^y D_2D_1g(\sigma,\tau)d\tau d\sigma\right|\right]dydx,\end{aligned}\tag{4.2.32}$$

for all $(x,y)\in\Delta$.

Proof. From the hypotheses, it is easy to observe that, the following identities hold:

$$f(x,y)-f(s,y)-f(x,t)+f(s,t)=\int_s^x\int_t^y D_2D_1f(\sigma,t)d\tau d\sigma,\tag{4.2.33}$$

and

$$g(x,y)-g(s,y)-g(x,t)+g(s,t)=\int_s^x\int_t^y D_2D_1g(\sigma,\tau)d\tau d\sigma,\tag{4.2.34}$$

for $(x,y),(s,t)\in\Delta$. Multiplying both sides of (4.2.33) and (4.2.34) by $g(x,y)$ and $f(x,y)$ respectively and adding the resulting identities, we have

$$\begin{aligned}&2f(x,y)g(x,y)-g(x,y)[f(s,y)+f(x,t)-f(s,t)]-f(x,y)[g(s,y)+g(x,t)-g(s,t)]\\&=g(x,y)\int_s^x\int_t^y D_2D_1f(\sigma,t)d\tau d\sigma+f(x,y)\int_s^x\int_t^y D_2D_1g(\sigma,t)d\tau d\sigma.\end{aligned}\tag{4.2.35}$$

Integrating both sides of (4.2.35) with respect to (s,t) over Δ and rewriting, we have

$$\begin{aligned}&f(x,y)g(x,y)-\frac{1}{2}g(x,y)\left[\frac{1}{b-a}\int_a^b f(s,y)ds+\frac{1}{d-c}\int_c^d f(x,t)dt\right.\\&\left.\qquad-\frac{1}{(b-a)(d-c)}\int_a^b\int_c^d f(s,t)dtds\right]\\&-\frac{1}{2}f(x,y)\left[\frac{1}{b-a}\int_a^b g(s,y)ds+\frac{1}{d-c}\int_c^d g(x,t)dt-\frac{1}{(b-a)(d-c)}\int_a^b\int_c^d g(s,t)dtds\right]\\&=\frac{1}{2(b-a)(d-c)}\int_a^b\int_c^d\left[g(x,y)\int_s^x\int_t^y D_2D_1f(\sigma,\tau)d\tau d\sigma\right.\\&\left.\qquad+f(x,y)\int_s^x\int_t^y D_2D_1g(\sigma,\tau)d\tau d\sigma\right]dydx.\end{aligned}\tag{4.2.36}$$

From (4.2.36) and using the properties of modulus, it is easy to observe that the required inequality in (4.2.32) holds. The proof is complete.

Remark 4.2.3. By taking $g(x,y) = 1$ and hence $D_2D_1g(x,y) = 0$ in Theorem 4.2.4, we get

$$\left| f(x,y) - \left[\frac{1}{b-a} \int_a^b f(s,y)ds + \frac{1}{d-c} \int_c^d f(x,t)dt - \frac{1}{(b-a)(d-c)} \int_a^b \int_c^d f(s,t)dtds \right] \right|$$
$$\leqslant \frac{1}{(b-a)(d-c)} \int_a^b \int_c^d \left| \int_s^x \int_t^y D_2D_1f(\sigma,\tau)d\tau d\sigma \right| dydx, \tag{4.2.37}$$

for all $(x,y) \in \Delta$. Further, if we assume that $D_2D_1f(x,y)$ is bounded on $(a,b) \times (c,d)$, i.e.,

$$\|D_2D_1f\|_\infty = \sup_{(x,y)\in(a,b)\times(c,d)} |D_2D_1f(x,y)| < \infty,$$

then after rewriting (4.2.37) and by elementary calculations, we get the inequality (4.2.1) given in Theorem 4.2.1.

In another paper [125], Pachpatte has given the following generalization of Theorem 4.2.4, involving three functions.

Theorem 4.2.5. Let $f,\ g,\ h : \Delta \to \mathbb{R}$ be continuous functions such that $D_2D_1f(x,y)$, $D_2D_1g(x,y)$, $D_2D_1h(x,y)$ exist and are continuous on Δ. Then

$$|A(f,g,h;F,G,H;l)(x,y)| \leqslant \frac{1}{3\Delta} B(|f|,|g|,|h|;|I|)(x,y), \tag{4.2.38}$$

for all $(x,y) \in \Delta$, where

$$A(f,g,h;F,G,H;l)(x,y) = f(x,y)g(x,y)h(x,y)$$
$$-\frac{1}{3l}\left[g(x,y)h(x,y)\left\{ F(x,y) - \int_a^b \int_c^d f(s,t)dtds \right\} \right.$$
$$+h(x,y)f(x,y)\left\{ G(x,y) - \int_a^b \int_c^d g(s,t)dtds \right\}$$
$$\left. +f(x,y)g(x,y)\left\{ H(x,y) - \int_a^b \int_c^d h(s,t)dtds \right\} \right],$$

$$B(f,g,h;I)(x,y) = g(x,y)h(x,y) \int_a^b \int_c^d I[f]dtds$$
$$+h(x,y)f(x,y) \int_a^b \int_c^d I[g]dtds + f(x,y)g(x,y) \int_a^b \int_c^d I[h]dtds,$$

in which $l = (b-a)(d-c)$,

$$F(x,y) = (c-d) \int_a^b f(s,y)ds + (b-a) \int_c^d f(x,t)dt,$$
$$G(x,y) = (c-d) \int_a^b g(s,y)ds + (b-a) \int_c^d g(x,t)dt,$$
$$H(x,y) = (c-d) \int_a^b h(s,y)ds + (b-a) \int_c^d h(x,t)dt,$$

and for a suitable function $p : \Delta \to \mathbb{R}$, the notation

$$I[p] = \int_s^x \int_t^y D_2D_1p(\sigma,\tau)d\tau d\sigma,$$

is set to simplify the presentation.

Proof. From the hypotheses, it is easy to observe that the following identities hold:

$$f(x,y)-f(s,y)-f(x,t)+f(s,t)=I[f], \tag{4.2.39}$$

$$g(x,y)-g(s,y)-g(x,t)+g(s,t)=I[g], \tag{4.2.40}$$

$$h(x,y)-h(s,y)-h(x,t)+h(s,t)=I[h], \tag{4.2.41}$$

for (x,y), $(s,t)\in\Delta$. Multiplying both sides of (4.2.39), (4.2.40) and (4.2.41) by $g(x,y)h(x,y)$, $h(x,y)f(x,y)$ and $f(x,y)g(x,y)$ respectively and adding the resulting identities, we get

$$3f(x,y)g(x,y)h(x,y)-g(x,y)h(x,y)[f(s,y)+f(x,t)-f(s,t)]$$

$$-h(x,y)f(x,y)[g(s,y)+g(x,t)-g(s,t)]-f(x,y)g(x,y)[h(s,y)+h(x,t)-h(s,t)]$$

$$=g(x,y)h(x,y)I[f]+h(x,y)f(x,y)I[g]+f(x,y)g(x,y)I[h]. \tag{4.2.42}$$

Integrating both sides of (4.2.42) with respect to (s,t) over Δ and rewriting, we have

$$A(f,g,h;F,G,H;l)(x,y)=\frac{1}{3l}B(f,g,h;I)(x,y). \tag{4.2.43}$$

From (4.2.43) and using the properties of modulus, we get the desired inequality in (4.2.38). The proof is complete.

Remark 4.2.4. If we take $h(x,y)=1$ and hence $D_2D_1h(x,y)=0$ in Theorem 4.2.5, then by elementary calculations, we get

$$\Bigg|f(x,y)g(x,y)-\frac{1}{2l}\left[g(x,y)\left\{F(x,y)-\int_a^b\int_c^d f(s,t)dtds\right\}\right.$$

$$\left.+f(x,y)\left\{G(x,y)-\int_a^b\int_c^d g(s,t)dtds\right\}\right]\Bigg|$$

$$\leqslant\frac{1}{2l}\left[|g(x,y)|\int_a^b\int_c^d|I[f]|dtds+|f(x,y)|\int_a^b\int_c^d|I[g]|dtds\right], \tag{4.2.44}$$

which in turn is the inequality given in Theorem 4.2.4.

Remark 4.2.5. Integrating both sides of (4.2.43) with respect to (x,y) over Δ and rewriting, we get

$$T(f,g,h;F,G,H;l)=\frac{1}{3l^2}\int_a^b\int_c^d B(f,g,h;I)(x,y)dydx, \tag{4.2.45}$$

where

$$T(f,g,h;F,G,H;l)=\frac{1}{l}\int_a^b\int_c^d f(x,y)g(x,y)h(x,y)dydx-\frac{1}{3l^2}\int_a^b\int_c^d[g(x,y)h(x,y)F(x,y)$$

$$+h(x,y)f(x,y)G(x,y)+f(x,y)g(x,y)H(x,y)]\,dydx$$
$$+\frac{1}{3}\left[\left(\frac{1}{l}\int_a^b\int_c^d g(x,y)h(x,y)dydx\right)\left(\frac{1}{l}\int_a^b\int_c^d f(x,y)dydx\right)\right.$$
$$+\left(\frac{1}{l}\int_a^b\int_c^d h(x,y)f(x,y)dydx\right)\left(\frac{1}{l}\int_a^b\int_c^d g(x,y)dydx\right)$$
$$\left.+\left(\frac{1}{l}\int_a^b\int_c^d f(x,y)g(x,y)dydx\right)\left(\frac{1}{l}\int_a^b\int_c^d h(x,y)dydx\right)\right].$$

From (4.2.45) and using the properties of modulus and integrals, we get

$$|T(f,g,h;F,G,H;l)| \leqslant \frac{1}{3l^2}\int_a^b\int_c^d B(|f|,|g|,|h|;|I|)(x,y)dydx. \tag{4.2.46}$$

We note that the inequality obtained in (4.2.46) is similar to the Grüss-type inequalities given in [89].

4.3 Ostrowski-type inequalities in three variables

This section is devoted to some basic Ostrowski-type inequalities involving functions of three independent variables, recently investigated in [83,126,148].

The following Ostrowski-type inequality is proved by Pachpatte in [83].

Theorem 4.3.1. Let $\Omega = [a,k]\times[b,m]\times[c,n]$; $a,\ b,\ c,\ k,\ m,\ n \in \mathbb{R}$ and $f:\Omega\to\mathbb{R}$ be a continuous function on Ω, $D_3D_2D_1 f(r,s,t) = \frac{\partial^3 f(r,s,t)}{\partial t\partial s\partial r}$ exists and is continuous on Ω. Then

$$\left|\int_a^k\int_b^m\int_c^n f(r,s,t)dtdsdr - \frac{1}{8}(k-a)(m-b)(n-c)[f(a,b,c)+f(k,m,n)]\right.$$
$$+\frac{1}{4}(m-b)(n-c)\int_a^k [f(r,b,c)+f(r,m,n)+f(r,m,c)+f(r,b,n)]dr$$
$$+\frac{1}{4}(k-a)(n-c)\int_b^m [f(a,s,c)+f(k,s,n)+f(a,s,n)+f(k,s,c)]ds$$
$$+\frac{1}{4}(k-a)(m-b)\int_c^n [f(a,b,t)+f(k,m,t)+f(k,b,t)+f(a,m,t)]\,dt$$
$$-\frac{1}{2}(k-a)\int_b^m\int_c^n [f(a,s,t)+f(k,s,t)]\,dtds$$
$$-\frac{1}{2}(m-b)\int_a^k\int_c^n [f(r,b,t)+f(r,m,t)]\,dtdr$$
$$\left.-\frac{1}{2}(n-c)\int_a^k\int_b^m [f(r,s,c)+f(r,s,n)]\,dsdr\right|$$
$$\leqslant \frac{1}{8}(k-a)(m-b)(n-c)\int_a^k\int_b^m\int_c^n |D_3D_2D_1 f(r,s,t)|dtdsdr. \tag{4.3.1}$$

Proof. As in the proof of Theorem 2.3.4, we have the following identity:

$$f(r,s,t) - L(f(r,s,t)) = \frac{1}{8}B(D_3D_2D_1f(r,s,t)), \tag{4.3.2}$$

for $(r,s,t) \in \Omega$, where $L(f(r,s,t))$ and $B(D_3D_2D_1f(r,s,t))$ are as given in Section 2.3. Integrating both sides of (4.3.2) over Ω and by elementary calculations, we get

$$\begin{aligned}
&\int_a^k \int_b^m \int_c^n f(r,s,t)dtdsdr - \frac{1}{8}(k-a)(m-b)(n-c)[f(a,b,c)+f(k,m,n)] \\
&+\frac{1}{4}(m-b)(n-c)\int_a^k [f(r,b,c)+f(r,m,n)+f(r,m,c)+f(r,b,n)]dr \\
&+\frac{1}{4}(k-a)(n-c)\int_b^m [f(a,s,c)+f(k,s,n)+f(a,s,n)+f(k,s,c)]ds \\
&+\frac{1}{4}(k-a)(m-b)\int_c^n [f(a,b,t)+f(k,m,t)+f(k,b,t)+f(a,m,t)]dt \\
&-\frac{1}{2}(k-a)\int_b^m \int_c^n [f(a,s,t)+f(k,s,t)]dtds \\
&-\frac{1}{2}(m-b)\int_a^k \int_c^n [f(r,b,t)+f(r,m,t)]dtdr \\
&-\frac{1}{2}(n-c)\int_a^k \int_b^m [f(r,s,c)+f(r,s,n)]dsdr \\
&=\frac{1}{8}\int_a^k \int_b^m \int_c^n B(D_3D_2D_1f(r,s,t))dtdsdr.
\end{aligned} \tag{4.3.3}$$

Using the properties of modulus and integrals, we observe that

$$|B(D_3D_2D_1f(r,s,t))| \leqslant \int_a^k \int_b^m \int_c^n |D_3D_2D_1f(u,v,w)|dwdvdu. \tag{4.3.4}$$

Now, from (4.3.3) and (4.3.4), we easily get the required inequality in (4.3.1) and the proof is complete.

Remark 4.3.1. From (4.3.2) and using (4.3.4), it is easy to observe that the following inequality also holds

$$|f(r,s,t) - L(f(r,s,t))| \leqslant \frac{1}{8}\int_a^k \int_b^m \int_c^n |D_3D_2D_1f(u,v,w)|dwdvdu. \tag{4.3.5}$$

for $(r,s,t) \in \Omega$.

We give below an integral identity proved in [148], which is essential for our proof of the next theorem.

Lemma 4.3.1. Let $H = [a_1,b_1] \times [a_2,b_2] \times [a_3,b_3]$ ($a_i < b_i$, $i = 1,2,3$) and $f : H \to \mathbb{R}$ be a continuous mapping such that the partial derivatives $\frac{\partial^{i+j+k} f(\cdot,\cdot,\cdot)}{\partial x^i \partial y^j \partial z^k}$; $i = 0,1,\ldots,n-1$; $j = 0,1,\ldots,m-1$; $k = 0,1,\ldots,p-1$; exist and are continuous on H. Also, let

$$P_n(x,r) = \begin{cases} \dfrac{(r-a_1)^n}{n!}; & r \in [a_1,x], \\ \dfrac{(r-b_1)^n}{n!}; & r \in (x,b_1], \end{cases} \tag{4.3.6}$$

$$Q_m(y,s) = \begin{cases} \dfrac{(s-a_2)^m}{m!}; & s \in [a_2,y], \\ \dfrac{(s-b_2)^m}{m!}; & s \in (y,b_2], \end{cases} \tag{4.3.7}$$

$$S_p(z,t) = \begin{cases} \dfrac{(t-a_3)^p}{p!}; & t \in [a_3,z], \\ \dfrac{(t-b_3)^p}{p!}; & t \in (z,b_3], \end{cases} \tag{4.3.8}$$

then for all $(x,y,z) \in H$, we have the identity:

$$\begin{aligned} V(x,y,z) &:= \int_{a_1}^{b_1}\int_{a_2}^{b_2}\int_{a_3}^{b_3} f(r,s,t)\,dt\,ds\,dr - \sum_{i=0}^{n-1}\sum_{j=0}^{m-1}\sum_{k=0}^{p-1} X_i(x)Y_j(y)Z_k(z)\frac{\partial^{i+j+k} f(x,y,z)}{\partial x^i \partial y^j \partial z^k} \\ &\quad +(-1)^p \sum_{i=0}^{n-1}\sum_{j=0}^{m-1} X_i(x)Y_j(y)\int_{a_3}^{b_3} S_p(z,t)\frac{\partial^{i+j+p} f(x,y,t)}{\partial x^i \partial y^j \partial t^p}dt \\ &\quad +(-1)^m \sum_{i=0}^{n-1}\sum_{k=0}^{p-1} X_i(x)Z_k(z)\int_{a_2}^{b_2} Q_m(y,s)\frac{\partial^{i+m+k} f(x,s,z)}{\partial x^i \partial s^m \partial z^k}ds \\ &\quad +(-1)^n \sum_{j=0}^{m-1}\sum_{k=0}^{p-1} Y_j(y)Z_k(z)\int_{a_1}^{b_1} P_n(x,r)\frac{\partial^{n+j+k} f(r,y,z)}{\partial r^n \partial y^j \partial z^k}dr \\ &\quad -(-1)^{m+p} \sum_{i=0}^{n-1} X_i(x)\int_{a_2}^{b_2}\int_{a_3}^{b_3} Q_m(y,s)S_p(z,t)\frac{\partial^{i+m+p} f(x,s,t)}{\partial x^i \partial s^m \partial t^p}dt\,ds \\ &\quad -(-1)^{n+p} \sum_{j=0}^{m-1} Y_j(y)\int_{a_1}^{b_1}\int_{a_3}^{b_3} P_n(x,r)S_p(z,t)\frac{\partial^{n+j+p} f(r,y,t)}{\partial r^n \partial y^j \partial t^p}dt\,dr \\ &\quad -(-1)^{n+m} \sum_{k=0}^{p-1} Z_k(z)\int_{a_1}^{b_1}\int_{a_2}^{b_2} P_n(x,r)Q_m(y,s)\frac{\partial^{n+m+k} f(r,s,z)}{\partial r^n \partial s^m \partial z^k}ds\,dr \\ &= -(-1)^{n+m+p}\int_{a_1}^{b_1}\int_{a_2}^{b_2}\int_{a_3}^{b_3} P_n(x,r)Q_m(y,s)S_p(z,t)\frac{\partial^{n+m+p} f(r,s,t)}{\partial r^n \partial s^m \partial t^p}dt\,ds\,dr, \end{aligned} \tag{4.3.9}$$

where

$$X_i(x) = \frac{(b_1-x)^{i+1} + (-1)^i (x-a_1)^{i+1}}{(i+1)!}, \tag{4.3.10}$$

$$Y_j(y) = \frac{(b_2-y)^{j+1} + (-1)^j (y-a_2)^{j+1}}{(j+1)!}, \tag{4.3.11}$$

$$Z_k(z) = \frac{(b_3-z)^{k+1} + (-1)^k (z-a_3)^{k+1}}{(k+1)!}. \tag{4.3.12}$$

Proof. As in the proof of Lemma 4.2.1, we have an identity

$$\int_{a_1}^{b_1} g(r)\,dr = \sum_{i=0}^{n-1} X_i(x) g^{(I)}(x) + (-1)^n \int_{a_1}^{b_1} P_n(x,r) g^{(n)}(r)\,dr. \tag{4.3.13}$$

Now, for the partial mapping $f(\cdot,s,t)$ we have

$$\int_{a_1}^{b_1} f(r,s,t)dr = \sum_{i=0}^{n-1} X_i(x)\frac{\partial^i f}{\partial x^i} + (-1)^n \int_{a_1}^{b_1} P_n(x,r)\frac{\partial^n f}{\partial r^n}dr, \tag{4.3.14}$$

for every $r \in [a_1,b_1]$, $s \in [a_2,b_2]$ and $t \in [a_3,b_3]$. Now, integrate (4.3.14) over $s \in [a_2,b_2]$, to get

$$\int_{a_1}^{b_1}\int_{a_2}^{b_2} f(r,s,t)dsdr = \sum_{i=0}^{n-1} X_i(x)\int_{a_2}^{b_2}\frac{\partial^i f}{\partial x^i}ds + (-1)^n \int_{a_1}^{b_1} P_n(x,r)\left(\int_{a_2}^{b_2}\frac{\partial^n f}{\partial r^n}ds\right)dr, \tag{4.3.15}$$

for all $x \in [a_1,b_1]$.

From (4.3.13) for the partial mapping $\frac{\partial^i f}{\partial x^i}$, we have

$$\begin{aligned}\int_{a_2}^{b_2}\frac{\partial^i}{\partial x^i} f(x,s,t)ds &= \sum_{j=0}^{m-1} Y_j(y)\frac{\partial^j}{\partial y^j}\left(\frac{\partial^i f}{\partial x^i}\right) + (-1)^m \int_{a_2}^{b_2} Q_m(y,s)\frac{\partial^m}{\partial s^m}\left(\frac{\partial^i f}{\partial x^i}\right)ds \\ &= \sum_{j=0}^{m-1} Y_j(y)\frac{\partial^{i+j} f}{\partial x^i \partial y^j} + (-1)^m \int_{a_2}^{b_2} Q_m(y,s)\frac{\partial^{i+m} f}{\partial x^i \partial s^m}ds. \end{aligned}\tag{4.3.16}$$

Also, from (4.3.13), we have

$$\int_{a_2}^{b_2}\frac{\partial^n f}{\partial r^n}ds = \sum_{j=0}^{m-1} Y_j(y)\frac{\partial^{j+n} f}{\partial y^j \partial r^n} + (-1)^m \int_{a_2}^{b_2} Q_m(y,s)\frac{\partial^m}{\partial s^m}\left(\frac{\partial^n f}{\partial r^n}\right)ds. \tag{4.3.17}$$

From (4.3.16), (4.3.17) and (4.3.15), we get

$$\begin{aligned}\int_{a_1}^{b_1}\int_{a_2}^{b_2} f(r,s,t)dsdr &= \sum_{i=0}^{n-1} X_i(x)\left[\sum_{j=0}^{m-1} Y_j(y)\frac{\partial^{i+j} f}{\partial x^i \partial y^j} + (-1)^m \int_{a_2}^{b_2} Q_m(y,s)\frac{\partial^{i+m} f}{\partial x^i \partial s^m}ds\right] \\ &\quad + (-1)^n \int_{a_1}^{b_1} P_n(x,r)\left[\sum_{j=0}^{m-1} Y_j(y)\frac{\partial^{j+n} f}{\partial y^j \partial r^n} + (-1)^m \int_{a_2}^{b_2} Q_m(y,s)\left(\frac{\partial^n f}{\partial r^n}\right)ds\right]dr \\ &= \sum_{i=0}^{n-1} X_i(x)\sum_{j=0}^{m-1} Y_j(y)\frac{\partial^{i+j} f}{\partial x^i \partial y^j} + (-1)^m \sum_{i=0}^{n-1} X_i(x)\int_{a_2}^{b_2} Q_m(y,s)\frac{\partial^{i+m} f}{\partial x^i \partial s^m}ds \\ &\quad + (-1)^n \sum_{j=0}^{m-1} Y_j(y)\int_{a_1}^{b_1} P_n(x,r)\frac{\partial^{j+n} f}{\partial y^j \partial r^n}dr \\ &\quad + (-1)^{n+m}\int_{a_1}^{b_1}\int_{a_2}^{b_2} P_n(x,r)Q_m(y,s)\frac{\partial^{n+m} f}{\partial s^m \partial r^n}dsdr. \end{aligned}\tag{4.3.18}$$

Now, integrate (4.3.18) over $t \in [a_3, b_3]$ to get

$$\int_{a_1}^{b_1}\int_{a_2}^{b_2}\int_{a_3}^{b_3} f(r,s,t)dtdsdr = \sum_{i=0}^{n-1}\sum_{j=0}^{m-1} X_i(x)Y_j(y)\int_{a_3}^{b_3}\frac{\partial^{i+j}f}{\partial x^i \partial y^j}dt$$

$$+(-1)^m \sum_{i=0}^{n-1} X_i(x)\int_{a_2}^{b_2} Q_m(y,s)\left(\int_{a_3}^{b_3}\frac{\partial^{i+m}f}{\partial x^i \partial s^m}dt\right)ds$$

$$+(-1)^n \sum_{j=0}^{m-1} Y_j(y)\int_{a_1}^{b_1} P_n(x,r)\left(\int_{a_3}^{b_3}\frac{\partial^{j+n}f}{\partial y^j \partial r^n}dt\right)dr$$

$$+(-1)^{n+m}\int_{a_1}^{b_1}\int_{a_2}^{b_2} P_n(x,r)Q_m(y,s)\left(\int_{a_3}^{b_3}\frac{\partial^{n+m}f}{\partial s^m \partial r^n}dt\right)dsdr. \tag{4.3.19}$$

Also, from (4.3.13) we have

$$\int_{a_3}^{b_3}\frac{\partial^{i+j}f}{\partial x^i \partial y^j}dt = \sum_{k=0}^{p-1} Z_k(z)\frac{\partial^k}{\partial z^k}\left(\frac{\partial^{i+j}f}{\partial x^i \partial y^j}\right)$$

$$+(-1)^p\int_{a_3}^{b_3} S_p(z,t)\frac{\partial^p}{\partial t^p}\left(\frac{\partial^{i+j}f}{\partial x^i \partial y^j}\right)dt, \tag{4.3.20}$$

$$\int_{a_3}^{b_3}\frac{\partial^{i+m}f}{\partial x^i \partial s^m}dt = \sum_{k=0}^{p-1} Z_k(z)\frac{\partial^k}{\partial z^k}\left(\frac{\partial^{i+m}f}{\partial x^i \partial s^m}\right)$$

$$+(-1)^p\int_{a_3}^{b_3} S_p(z,t)\frac{\partial^p}{\partial t^p}\left(\frac{\partial^{i+m}f}{\partial x^i \partial s^m}\right)dt, \tag{4.3.21}$$

$$\int_{a_3}^{b_3}\frac{\partial^{j+n}f}{\partial y^j \partial r^n}dt = \sum_{k=0}^{p-1} Z_k(z)\frac{\partial^k}{\partial z^k}\left(\frac{\partial^{j+n}f}{\partial y^j \partial r^n}\right)$$

$$+(-1)^p\int_{a_3}^{b_3} S_p(z,t)\frac{\partial^p}{\partial t^p}\left(\frac{\partial^{j+n}f}{\partial y^j \partial r^n}\right)dt, \tag{4.3.22}$$

$$\int_{a_3}^{b_3}\frac{\partial^{n+m}f}{\partial s^m \partial r^n}dt = \sum_{k=0}^{p-1} Z_k(z)\frac{\partial^k}{\partial z^k}\left(\frac{\partial^{n+m}f}{\partial r^n \partial s^m}\right)$$

$$+(-1)^p\int_{a_3}^{b_3} S_p(z,t)\frac{\partial^p}{\partial t^p}\left(\frac{\partial^{n+m}f}{\partial r^n \partial s^m}\right)dt. \tag{4.3.23}$$

Using (4.3.20)–(4.3.23) in (4.3.19), we get the required identity in (4.3.9).

The following result deals with the Ostrowski-type inequality investigated by Sofo in [148].

Theorem 4.3.2. Assume that the hypotheses of Lemma 4.3.1 hold. Then we have the inequality

$$|V(x,y,z)| \leqslant \begin{cases} \left\| \dfrac{\partial^{n+m+p} f}{\partial r^n \partial s^n \partial t^p} \right\|_\infty N_1(x,y,z), & \text{if } \dfrac{\partial^{n+m+p} f(r,s,t)}{\partial r^n \partial s^n \partial t^p} \in L_\infty(H); \\ \left\| \dfrac{\partial^{n+m+p} f}{\partial r^n \partial s^n \partial t^p} \right\|_\alpha N_2(x,y,z), & \text{if } \dfrac{\partial^{n+m+p} f(r,s,t)}{\partial r^n \partial s^n \partial t^p} \in L_\alpha(H); \\ & \quad \dfrac{1}{\alpha} + \dfrac{1}{\beta} = 1, \ \alpha > 1; \\ \left\| \dfrac{\partial^{n+m+p} f}{\partial r^n \partial s^n \partial t^p} \right\|_1 N_3(x,y,z), & \text{if } \dfrac{\partial^{n+m+p} f(r,s,t)}{\partial r^n \partial s^n \partial t^p} \in L_1(H); \end{cases} \tag{4.3.24}$$

for all $(x,y,z) \in H$, where

$$\left\| \frac{\partial^{n+m+p} f}{\partial r^n \partial s^n \partial t^p} \right\|_\infty = \sup_{(r,s,t)\in H} \left| \frac{\partial^{n+m+p} f(r,s,t)}{\partial r^n \partial s^n \partial t^p} \right| < \infty,$$

$$\left\| \frac{\partial^{n+m+p} f}{\partial r^n \partial s^n \partial t^p} \right\|_\alpha = \left(\int_{a_1}^{b_1} \int_{a_2}^{b_2} \int_{a_3}^{b_3} \left| \frac{\partial^{n+m+p} f(r,s,t)}{\partial r^n \partial s^n \partial t^p} \right|^\alpha dtdsdr \right)^{\frac{1}{\alpha}} < \infty,$$

and

$$N_1(x,y,z) = \left[\frac{(x-a_1)^{n+1} + (b_1-x)^{n+1}}{(n+1)!} \right]$$

$$\times \left[\frac{(y-a_2)^{m+1} + (b_2-y)^{m+1}}{(m+1)!} \right] \left[\frac{(z-a_3)^{p+1} + (b_3-z)^{p+1}}{(p+1)!} \right],$$

$$N_2(x,y,z) = \frac{1}{n!m!p!} \left[\frac{(x-a_1)^{n\beta+1} + (b_1-x)^{n\beta+1}}{n\beta+1} \right]^{\frac{1}{\beta}}$$

$$\times \left[\frac{(y-a_2)^{m\beta+1} + (b_2-y)^{m\beta+1}}{m\beta+1} \right]^{\frac{1}{\beta}} \left[\frac{(z-a_3)^{p\beta+1} + (b_3-z)^{p\beta+1}}{p\beta+1} \right]^{\frac{1}{\beta}},$$

$$N_3(x,y,z) = \frac{1}{8} \frac{1}{n!m!p!} \left[(x-a_1)^n + (b_1-x)^n + |(x-a_1)^n - (b_1-x)^n| \right]$$

$$\times \left[(y-a_2)^m + (b_2-y)^m + |(y-a_2)^m - (b_2-y)^m| \right]$$

$$\times \left[(z-a_3)^p + (b_3-z)^p + |(z-a_3)^p - (b_3-z)^p| \right].$$

Proof. From the hypotheses, the identity (4.3.9) holds. From (4.3.9), we have

$$|V(x,y,z)| \leqslant \int_{a_1}^{b_1}\int_{a_2}^{b_2}\int_{a_3}^{b_3} |P_n(x,r)Q_m(y,s)S_p(z,t)| \left|\frac{\partial^{n+m+p} f(r,s,t)}{\partial r^n \partial s^m \partial t^p}\right| dtdsdr$$

$$= U(x,y,z) \quad \text{(say)}. \tag{4.3.25}$$

Now, using Hölder's inequality and property of the modulus and integral, we have

$$U(x,y,z) \leqslant$$

$$\begin{cases} \left\|\frac{\partial^{n+m+p} f}{\partial r^n \partial s^m \partial t^p}\right\|_\infty \int_{a_1}^{b_1}\int_{a_2}^{b_2}\int_{a_3}^{b_3} |P_n(x,r)Q_m(y,s)S_p(z,t)|\, dtdsdr; \\ \left\|\frac{\partial^{n+m+p} f}{\partial r^n \partial s^m \partial t^p}\right\|_\alpha \left(\int_{a_1}^{b_1}\int_{a_2}^{b_2}\int_{a_3}^{b_3} |P_n(x,r)Q_m(y,s)S_p(z,t)|^\beta dtdsdr\right)^{\frac{1}{\beta}}, \\ \qquad \frac{1}{\alpha}+\frac{1}{\beta}=1,\ \alpha>1; \\ \left\|\frac{\partial^{n+m+p} f}{\partial r^n \partial s^m \partial t^p}\right\|_1 \sup_{(r,s,t)\in H} |P_n(x,r)Q_m(y,s)S_p(z,t)|. \end{cases} \tag{4.3.26}$$

Using (4.3.6), (4.3.7) and (4.3.8), we have

$$\int_{a_1}^{b_1}\int_{a_2}^{b_2}\int_{a_3}^{b_3} |P_n(x,r)Q_m(y,s)S_p(z,t)|\, dtdsdr$$

$$= \int_{a_1}^{b_1} |P_n(x,r)|dr \int_{a_2}^{b2} |Q_m(y,s)|ds \int_{a_3}^{b_3} |S_p(z,t)|dt$$

$$= \left[\int_{a_1}^{x} \frac{(r-a_1)^n}{n!}dr + \int_{x}^{b_1} \frac{(b_1-r)^n}{n!}dr\right]$$

$$\times \left[\int_{a_2}^{y} \frac{(s-a_2)^m}{m!}ds + \int_{y}^{b_2} \frac{(b_2-s)^m}{m!}ds\right]\left[\int_{a_3}^{z} \frac{(t-a_3)^p}{p!}dt + \int_{z}^{b_3} \frac{(b_3-t)^p}{p!}dt\right]$$

$$= N_1(x,y,z). \tag{4.3.27}$$

Further, by using (4.3.6), (4.3.7) and (4.3.8), we have

$$\left(\int_{a_1}^{b_1}\int_{a_2}^{b_2}\int_{a_3}^{b_3} |P_n(x,r)Q_m(y,s)S_p(z,t)|^\beta dtdsdr\right)^{\frac{1}{\beta}}$$

$$= \left(\int_{a_1}^{b_1} |P_n(x,r)|^\beta dr\right)^{\frac{1}{\beta}} \left(\int_{a_2}^{b2} |Q_m(y,s)|^\beta ds\right)^{\frac{1}{\beta}} \left(\int_{a_3}^{b_3} |S_p(z,t)|^\beta dt\right)^{\frac{1}{\beta}}$$

$$= \frac{1}{n!m!p!}\left[\int_{a_1}^{x} (r-a_1)^{n\beta}dr + \int_{x}^{b_1} (b_1-r)^{n\beta}dr\right]^{\frac{1}{\beta}}$$

$$\times\left[\int_{a_2}^{y}(s-a_2)^{m\beta}ds+\int_{y}^{b_2}(b_2-s)^{m\beta}ds\right]^{\frac{1}{\beta}}\left[\int_{a_3}^{z}(t-a_3)^{p\beta}dt+\int_{z}^{b_3}(b_3-t)^{p\beta}dt\right]^{\frac{1}{\beta}}$$

$$=N_2(x,y,z). \tag{4.3.28}$$

Finally, using (4.3.6), (4.3.7) and (4.3.8), we have

$$\sup_{(r,s,t)\in H}|P_n(x,r)Q_m(y,s)S_p(z,t)|=\sup_{r\in[a_1,b_1]}|P_n(x,r)|\sup_{s\in[a_2,b_2]}|Q_m(y,s)|\sup_{t\in[a_3,b_3]}|S_p(z,t)|$$

$$=\max\left\{\frac{(x-a_1)^n}{n!},\frac{(b_1-x)^n}{n!}\right\}\max\left\{\frac{(y-a_2)^m}{m!},\frac{(b_2-y)^m}{m!}\right\}\max\left\{\frac{(z-a_3)^p}{p!},\frac{(b_3-z)^p}{p!}\right\}$$

$$=N_3(x,y,z), \tag{4.3.29}$$

here we have used the fact that

$$\max\{A,B\}=\frac{a+b}{2}+\left|\frac{b-a}{2}\right|.$$

The required inequality in (4.3.24) follows from (4.3.25)–(4.3.29) and the proof is complete.

In our further discussion, we make use of the following notation to simplify the details of presentation. The partial derivatives of a function $e=e(x,y,z):H\to\mathbb{R}$ are denoted by $D_1e=\frac{\partial e}{\partial x}$, $D_2e=\frac{\partial e}{\partial y}$, $D_3e=\frac{\partial e}{\partial z}$, $D_1D_2e=\frac{\partial^2 e}{\partial x\partial y}$, $D_2D_3e=\frac{\partial^2 e}{\partial y\partial z}$, $D_3D_1e=\frac{\partial^2 e}{\partial z\partial x}$ and $D_3D_2D_1e=\frac{\partial^3 e}{\partial z\partial y\partial x}$. We denote by $F(H)$ the class of continuous functions $e:H\to\mathbb{R}$ for which D_1e, D_2e, D_3e, D_1D_2e, D_2D_3e, D_3D_1e, $D_3D_2D_1e$ exist and are continuous on H. For (x,y,z), $(r,s,t)\in H$ and some suitable functions $p,f,\ g,\ h:H\to\mathbb{R}$, we set

$$\bar{l}=(b_1-a_1)(b_2-a_2)(b_3-a_3),$$

$$\bar{I}[p]=\int_r^x\int_s^y\int_t^z D_3D_2D_1p(u,v,w)dwdvdu,$$

$$J[p]=(b_2-a_2)(b_3-a_3)\int_{a_1}^{b_1}p(r,y,z)dr$$

$$+(b_1-a_1)(b_3-a_3)\int_{a_2}^{b_2}p(x,s,z)ds+(b_1-a_1)(b_2-a_2)\int_{a_3}^{b_3}p(x,y,t)dt,$$

$$L[p]=(b_3-a_3)\int_{a_1}^{b_1}\int_{a_2}^{b_2}p(r,s,z)dsdr$$

$$+(b_2-a_2)\int_{a_1}^{b_1}\int_{a_3}^{b_3}p(r,y,t)dtdr+(b_1-a_1)\int_{a_2}^{b_2}\int_{a_3}^{b_3}p(x,s,t)dtds,$$

$$\bar{A}(f,g,h;J,L;\bar{l})(x,y,z)=f(x,y,z)g(x,y,z)h(x,y,z)$$

$$-\frac{1}{3\bar{l}}\left[g(x,y,z)h(x,y,z)\left\{J[f]-L[f]+\int_{a_1}^{b_1}\int_{a_2}^{b_2}\int_{a_3}^{b_3}f(r,s,t)dtdsdr\right\}\right.$$

$$+h(x,y,z)f(x,y,z)\left\{J[g]-L[g]+\int_{a_1}^{b_1}\int_{a_2}^{b_2}\int_{a_3}^{b_3}g(r,s,t)dtdsdr\right\}$$

$$\left.+f(x,y,z)g(x,y,z)\left\{J[h]-L[h]+\int_{a_1}^{b_1}\int_{a_2}^{b_2}\int_{a_3}^{b_3}h(r,s,t)dtdsdr\right\}\right],$$

$$\overline{B}(f,g,h;\overline{I})(x,y,z)=g(x,y,z)h(x,y,z)\int_{a_1}^{b_1}\int_{a_2}^{b_2}\int_{a_3}^{b_3}\overline{I}[f]dtdsdr$$

$$+h(x,y,z)f(x,y,z)\int_{a_1}^{b_1}\int_{a_2}^{b_2}\int_{a_3}^{b_3}\overline{I}[g]dtdsdr+f(x,y,z)g(x,y,z)\int_{a_1}^{b_1}\int_{a_2}^{b_2}\int_{a_3}^{b_3}\overline{I}[h]dtdsdr.$$

In [126], Pachpatte has established the following Ostrowski-type inequality for triple integrals.

Theorem 4.3.3. Let $f\in F(H)$. Then

$$\left|f(x,y,z)-\left[\frac{1}{b_1-a_1}\int_{a_1}^{b_1}f(r,y,z)dr+\frac{1}{b_2-a_2}\int_{a_2}^{b_2}f(x,s,z)ds\right.\right.$$

$$\left.+\frac{1}{b_3-a_3}\int_{a_3}^{b_3}f(x,y,t)dt\right]+\left[\frac{1}{(b_1-a_1)(b_2-a_2)}\int_{a_1}^{b_1}\int_{a_2}^{b_2}f(r,s,z)dsdr\right.$$

$$\left.+\frac{1}{(b_1-a_1)(b_3-a_3)}\int_{a_1}^{b_1}\int_{a_3}^{b_3}f(r,y,t)dtdr+\frac{1}{(b_2-a_2)(b_3-a_3)}\int_{a_2}^{b_2}\int_{a_3}^{b_3}f(x,s,t)dtds\right]$$

$$\left.-\frac{1}{\bar{l}}\int_{a_1}^{b_1}\int_{a_2}^{b_2}\int_{a_3}^{b_3}f(r,s,t)dtdsdr\right|\leqslant\frac{1}{\bar{l}}\int_{a_1}^{b_1}\int_{a_2}^{b_2}\int_{a_3}^{b_3}\left|\overline{I}[f]\right|dtdsdr, \tag{4.3.30}$$

for all $(x,y,z)\in H$.

Proof. First, we prove the following identity

$$\overline{I}[f]=f(x,y,z)-[f(r,y,z)+f(x,s,z)+f(x,y,t)]$$

$$+[f(r,s,z)+f(r,y,t)+f(x,s,t)]-f(r,s,t), \tag{4.3.31}$$

for (x,y,z), $(r,s,t)\in H$, where

$$\overline{I}[f]=\int_r^x\int_s^y\int_t^z D_3D_2D_1f(u,v,w)dwdvdu. \tag{4.3.32}$$

From (4.3.32), it is easy to observe that

$$\overline{I}[f]=\int_r^x\int_s^y D_2D_1f(u,v,z)dvdu-\int_r^x\int_s^y D_2D_1f(u,v,t)dvdu=I_1[f]-I_2[f]. \tag{4.3.33}$$

By simple computation, we have

$$I_1[f] = \int_r^x \int_s^y D_2 D_1 f(u,v,z)\,dv\,du$$
$$= \int_r^x D_1 f(u,y,z)\,du - \int_r^x D_1 f(u,s,z)\,du$$
$$= f(x,y,z) - f(r,y,z) - f(x,s,z) + f(r,s,z). \tag{4.3.34}$$

Similarly, we have

$$I_2[f] = \int_r^x \int_s^y D_2 D_1 f(u,v,t)\,dv\,du$$
$$= f(x,y,t) - f(r,y,t) - f(x,s,t) + f(r,s,t). \tag{4.3.35}$$

Using (4.3.34) and (4.3.35) in (4.3.33), we get (4.3.31).

Integrating both sides of (4.3.31) with respect to (r,s,t) over H and rewriting, we get

$$f(x,y,z) - \left[\frac{1}{b_1-a_1}\int_{a_1}^{b_1} f(r,y,z)\,dr + \frac{1}{b_2-a_2}\int_{a_2}^{b_2} f(x,s,z)\,ds\right.$$
$$\left. + \frac{1}{b_3-a_3}\int_{a_3}^{b_3} f(x,y,t)\,dt\right] + \left[\frac{1}{(b_1-a_1)(b_2-a_2)}\int_{a_1}^{b_1}\int_{a_2}^{b_2} f(r,s,z)\,ds\,dr\right.$$
$$\left. + \frac{1}{(b_1-a_1)(b_3-a_3)}\int_{a_1}^{b_1}\int_{a_3}^{b_3} f(r,y,t)\,dt\,dr + \frac{1}{(b_2-a_2)(b_3-a_3)}\int_{a_2}^{b_2}\int_{a_3}^{b_3} f(x,s,t)\,dt\,ds\right]$$
$$-\frac{1}{\bar{l}}\int_{a_1}^{b_1}\int_{a_2}^{b_2}\int_{a_3}^{b_3} f(r,s,t)\,dt\,ds\,dr = \frac{1}{\bar{l}}\int_{a_1}^{b_1}\int_{a_2}^{b_2}\int_{a_3}^{b_3} \left\{\bar{I}[f]\right\} dt\,ds\,dr, \tag{4.3.36}$$

for $(x,y,z) \in H$. From (4.3.36) and using the properties of modulus and integrals, we get the required inequality in (4.3.30). The proof is complete.

The following corollary holds.

Corollary 4.3.1. Let f be as in Theorem 4.3.3, then

$$\left| f\left(\frac{a_1+b_1}{2}, \frac{a_2+b_2}{2}, \frac{a_3+b_3}{2}\right) - \left[\frac{1}{b_1-a_1}\int_{a_1}^{b_1} f\left(r, \frac{a_2+b_2}{2}, \frac{a_3+b_3}{2}\right) dr\right.\right.$$
$$\left. + \frac{1}{b_2-a_2}\int_{a_2}^{b_2} f\left(\frac{a_1+b_1}{2}, s, \frac{a_3+b_3}{2}\right) ds + \frac{1}{b_3-a_3}\int_{a_3}^{b_3} f\left(\frac{a_1+b_1}{2}, \frac{a_2+b_2}{2}, t\right) dt\right]$$
$$+ \left[\frac{1}{(b_1-a_1)(b_2-a_2)}\int_{a_1}^{b_1}\int_{a_2}^{b_2} f\left(r, s, \frac{a_3+b_3}{2}\right) ds\,dr\right.$$
$$+ \frac{1}{(b_1-a_1)(b_3-a_3)}\int_{a_1}^{b_1}\int_{a_3}^{b_3} f\left(r, \frac{a_2+b_2}{2}, t\right) dt\,dr$$

$$+\frac{1}{(b_2-a_2)(b_3-a_3)}\int_{a_2}^{b_2}\int_{a_3}^{b_3} f\left(\frac{a_1+b_1}{2},s,t\right)dtds\Bigg]$$

$$-\frac{1}{\bar{l}}\int_{a_1}^{b_1}\int_{a_2}^{b_2}\int_{a_3}^{b_3} f(r,s,t)dtdsdr\Bigg| \leqslant \frac{\bar{l}}{64}\|D_3D_2D_1f\|_\infty, \tag{4.3.37}$$

where

$$\|D_3D_2D_1f\|_\infty = \sup_{(u,v,w)\in H} |D_3D_2D_1f(u,v,w)| < \infty.$$

By taking $x=\frac{a_1+b_1}{2}$, $y=\frac{a_2+b_2}{2}$, $z=\frac{a_3+b_3}{2}$ in (4.3.30) and simple computation, we get the desired inequality in (4.3.37).

Before concluding this section, we shall give the following generalization of Theorem 4.3.3 recently investigated by Pachpatte in [134].

Theorem 4.3.4. Let $f,\ g,\ h \in F(H)$. Then

$$\left|\bar{A}\left(f,g,h;J,L;\bar{l}\right)(x,y,z)\right| \leqslant \frac{1}{3\bar{l}}\bar{B}\left(|f|,|g|,|h|;|\bar{I}|\right)(x,y,z), \tag{4.3.38}$$

for all $(x,y,z)\in H$.

Proof. From the hypotheses, we have the following identities (see, the proof of Theorem 4.3.3):

$$\bar{I}[f] = f(x,y,z) - [f(r,y,z)+f(x,s,z)+f(x,y,t)]$$

$$+[f(r,s,z)+f(r,y,t)+f(x,s,t)] - f(r,s,t), \tag{4.3.39}$$

$$\bar{I}[g] = g(x,y,z) - [g(r,y,z)+g(x,s,z)+g(x,y,t)]$$

$$+[g(r,s,z)+g(r,y,t)+g(x,s,t)] - g(r,s,t), \tag{4.3.40}$$

$$\bar{I}[h] = h(x,y,z) - [h(r,y,z)+h(x,s,z)+h(x,y,t)]$$

$$+[h(r,s,z)+h(r,y,t)+h(x,s,t)] - h(r,s,t), \tag{4.3.41}$$

for all $(x,y,z),(r,s,t)\in H$.

Multiplying both sides of (4.3.39), (4.3.40) and (4.3.41) by $g(x,y,z)h(x,y,z)$, $h(x,y,z)f(x,y,z)$ and $f(x,y,z)g(x,y,z)$ respectively and adding the resulting identities, we get

$$3f(x,y,z)g(x,y,z)h(x,y,z) - g(x,y,z)h(x,y,z)\left\{[f(r,y,z)+f(x,s,z)+f(x,y,t)]\right.$$

$$\left.-[f(r,s,z)+f(r,y,t)+f(x,s,t)]+f(r,s,t)\right\}$$

$$-h(x,y,z)f(x,y,z)\{[g(r,y,z)+g(x,s,z)+g(x,y,t)]$$

$$-[g(r,s,z)+g(r,y,t)+g(x,s,t)]+g(r,s,t)\}$$

$$-f(x,y,z)g(x,y,z)\{[h(r,y,z)+h(x,s,z)+h(x,y,t)]$$

$$-[h(r,s,z)+h(r,y,t)+h(x,s,t)]+h(r,s,t)\}$$

$$= g(x,y,z)h(x,y,z)\overline{I}[f]+h(x,y,z)f(x,y,z)\overline{I}[g]+f(x,y,z)g(x,y,z)\overline{I}[h]. \tag{4.3.42}$$

Integrating both sides of (4.3.42) with respect to (r,s,t) over H and rewriting, we have

$$\overline{A}\left(f,g,h;J,L;\overline{l}\right)(x,y,z) = \frac{1}{3\overline{l}}\overline{B}(f,g,h;\overline{I})(x,y,z). \tag{4.3.43}$$

From (4.3.43) and using the properties of modulus, we get the desired inequality in (4.3.38). The proof is complete.

Remark 4.3.2 If we take $h(x,y,z)=1$ and hence $\overline{I}[h]=0$ in Theorem 4.3.4, then by elementary calculation, we get

$$\Bigg|f(x,y,z)g(x,y,z)-\frac{1}{2\overline{l}}\Bigg[g(x,y,z)\left\{J[f]-L[f]+\int_{a_1}^{b_1}\int_{a_2}^{b_2}\int_{a_3}^{b_3}f(r,s,t)dtdsdr\right\}$$

$$+\,f(x,y,z)\left\{J[g]-L[g]+\int_{a_1}^{b_1}\int_{a_2}^{b_2}\int_{a_3}^{b_3}g(r,s,t)dtdsdr\right\}\Bigg]\Bigg|$$

$$\leqslant \frac{1}{2\overline{l}}\Bigg[|g(x,y,z)|\int_{a_1}^{b_1}\int_{a_2}^{b_2}\int_{a_3}^{b_3}\left|\overline{I}[f]\right|dtdsdr$$

$$+|f(x,y,z)|\int_{a_1}^{b_1}\int_{a_2}^{b_2}\int_{a_3}^{b_3}\left|\overline{I}[g]\right|dtdsdr\Bigg], \tag{4.3.44}$$

for all $(x,y,z)\in H$. Further, by taking $g(x,y,z)=1$ and hence $\overline{I}[g]=0$ in (4.3.44) and by simple computation, we get (4.3.30).

Remark 4.3.3. Integrating both sides of (4.3.43) with respect to (x,y,z) over H and rewriting, we have

$$\overline{T}(f,g,h;J,L;\overline{l}) = \frac{1}{3\overline{l}^2}\int_{a_1}^{b_1}\int_{a_2}^{b_2}\int_{a_3}^{b_3}\overline{B}(f,g,h;\overline{I})(x,y,z)dzdydx, \tag{4.3.45}$$

where

$$\overline{T}(f,g,h;J,L;\overline{l}) = \frac{1}{\overline{l}}\int_{a_1}^{b_1}\int_{a_2}^{b_2}\int_{a_3}^{b_3}f(x,y,z)g(x,y,z)h(x,y,z)dzdydx$$

$$-\frac{1}{3\overline{l}^2}\int_{a_1}^{b_1}\int_{a_2}^{b_2}\int_{a_3}^{b_3}[g(x,y,z)h(x,y,z)(J[f]-L[f])$$

$$+h(x,y,z)f(x,y,z)(J[g]-L[g]) + f(x,y,z)g(x,y,z)(J[h]-L[h])]\,dzdydx$$

$$-\frac{1}{3}\left[\left(\frac{1}{\bar{l}}\int_{a_1}^{b_1}\int_{a_2}^{b_2}\int_{a_3}^{b_3} g(x,y,z)h(x,y,z)dzdydx\right)\left(\frac{1}{\bar{l}}\int_{a_1}^{b_1}\int_{a_2}^{b_2}\int_{a_3}^{b_3} f(x,y,z)dzdydx\right)\right.$$

$$+\left(\frac{1}{\bar{l}}\int_{a_1}^{b_1}\int_{a_2}^{b_2}\int_{a_3}^{b_3} h(x,y,z)f(x,y,z)dzdydx\right)\left(\frac{1}{\bar{l}}\int_{a_1}^{b_1}\int_{a_2}^{b_2}\int_{a_3}^{b_3} g(x,y,z)dzdydx\right)$$

$$\left.+\left(\frac{1}{\bar{l}}\int_{a_1}^{b_1}\int_{a_2}^{b_2}\int_{a_3}^{b_3} f(x,y,z)g(x,y,z)dzdydx\right)\left(\frac{1}{\bar{l}}\int_{a_1}^{b_1}\int_{a_2}^{b_2}\int_{a_3}^{b_3} h(x,y,z)dzdydx\right)\right].$$

From (4.3.45) and using the properties of modulus and integrals, we get the following Grüss-type inequality

$$\left|\overline{T}(f,g,h;J,L;\bar{l})\right| \leqslant \frac{1}{3\bar{l}^2}\int_{a_1}^{b_1}\int_{a_2}^{b_2}\int_{a_3}^{b_3}\overline{B}(|f|,|g|,|h|;|\bar{l}|)(x,y,z)dzdydx. \tag{4.3.46}$$

For similar results see [89].

4.4 Ostrowski-type inequalities in several variables

In this section, we offer some multivariate inequalities related to the Ostrowski's inequality, investigated in [44,76,87].

In 1975, Milovanović [76] first proved the following multivariate version of the Ostrowski's inequality given in (7).

Theorem 4.4.1. Let $D = \{(x_1,\ldots,x_n) : a_i < x_i < b_i\ (i = 1,\ldots,n)\}$ and let $\overline{D}$ be the closure of D. Let $f : \mathbb{R}^n \to \mathbb{R}$ be a differentiable function defined on $\overline{D}$ and let $\left|\frac{\partial f}{\partial x_i}\right| \leqslant M_i$ ($M_i > 0$; $i = 1,\ldots,n$) in D. Then, for every $x = (x_1,\ldots,x_n) \in \overline{D}$,

$$\left| f(x) - \frac{1}{\prod\limits_{i=1}^{n}(b_i-a_i)}\int_D f(y)dy\right| \leqslant \sum_{i=1}^{n}\left(\frac{1}{4}+\frac{\left(x_i-\frac{a_i+b_i}{2}\right)^2}{(b_i-a_i)^2}\right)(b_i-a_i)M_i, \tag{4.4.1}$$

where, $\int_D f(y)dy = \int_{a_1}^{b_1}\cdots\int_{a_n}^{b_n} f(y_1,\ldots,y_n)dy_n\cdots dy_1$.

Proof. Let $x = (x_1,\ldots,x_n)$ and $y = (y_1,\ldots,y_n)$ $(x \in \overline{D}, y \in D)$. From the n-dimensional version of the mean value Theorem (see [146, p. 174], [76, p. 121]), we have

$$f(x) - f(y) = \sum_{i=1}^{n}\frac{\partial f(c)}{\partial x_i}(x_i - y_i), \tag{4.4.2}$$

where, $c = (y_1 + \alpha(x_1 - y_1),\ldots,y_n + \alpha(x_n - y_n))$ $(0 < \alpha < 1)$.

Integrating both sides of (4.4.2) with respect to y over D, we obtain

$$f(x)\operatorname{mes}D-\int_D f(y)dy=\sum_{i=1}^{n}\int_D\frac{\partial f(c)}{\partial x_i}(x_i-y_i)dy, \tag{4.4.3}$$

where $\operatorname{mes}D=\prod_{i=1}^{n}(b_i-a_i)$. From (4.4.3), it follows that

$$\begin{aligned}\left|f(x)\operatorname{mes}D-\int_D f(y)dy\right| &\leqslant \sum_{i=1}^{n}\left|\int_D\frac{\partial f(c)}{\partial x_i}(x_i-y_i)dy\right|\\ &\leqslant \sum_{i=1}^{n}\int_D\left|\frac{\partial f(c)}{\partial x_i}\right||x_i-y_i|dy\leqslant\sum_{i=1}^{n}M_i\int_D|x_i-y_i|dy.\end{aligned} \tag{4.4.4}$$

Since

$$\int_{a_i}^{b_i}|x_i-y_i|dy_i=\frac{1}{4}(b_i-a_i)^2+\left(x_i-\frac{a_i+b_i}{2}\right)^2,$$

we have

$$\begin{aligned}\int_D|x_i-y_i|dy &= \frac{\operatorname{mes}D}{b_i-a_i}\int_{a_i}^{b_i}|x_i-y_i|dy_i\\ &= (\operatorname{mes}D)(b_i-a_i)\left(\frac{1}{4}+\frac{\left(x_i-\frac{a_i+b_i}{2}\right)^2}{(b_i-a_i)^2}\right).\end{aligned} \tag{4.4.5}$$

Using (4.4.5) in (4.4.4) and the fact that $\operatorname{mes}D>0$, we get the required inequality in (4.4.1). The proof is complete.

In [44], Dragomir, Barnett and Cerone established the following Ostrowski-type inequality for multivariate mappings of the r-Hölder type.

Theorem 4.4.2. Let $B=\prod_{i=1}^{n}[a_i,b_i]\subset\mathbb{R}^n$. Assume that the mapping $f:B\to\mathbb{R}$ satisfies the following r-Hölder type condition

$$|f(x)-f(y)|\leqslant\sum_{i=1}^{n}L_i|x_i-y_i|^{r_i}\quad(L_i\geqslant 0, i=1,\ldots,n), \tag{4.4.6}$$

for all $x=(x_1,\ldots,x_n)$, $y=(y_1,\ldots,y_n)$ in B, where $r_i\in(0,1]$, $i=1,\ldots,n$. Then we have the inequality

$$\begin{aligned}\left|f(x)-\frac{1}{\prod_{i=1}^{n}(b_i-a_i)}\int_B f(y)dy\right| &\leqslant\sum_{i=1}^{n}\frac{L_i}{r_i+1}\left[\left(\frac{x_i-a_i}{b_i-a_i}\right)^{r_i+1}+\left(\frac{b_i-x_i}{b_i-a_i}\right)^{r_i+1}\right](b_i-a_i)^{r_i}\\ &\leqslant\sum_{i=1}^{n}\frac{L_i}{r_i+1}(b_i-a_i)^{r_i},\end{aligned} \tag{4.4.7}$$

for all $x\in B$, where $\int_B f(y)dy=\int_{a_1}^{b_1}\cdots\int_{a_n}^{b_n}f(y_1,\ldots,y_n)dy_n\cdots dy_1$.

Proof. Using the properties of modulus and (4.4.6), we observe that

$$\left| f(x)\int_B dy - \int_B f(y)dy \right| \leqslant \int_B |f(x)-f(y)|dy$$
$$\leqslant \sum_{i=1}^{n} L_i \int_{a_1}^{b_1} \cdots \int_{a_n}^{b_n} |x_i - y_i|^{r_i} dy_n \cdots dy_1. \tag{4.4.8}$$

As

$$\int_B dy = \int_{a_1}^{b_1} \cdots \int_{a_n}^{b_n} dy_n \cdots dy_1 = \prod_{i=1}^{n}(b_i - a_i),$$

and

$$\int_{a_1}^{b_1} \cdots \int_{a_n}^{b_n} |x_i - y_i|^{r_i} dy_n \cdots dy_1 = \prod_{\substack{j=1\\ j\neq i}}^{n}(b_j - a_j) \int_{a_i}^{b_i} |x_i - y_i|^{r_i} dy_i$$

$$= \prod_{\substack{j=1\\ j\neq i}}^{n}(b_j - a_j) \left[\frac{(b_i - x_i)^{r_i+1} + (x_i - a_i)^{r_i+1}}{r_i + 1} \right]$$

$$= \prod_{j=1}^{n}(b_j - a_j) \frac{1}{r_i+1} \left[\left(\frac{x_i - a_i}{b_i - a_i} \right)^{r_i+1} + \left(\frac{b_i - x_i}{b_i - a_i} \right)^{r_i+1} \right] (b_i - a_i)^{r_i},$$

then dividing (4.4.8) by $\prod_{j=1}^{n}(b_j - a_j)$, we get the first part of (4.4.7).

Using the elementary inequality

$$(y-\alpha)^{p+1} + (\beta - y)^{p+1} \leqslant (\beta - \alpha)^{p+1},$$

for all $\alpha \leqslant y \leqslant \beta$ and $p > 0$, we get

$$\left(\frac{x_i - a_i}{b_i - a_i} \right)^{r_i+1} + \left(\frac{b_i - x_i}{b_i - a_i} \right)^{r_i+1} \leqslant 1, \quad i = 1, \ldots, n,$$

and the last part of (4.4.7) is also proved.

The following Corollaries hold.

Corollary 4.4.1. Under the assumptions of Theorem 4.4.2, we have the mid-point inequality

$$\left| f\left(\frac{a_1+b_1}{2}, \ldots, \frac{a_n+b_n}{2} \right) - \frac{1}{\prod_{i=1}^{n}(b_i - a_i)} \int_B f(y)dy \right| \leqslant \sum_{i=1}^{n} \frac{L_i (b_i - a_i)^{r_i}}{2^{r_i}(r_i + 1)}, \tag{4.4.9}$$

which is the best inequality we can get from (4.4.7).

Proof. Note that the mapping $h_p(y) : [\alpha,\beta] \to \mathbb{R}$, $h_p(y) = (y-\alpha)^{p+1} + (\beta - y)^{p+1}$ $(p > 0)$ has its infimum at $y_0 = \frac{\alpha+\beta}{2}$ and

$$\inf_{y\in[\alpha,\beta]} h_p(y) = \frac{(\beta-\alpha)^{p+1}}{2^p}.$$

Consequently, the best inequality we can get from (4.4.7) is the one for which $x_i = \frac{a_i+b_i}{2}$, giving the desired inequality (4.4.9).

Corollary 4.4.2. Under the assumptions of Theorem 4.4.2, we have the trapezoid-type inequality

$$\left| \frac{f(a_1,\dots,a_n)+f(b_1,\dots,b_n)}{2} - \frac{1}{\prod\limits_{i=1}^{n}(b_i-a_i)} \int_B f(y)dy \right| \leqslant \sum_{i=1}^{n} \frac{L_i(b_i-a_i)^{r_i}}{r_i+1}. \tag{4.4.10}$$

Proof. Put in (4.4.7), $x = (x_1,\dots,x_n) = (a_1,\dots,a_n)$ and then $x = (x_1,\dots,x_n) = (b_1,\dots,b_n)$, add the obtained inequalities and use the triangle inequality, to get (4.4.10).

An important particular case is one for which the mapping f is Lipschitzian, i.e.,

$$|f(x)-f(y)| \leqslant \sum_{i=1}^{n} L_i|x_i-y_i|, \tag{4.4.11}$$

for all $x,\ y \in B$ and $L_i \geqslant 0$, $i = 1,\dots,n$.

The following Corollary holds.

Corollary 4.4.3. Let f be a Lipschitzian mapping with constants L_i. Then we have

$$\left| f(x) - \frac{1}{\prod\limits_{i=1}^{n}(b_i-a_i)} \int_B f(y)dy \right| \leqslant \sum_{i=1}^{n} L_i \left[\frac{1}{4} + \left(\frac{x-\frac{a_i+b_i}{2}}{b_i-a_i} \right)^2 \right] (b_i-a_i), \tag{4.4.12}$$

for all $x \in B$. The constant $\frac{1}{4}$, in all the brackets, is the best possible.

Proof. Choose $r_i = 1$ $(i = 1,\dots,n)$ in (4.4.7), to get

$$\left| f(x) - \frac{1}{\prod\limits_{i=1}^{n}(b_i-a_i)} \int_B f(y)dy \right| \leqslant \frac{1}{2}\sum_{i=1}^{n} L_i \left[\left(\frac{x_i-a_i}{b_i-a_i} \right)^2 + \left(\frac{b_i-x_i}{b_i-a_i} \right)^2 \right] (b_i-a_i).$$

A simple computation shows that

$$\frac{1}{2}\left[\left(\frac{x_i-a_i}{b_i-a_i} \right)^2 + \left(\frac{b_i-x_i}{b_i-a_i} \right)^2 \right] = \frac{1}{4} + \left(\frac{x-\frac{a_i+b_i}{2}}{b_i-a_i} \right)^2,$$

for $i = 1,\dots,n$, giving the desired inequality (4.4.12).

To prove the sharpness of the constants $\frac{1}{4}$, assume that the inequality (4.4.12) holds for some constants $c_i > 0$, i.e.,

$$\left| f(x) - \frac{1}{\prod_{i=1}^{n}(b_i - a_i)} \int_B f(y)dy \right| \leqslant \sum_{i=1}^{n} L_i \left[c_i + \left(\frac{x - \frac{a_i+b_i}{2}}{b_i - a_i} \right)^2 \right] (b_i - a_i), \tag{4.4.13}$$

for all $x \in B$.

Choose $f(x) = x_i (i = 1, \ldots, n)$. Then by (4.4.13), we get

$$\left| x_i - \frac{a_i + b_i}{2} \right| \leqslant \left[c_i + \left(\frac{x - \frac{a_i+b_i}{2}}{b_i - a_i} \right)^2 \right] (b_i - a_i),$$

for all $x_i \in [a_i, b_i]$. Put $x_i = a_i$, to get

$$\frac{b_i - a_i}{2} \leqslant \left(c_i + \frac{1}{4} \right) (b_i - a_i),$$

from which we deduce $c_i \geqslant \frac{1}{4}$, and the sharpness of $\frac{1}{4}$ is proved.

The multivariate Ostrowski-type inequality investigated by Pachpatte in [87], is embodied in the following theorem.

Theorem 4.4.3. Let the set B and the n-fold integral be as defined in Theorem 4.4.2. Let $f : B \to \mathbb{R}$ be a differentiable function and

$$v_i(x_1, \ldots, x_{i-1}, x_{i+1}, \ldots, x_n) = f(x)\big|_{x_i = a_i} + f(x)\big|_{x_i = b_i},$$

for $i = 1, \ldots, n$. Then

$$\left| \int_B f(x)dx - \frac{1}{2n} \sum_{i=1}^{n} \int_B v_i(x_1, \ldots, x_{i-1}, x_{i+1}, \ldots, x_n)dx \right|$$

$$\leqslant \frac{1}{2n} \left(\sum_{i=1}^{n} (b_i - a_i) \int_B \left| \frac{\partial f(x)}{\partial x_i} \right| dx \right). \tag{4.4.14}$$

Proof. For $x \in B$ it is easy to observe that the following identities hold (see [108, p. 392]):

$$nf(x) = f(a_1, x_2, \ldots, x_n) + \cdots + f(x_1, \ldots, x_{n-1}, a_n)$$

$$+ \int_{a_1}^{x_1} \frac{\partial}{\partial t_1} f(t_1, x_2, \ldots, x_n)dt_1 + \cdots + \int_{a_n}^{x_n} \frac{\partial}{\partial t_n} f(x_1, x_2, \ldots, x_{n-1}, t_n)dt_n, \tag{4.4.15}$$

and

$$nf(x) = f(b_1, x_2, \ldots, x_n) + \cdots + f(x_1, \ldots, x_{n-1}, b_n)$$

$$- \int_{x_1}^{b_1} \frac{\partial}{\partial t_1} f(t_1, x_2, \ldots, x_n)dt_1 - \cdots - \int_{x_n}^{b_n} \frac{\partial}{\partial t_n} f(x_1, x_2, \ldots, x_{n-1}, t_n)dt_n. \tag{4.4.16}$$

From (4.4.15) and (4.4.16), we get

$$f(x) = \frac{1}{2n}\left\{\sum_{i=1}^{n} v_i(x_1,\dots,x_{i-1},x_{i+1},\dots,x_n)\right.$$

$$+\left[\int_{a_1}^{x_1}\frac{\partial}{\partial t_1}f(t_1,x_2,\dots,x_n)dt_1 - \int_{x_1}^{b_1}\frac{\partial}{\partial t_1}f(t_1,x_2,\dots,x_n)dt_1\right]+\cdots$$

$$\vdots$$

$$\left.+\left[\int_{a_n}^{x_n}\frac{\partial}{\partial t_n}f(x_1,\dots,x_{n-1},t_n)dt_n - \int_{x_n}^{b_n}\frac{\partial}{\partial t_n}f(x_1,\dots,x_{n-1},t_n)dt_n\right]\right\}. \tag{4.4.17}$$

Integrating both sides of (4.4.17) over B and by making elementary calculations, we get the desired inequality in (4.4.14) and the proof is complete.

4.5 More Ostrowski-type inequalities in several variables

The main goal of this section is to present certain Ostrowski-type inequalities involving two functions of several independent variables.

Let $D = \{(x_1,\dots,x_n) : a_i < x_i < b_i\ (i = 1,\dots,n)\}$ and $\overline{D}$ be the closure of D. For $x_i \in \mathbb{R}$, $x = (x_1,\dots,x_n)$ is a variable point in D and $dx = dx_1\cdots dx_n$. The first order partial derivatives of a function $u(x) : D \to \mathbb{R}$ are denoted by $\frac{\partial u(x)}{\partial x_i}$ $(i = 1,\dots,n)$ and said to be bounded if

$$\left\|\frac{\partial u}{\partial x_i}\right\|_\infty = \sup_{x\in D}\left|\frac{\partial u(x)}{\partial x_i}\right| < \infty.$$

For any integrable function $u(x) : D \to \mathbb{R}$ we denote by $\int_D u(x)dx$ the n-fold integral $\int_{a_1}^{b_1}\cdots\int_{a_n}^{b_n} u(x_1,\dots,x_n)dx_1\cdots dx_n$.

The first Theorem deals with the Ostrowski-type inequality involving two functions of several variables established by Pachpatte in [92].

Theorem 4.5.1. Let $f,\ g : \mathbb{R}^n \to \mathbb{R}$ be continuous functions on $\overline{D}$ and differentiable on D, with derivatives $\frac{\partial f(x)}{\partial x_i}$, $\frac{\partial g(x)}{\partial x_i}$ being bounded. Then for all $x \in \overline{D}$,

$$\left|f(x)g(x) - \frac{1}{2M}\left[g(x)\int_D f(y)dy + f(x)\int_D g(y)dy\right]\right|$$

$$\leqslant \frac{1}{2M}\sum_{i=1}^{n}\left[|g(x)|\left\|\frac{\partial f}{\partial x_i}\right\|_\infty + |f(x)|\left\|\frac{\partial g}{\partial x_i}\right\|_\infty\right]E_i(x), \tag{4.5.1}$$

where, $M = \operatorname{mes} D = \prod_{i=1}^{n}(b_i - a_i)$ and $E_i(x) = \int_D |x_i - y_i|dy$.

Proof. Let $x = (x_1, \ldots, x_n)$, $y = (y_1, \ldots, y_n)$ $\left(x \in \overline{D}, y \in D\right)$. From the n-dimensional version of the mean value Theorem (see [146, p. 174] or [76, p. 121]), we have

$$f(x) - f(y) = \sum_{i=1}^{n} \frac{\partial f(c_1)}{\partial x_i}(x_i - y_i), \tag{4.5.2}$$

$$g(x) - g(y) = \sum_{i=1}^{n} \frac{\partial g(c_2)}{\partial x_i}(x_i - y_i), \tag{4.5.3}$$

where, $c_i = (y_1 + \alpha_i(x_1 - y_1), \ldots, y_n + \alpha_i(x_n - y_n))$ $(0 < \alpha_i < 1)$, $i = 1, 2$. Multiplying both sides of (4.5.2) and (4.5.3) by $g(x)$ and $f(x)$ respectively and adding, we get

$$2f(x)g(x) - g(x)f(y) - f(x)g(y)$$

$$= g(x)\sum_{i=1}^{n} \frac{\partial f(c_1)}{\partial x_i}(x_i - y_i) + f(x)\sum_{i=1}^{n} \frac{\partial g(c_2)}{\partial x_i}(x_i - y_i). \tag{4.5.4}$$

Integrating both sides of (4.5.4) with respect to y over D, using the fact that $\operatorname{mes} D > 0$ and rewriting, we have

$$f(x)g(x) - \frac{1}{2M}g(x)\int_D f(y)dy - \frac{1}{2M}f(x)\int_D g(y)dy$$

$$= \frac{1}{2M}\left[g(x)\int_D \sum_{i=1}^{n} \frac{\partial f(c_1)}{\partial x_i}(x_i - y_i)dy + f(x)\int_D \sum_{i=1}^{n} \frac{\partial g(c_2)}{\partial x_i}(x_i - y_i)dy\right]. \tag{4.5.5}$$

From (4.5.5) and using the properties of modulus, we have

$$\left| f(x)g(x) - \frac{1}{2M}\left[g(x)\int_D f(y)dy + f(x)\int_D g(y)dy\right]\right|$$

$$\leqslant \frac{1}{2M}\left[|g(x)|\int_D \sum_{i=1}^{n} \left|\frac{\partial f(c_1)}{\partial x_i}\right| |x_i - y_i|dy + |f(x)|\int_D \sum_{i=1}^{n} \left|\frac{\partial g(c_2)}{\partial x_i}\right| |x_i - y_i|dy\right]$$

$$\leqslant \frac{1}{2M}\sum_{i=1}^{n}\left[|g(x)|\left\|\frac{\partial f}{\partial x_i}\right\|_\infty + |f(x)|\left\|\frac{\partial g}{\partial x_i}\right\|_\infty\right] E_i(x).$$

The proof is complete.

Remark 4.5.1. If we take $g(x) = 1$ and hence $\frac{\partial g(x)}{\partial x_i} = 0$ in Theorem 4.5.1, then the inequality (4.5.1) reduces to the inequality established by Milovanović in [76, Theorem 2], which in turn is a generalization of the well-known Ostrowski's inequality given in (7). The inequalities in the following Theorem are also established by Pachpatte in [92].

Theorem 4.5.2. Let $f(x), g(x), \frac{\partial f(x)}{\partial x_i}, \frac{\partial g(x)}{\partial x_i}$ be as in Theorem 4.5.1. Then for all $x \in \overline{D}$,

$$\left| f(x)g(x) - \frac{1}{M}\left[f(x)\int_D g(y)dy + g(x)\int_D f(y)dy \right] + \frac{1}{M}\int_D f(y)g(y)dy \right|$$

$$\leqslant \frac{1}{M}\int_D \left(\sum_{i=1}^{n} \left\| \frac{\partial f}{\partial x_i} \right\|_\infty |x_i - y_i| \right) \left(\sum_{i=1}^{n} \left\| \frac{\partial g}{\partial x_i} \right\|_\infty |x_i - y_i| \right) dy, \tag{4.5.6}$$

and

$$\left| f(x)g(x) - \frac{1}{M}\left[f(x)\int_D g(y)dy + g(x)\int_D f(y)dy \right] + \frac{1}{M^2}\left(\int_D f(y)dy \right)\left(\int_D g(y)dy \right) \right|$$

$$\leqslant \frac{1}{M^2}\left(\sum_{i=1}^{n} \left\| \frac{\partial f}{\partial x_i} \right\|_\infty E_i(x) \right) \left(\sum_{i=1}^{n} \left\| \frac{\partial g}{\partial x_i} \right\|_\infty E_i(x) \right), \tag{4.5.7}$$

where M and $E_i(x)$ are as defined in Theorem 4.5.1.

Proof. From the hypotheses, the identities (4.5.2) and (4.5.3) hold. Multiplying the left hand and right hand sides of (4.5.2) and (4.5.3), we get

$$f(x)g(x) - f(x)g(y) - g(x)f(y) + f(y)g(y)$$

$$= \left(\sum_{i=1}^{n} \frac{\partial f(c_1)}{\partial x_i}(x_i - y_i) \right) \left(\sum_{i=1}^{n} \frac{\partial g(c_2)}{\partial x_i}(x_i - y_i) \right). \tag{4.5.8}$$

Integrating both sides of (4.5.8) with respect to y over D and rewriting, we have

$$f(x)g(x) - \frac{1}{M}f(x)\int_D g(y)dy - \frac{1}{M}g(x)\int_D f(y)dy + \frac{1}{M}\int_D f(y)g(y)dy$$

$$= \frac{1}{M}\int_D \left(\sum_{i=1}^{n} \frac{\partial f(c_1)}{\partial x_i}(x_i - y_i) \right) \left(\sum_{i=1}^{n} \frac{\partial g(c_2)}{\partial x_i}(x_i - y_i) \right) dy. \tag{4.5.9}$$

From (4.5.9) and using the properties of the modulus, we have

$$\left| f(x)g(x) - \frac{1}{M}\left[f(x)\int_D g(y)dy + g(x)\int_D f(y)dy \right] + \frac{1}{M}\int_D f(y)g(y)dy \right|$$

$$\leqslant \frac{1}{M}\int_D \left(\sum_{i=1}^{n} \left| \frac{\partial f(c_1)}{\partial x_i} \right| |x_i - y_i| \right) \left(\sum_{i=1}^{n} \left| \frac{\partial g(c_2)}{\partial x_i} \right| |x_i - y_i| \right) dy$$

$$\leqslant \frac{1}{M}\int_D \left(\sum_{i=1}^{n} \left\| \frac{\partial f}{\partial x_i} \right\|_\infty |x_i - y_i| \right) \left(\sum_{i=1}^{n} \left\| \frac{\partial g}{\partial x_i} \right\|_\infty |x_i - y_i| \right) dy,$$

which is the required inequality in (4.5.6).

Integrating both sides of (4.5.2) and (4.5.3) with respect to y over D and rewriting, we get

$$f(x) - \frac{1}{M}\int_D f(y)dy = \frac{1}{M}\int_D \sum_{i=1}^{n} \frac{\partial f(c_1)}{\partial x_i}(x_i - y_i)dy, \tag{4.5.10}$$

and

$$g(x) - \frac{1}{M}\int_D g(y)dy = \frac{1}{M}\int_D \sum_{i=1}^{n} \frac{\partial g(c_2)}{\partial x_i}(x_i - y_i)dy, \tag{4.5.11}$$

respectively. Multiplying the left hand and right hand sides of (4.5.10) and (4.5.11), we get

$$f(x)g(x) - \frac{1}{M}\left[f(x)\int_D g(y)dy + g(x)\int_D f(y)dy\right] + \frac{1}{M^2}\left(\int_D f(y)dy\right)\left(\int_D g(y)dy\right)$$

$$= \frac{1}{M^2}\left(\int_D \sum_{i=1}^{n} \frac{\partial f(c_1)}{\partial x_i}(x_i - y_i)dy\right)\left(\int_D \sum_{i=1}^{n} \frac{\partial g(c_2)}{\partial x_i}(x_i - y_i)dy\right). \tag{4.5.12}$$

From (4.5.12) and using the properties of modulus, we have

$$\left|f(x)g(x) - \frac{1}{M}\left[f(x)\int_D g(y)dy + g(x)\int_D f(y)dy\right] + \frac{1}{M^2}\left(\int_D f(y)dy\right)\left(\int_D g(y)dy\right)\right|$$

$$\leqslant \frac{1}{M^2}\left(\int_D \sum_{i=1}^{n} \left|\frac{\partial f(c_1)}{\partial x_i}\right| |x_i - y_i|dy\right)\left(\int_D \sum_{i=1}^{n} \left|\frac{\partial g(c_2)}{\partial x_i}\right| |x_i - y_i|dy\right)$$

$$\leqslant \frac{1}{M^2}\left(\sum_{i=1}^{n} \left\|\frac{\partial f}{\partial x_i}\right\|_\infty E_i(x)\right)\left(\sum_{i=1}^{n} \left\|\frac{\partial g}{\partial x_i}\right\|_\infty E_i(x)\right).$$

This is the desired inequality in (4.5.7) and the proof is complete.

Remark 4.5.2. In [2,74], the authors have established a number of Ostrowski-type inequalities for functions of several variables by using different techniques. We note that the inequalities given in Theorems 4.5.1 and 4.5.2 are different from those established in [2,74], and the proofs are extremely simple. We also note that, by closely looking at the results obtained in [105,125,134], one can very easily extend the inequalities in Theorems 4.5.1 and 4.5.2 involving three functions of several independent variables. We omit the details. Another Ostrowski-type inequality for mappings of the Hölder type involving two functions of several independent variables is given in the following theorem.

Theorem 4.5.3. Let the set B be as defined in Theorem 4.4.2. Suppose that $f : B \to \mathbb{R}$ is of r-Hölder type, $g : B \to \mathbb{R}$ is of s-Hölder type, i.e.,

$$|f(x) - f(y)| \leqslant \sum_{i=1}^{n} H_i \,|\, x_i - y_i|^{r_i} \quad (H_i \geqslant 0,\ i = 1,\ldots,n),$$

$$|g(x) - g(y)| \leqslant \sum_{i=1}^{n} L_i \,|\, x_i - y_i|^{s_i} \quad (L_i \geqslant 0,\ i = 1,\ldots,n),$$

for all $x,\ y \in B$, where $r_i,\ s_i \in (0,1]$, $i = 1,\ldots,n$. Then we have the inequality

$$\left| f(x)g(x) - \frac{1}{M}\left[f(x)\int_B g(y)dy + g(x)\int_B f(y)dy \right] + \frac{1}{M}\int_B f(y)g(y)dy \right|$$

$$\leqslant \frac{1}{M}\int_B \left(\sum_{i=1}^{n} H_i |x_i - y_i|^{r_i} \right)\left(\sum_{i=1}^{n} L_i |x_i - y_i|^{s_i} \right) dy, \tag{4.5.13}$$

for all $x \in B$, where $M = \operatorname{mes} B = \prod_{i=1}^{n}(b_i - a_i)$.

Proof. From the assumptions, we have

$$|(f(x) - f(y))(g(x) - g(y))| \leqslant \left(\sum_{i=1}^{n} H_i |x_i - y_i|^{r_i} \right)\left(\sum_{i=1}^{n} L_i |x_i - y_i|^{s_i} \right). \tag{4.5.14}$$

Integrating both sides of (4.5.14) with respect to y over B, we get

$$\int_B |(f(x) - f(y))(g(x) - g(y))|\, dy \leqslant \int_B \left(\sum_{i=1}^{n} H_i |x_i - y_i|^{r_i} \right)\left(\sum_{i=1}^{n} L_i |x_i - y_i|^{s_i} \right) dy. \tag{4.5.15}$$

On the other hand, from the properties of integrals, we have

$$\int_B |(f(x) - f(y))(g(x) - g(y))|\, dy \geqslant \left| \int_B (f(x) - f(y))(g(x) - g(y)) dy \right|$$

$$= \left| f(x)g(x) - f(x)\int_B g(y)dy - g(x)\int_B f(y)dy + \int_B f(y)g(y)dy \right|. \tag{4.5.16}$$

Using (4.5.16) in (4.5.15) and the fact that $M = \operatorname{mes} B > 0$, we get the required inequality in (4.5.13). The proof is complete.

Remark 4.5.3. If we take $g = f$ and hence $L_i = H_i$, $s_i = r_i$, $i = 1,\ldots,n$ in Theorem 4.5.3, then we get

$$\left| f^2(x) - \frac{2}{M} f(x)\int_B f(y)dy + \frac{1}{M}\int_B f^2(y)dy \right| \leqslant \frac{1}{M}\int_B \left(\sum_{i=1}^{n} H_i |x_i - y_i|^{r_i} \right)^2 dy,$$

for all $x \in B$.

4.6 Discrete inequalities of the Ostrowski-type in several variables

In this section, we offer some discrete inequalities of Ostrowski-type involving functions of many independent variables, recently investigated by Pachpatte in [95,125,134].

In what follows, we use the following notation for simplicity of presentation.

Let $U = \{1,\ldots,k+1\}$, $V = \{1,\ldots,r+1\}$ $(k,r \in \mathbb{N})$ and $W = U \times V$. For some suitable functions $p,\ f,\ g,\ h : W \to \mathbb{R}$, we set

$$e = kr,$$

$$S[p] = \sum_{\sigma=s}^{m-1}\sum_{\tau=t}^{n-1} \Delta_2\Delta_1 p(\sigma,\tau),$$

$$\overline{F}(m,n) = r\sum_{s=1}^{k} f(s,n) + k\sum_{t=1}^{r} f(m,t),$$

$$\overline{G}(m,n) = r\sum_{s=1}^{k} g(s,n) + k\sum_{t=1}^{r} g(m,t),$$

$$\overline{H}(m,n) = r\sum_{s=1}^{k} h(s,n) + k\sum_{t=1}^{r} h(m,t),$$

$$P(f,g,h;\overline{F},\overline{G},\overline{H};e)(m,n) = f(m,n)g(m,n)h(m,n)$$

$$-\frac{1}{3e}\left[g(m,n)h(m,n)\left\{\overline{F}(m,n) - \sum_{s=1}^{k}\sum_{t=1}^{r} f(s,t)\right\}\right.$$

$$+h(m,n)f(m,n)\left\{\overline{G}(m,n) - \sum_{s=1}^{k}\sum_{t=1}^{r} g(s,t)\right\}$$

$$\left.+f(m,n)g(m,n)\left\{\overline{H}(m,n) - \sum_{s=1}^{k}\sum_{t=1}^{r} h(s,t)\right\}\right],$$

$$Q_0(f,g,h;S)(m,n) = g(m,n)h(m,n)\sum_{s=1}^{k}\sum_{t=1}^{r} S[f]$$

$$+h(m,n)f(m,n)\sum_{s=1}^{k}\sum_{t=1}^{r} S[g] + f(m,n)g(m,n)\sum_{s=1}^{k}\sum_{t=1}^{r} S[h],$$

Let $A = \{1,\ldots,a+1\}$, $B = \{1,\ldots,b+1\}$, $C = \{1,\ldots,c+1\}$ $(a,\ b,\ c \in \mathbb{N})$ and $E = A \times B \times C$. For some suitable functions $p,\ f,\ g,\ h : E \to \mathbb{R}$, we set

$$\overline{e} = abc,$$

$$\overline{S}[p] = \sum_{u=r}^{k-1}\sum_{v=s}^{m-1}\sum_{w=t}^{n-1}\Delta_3\Delta_2\Delta_1 p(u,v,w),$$

$$\overline{J}[p] = bc\sum_{r=1}^{a} p(r,m,n) + ca\sum_{s=1}^{b} p(k,s,n) + ab\sum_{t=1}^{c} p(k,m,t),$$

$$\overline{L}[p] = c\sum_{r=1}^{a}\sum_{s=1}^{b} p(r,s,n) + b\sum_{r=1}^{a}\sum_{t=1}^{c} p(r,m,t) + a\sum_{s=1}^{b}\sum_{t=1}^{c} p(k,s,t),$$

$$\overline{P}(f,g,h;\overline{J},\overline{L};\overline{e})(k,m,n) = f(k,m,n)g(k,m,n)h(k,m,n)$$

$$-\frac{1}{3\overline{e}}\left[g(k,m,n)h(k,m,n)\left\{\overline{J}[f] - \overline{L}[f] + \sum_{r=1}^{a}\sum_{s=1}^{b}\sum_{t=1}^{c} f(r,s,t)\right\}\right.$$

$$+h(k,m,n)f(k,m,n)\left\{\overline{J}[g] - \overline{L}[g] + \sum_{r=1}^{a}\sum_{s=1}^{b}\sum_{t=1}^{c} g(r,s,t)\right\}$$

$$\left.+f(k,m,n)g(k,m,n)\left\{\overline{J}[h] - \overline{L}[h] + \sum_{r=1}^{a}\sum_{s=1}^{b}\sum_{t=1}^{c} h(r,s,t)\right\}\right],$$

$$\overline{Q}(f,g,h;\overline{S})(k,m,n) = g(k,m,n)h(k,m,n)\sum_{r=1}^{a}\sum_{s=1}^{b}\sum_{t=1}^{c}\overline{S}[f]$$

$$+h(k,m,n)f(k,m,n)\sum_{r=1}^{a}\sum_{s=1}^{b}\sum_{t=1}^{c}\overline{S}[g] + f(k,m,n)g(k,m,n)\sum_{r=1}^{a}\sum_{s=1}^{b}\sum_{t=1}^{c}\overline{S}[h].$$

The discrete versions of Theorems 4.2.5 and 4.3.4 proved in [125] and [134] are embodied in the following theorems.

Theorem 4.6.1. Let the functions $f,\ g,\ h : W \to \mathbb{R}$ be such that $\Delta_2\Delta_1 f(m,n)$, $\Delta_2\Delta_1 g(m,n)$, $\Delta_2\Delta_1 h(m,n)$, exist on W. Then

$$\left|P(f,g,h;\overline{F},\overline{G},\overline{H};e)(m,n)\right| \leqslant \frac{1}{3e}Q_0(|f|,|g|,|h|;|S|)(m,n), \tag{4.6.1}$$

for all $(m,n) \in W$.

Proof. By simple computation we have the following identities:

$$f(m,n) - f(s,n) - f(m,t) + f(s,t) = S[f], \tag{4.6.2}$$

$$g(m,n) - g(s,n) - g(m,t) + g(s,t) = S[g], \tag{4.6.3}$$

$$h(m,n) - h(s,n) - h(m,t) + h(s,t) = S[h], \tag{4.6.4}$$

for (m,n), $(s,t) \in W$. Multiplying both sides of (4.6.2), (4.6.3) and (4.6.4) by $g(m,n)h(m,n)$, $h(m,n)f(m,n)$ and $f(m,n)g(m,n)$ respectively and adding the resulting identities, we have

$$3f(m,n)g(m,n)h(m,n) - g(m,n)h(m,n)[f(s,n) + f(m,t) - f(s,t)]$$

$$-h(m,n)f(m,n)[g(s,n) + g(m,t) - g(s,t)] - f(m,n)g(m,n)[h(s,n) + h(m,t) - h(s,t)]$$

$$= g(m,n)h(m,n)S[f] + h(m,n)f(m,n)S[g] + f(m,n)g(m,n)S[h]. \tag{4.6.5}$$

Summing both sides of (4.6.5) first with respect to t from 1 to r and then with respect to s from 1 to k and rewriting, we have

$$P(f,g,h;\overline{F},\overline{G},\overline{H};e)(m,n) = \frac{1}{3e}Q_0(f,g,h;S)(m,n). \tag{4.6.6}$$

From (4.6.6) and using the properties of modulus, we get the required inequality in (4.6.1). The proof is complete.

Remark 4.6.1. Taking $h(m,n) = 1$ and hence $\Delta_2\Delta_1 h(m,n) = 0$ in Theorem 4.6.1 and by simple computation, it is easy to see that the inequality (4.6.1) reduces to

$$\left| f(m,n)g(m,n) - \frac{1}{2e}\left[g(m,n)\left\{ \overline{F}(m,n) - \sum_{s=1}^{k}\sum_{t=1}^{r} f(s,t) \right\} \right.\right.$$

$$\left.\left. +f(m,n)\left\{ \overline{G}(m,n) - \sum_{s=1}^{k}\sum_{t=1}^{r} g(s,t) \right\} \right] \right|$$

$$\leqslant \frac{1}{2e}\left[|g(m,n)| \sum_{s=1}^{k}\sum_{t=1}^{r} |S[f]| + |f(m,n)| \sum_{s=1}^{k}\sum_{t=1}^{r} |S[g]| \right], \tag{4.6.7}$$

for all $(m,n) \in W$. Further, by taking $g(m,n) = 1$ and hence $\Delta_2\Delta_1 g(m,n) = 0$ in (4.6.7) and by simple computation, we get the following Ostrowski-type discrete inequality

$$\left| f(m,n) - \frac{1}{e}\left[\overline{F}(m,n) - \sum_{s=1}^{k}\sum_{t=1}^{r} f(s,t) \right] \right| \leqslant \frac{1}{e} \sum_{s=1}^{k}\sum_{t=1}^{r} |S[f]|, \tag{4.6.8}$$

for all $(m,n) \in W$.

Remark 4.6.2. Summing both sides of (4.6.6) first with respect to n from 1 to r and then with respect to m from 1 to k and rewriting, we have

$$M(f,g,h;\overline{F},\overline{G},\overline{H};e) = \frac{1}{3e^2}\sum_{m=1}^{k}\sum_{n=1}^{r} Q_0(f,g,h;S)(m,n), \tag{4.6.9}$$

where

$$M(f,g,h;\overline{F},\overline{G},\overline{H};e) = \frac{1}{e}\sum_{m=1}^{k}\sum_{n=1}^{r} f(m,n)g(m,n)h(m,n)$$

$$-\frac{1}{3e^2}\sum_{m=1}^{k}\sum_{n=1}^{r}\left[g(m,n)h(m,n)\overline{F}(m,n)\right.$$

$$\left.+h(m,n)f(m,n)\overline{G}(m,n)+f(m,n)g(m,n)\overline{H}(m,n)\right]$$

$$+\frac{1}{3}\left[\left(\frac{1}{e}\sum_{m=1}^{k}\sum_{n=1}^{r} g(m,n)h(m,n)\right)\left(\frac{1}{e}\sum_{m=1}^{k}\sum_{n=1}^{r} f(m,n)\right)\right.$$

$$+\left(\frac{1}{e}\sum_{m=1}^{k}\sum_{n=1}^{r} h(m,n)f(m,n)\right)\left(\frac{1}{e}\sum_{m=1}^{k}\sum_{n=1}^{r} g(m,n)\right)$$

$$\left.+\left(\frac{1}{e}\sum_{m=1}^{k}\sum_{n=1}^{r} f(m,n)g(m,n)\right)\left(\frac{1}{e}\sum_{m=1}^{k}\sum_{n=1}^{r} h(m,n)\right)\right].$$

From (4.6.9) and using the properties of modulus, we get the following Grüss-type discrete inequality

$$\left|M(f,g,h;\overline{F},\overline{G},\overline{H};e)\right| \leqslant \frac{1}{3e^2}\sum_{m=1}^{k}\sum_{n=1}^{r} Q_0(|f|,|g|,|h|;|S|)(m,n). \tag{4.6.10}$$

Theorem 4.6.2. Let the functions $f,\ g,\ h : E \to \mathbb{R}$ be such that $\Delta_3\Delta_2\Delta_1 f(k,m,n)$, $\Delta_3\Delta_2\Delta_1 g(k,m,n)$, $\Delta_3\Delta_2\Delta_1 h(k,m,n)$ exist on E. Then

$$\left|\overline{P}\left(f,g,h;\overline{J},\overline{L};\overline{e}\right)(k,m,n)\right| \leqslant \frac{1}{3\overline{e}}\overline{Q}\left(|f|,|g|,|h|;\left|\overline{S}\right|\right)(k,m,n), \tag{4.6.11}$$

for all $(k,m,n) \in E$.

Proof. We first prove the following identity:

$$\overline{S}[f] = f(k,m,n) - [f(r,m,n) + f(k,s,n) + f(k,m,t)]$$

$$+[f(r,s,n) + f(r,m,t) + f(k,s,t)] - f(r,s,t), \tag{4.6.12}$$

for $(k,m,n),(r,s,t) \in E$, where

$$\overline{S}[f] = \sum_{u=r}^{k-1}\sum_{v=s}^{m-1}\sum_{w=t}^{n-1} \Delta_3\Delta_2\Delta_1 f(u,v,w). \tag{4.6.13}$$

From (4.6.13), by simple calculation, we have

$$\overline{S}[f] = \sum_{u=r}^{k-1}\sum_{v=s}^{m-1}\left[\sum_{w=t}^{n-1}\{\Delta_2\Delta_1 f(u,v,w+1) - \Delta_2\Delta_1 f(u,v,w)\}\right]$$

$$= \sum_{u=r}^{k-1}\sum_{v=s}^{m-1} \Delta_2\Delta_1 f(u,v,n) - \sum_{u=r}^{k-1}\sum_{v=s}^{m-1} \Delta_2\Delta_1 f(u,v,t)$$

$$= \overline{S}_1[f] - \overline{S}_2[f] \text{ (say)}. \tag{4.6.14}$$

By simple computation, we have

$$\overline{S}_1[f] = \sum_{u=r}^{k-1}\sum_{v=s}^{m-1} \Delta_2\Delta_1 f(u,v,n)$$

$$= \sum_{u=r}^{k-1}\left[\sum_{v=s}^{m-1}\{\Delta_1 f(u,v+1,n) - \Delta_1 f(u,v,n)\}\right]$$

$$= \sum_{u=r}^{k-1}\Delta_1 f(u,m,n) - \sum_{u=r}^{k-1}\Delta_1 f(u,s,n)$$

$$= \sum_{u=r}^{k-1}\{f(u+1,m,n) - f(u,m,n)\} - \sum_{u=r}^{k-1}\{f(u+1,s,n) - f(u,s,n)\}$$

$$= f(k,m,n) - f(r,m,n) - f(k,s,n) + f(r,s,n). \tag{4.6.15}$$

Similarly, we have

$$\overline{S}_2[f] = \sum_{u=r}^{k-1}\sum_{v=s}^{m-1} \Delta_2\Delta_1 f(u,v,t) = f(k,m,t) - f(r,m,t) - f(k,s,t) + f(r,s,t) \tag{4.6.16}$$

Using (4.6.15) and (4.6.16) in (4.6.14), we get (4.6.12). Similarly, we have the following identities

$$\overline{S}[g] = g(k,m,n) - [g(r,m,n) + g(k,s,n) + g(k,m,t)]$$

$$+[g(r,s,n) + g(r,m,t) + g(k,s,t)] - g(r,s,t), \tag{4.6.17}$$

$$\overline{S}[h] = h(k,m,n) - [h(r,m,n) + h(k,s,n) + h(k,m,t)] \\ + [h(r,s,n) + h(r,m,t) + h(k,s,t)] - h(r,s,t), \tag{4.6.18}$$

for (k,m,n), $(r,s,t) \in E$.

Multiplying both sides of (4.6.12), (4.6.17) and (4.6.18) by $g(k,m,n)h(k,m,n)$, $h(k,m,n)f(k,m,n)$ and $f(k,m,n)g(k,m,n)$ respectively and adding the resulting identities, we have

$$\begin{aligned}
&3f(k,m,n)g(k,m,n)h(k,m,n) \\
&-g(k,m,n)h(k,m,n)\{[f(r,m,n)+f(k,s,n)+f(k,m,t)] \\
&-[f(r,s,n)+f(r,m,t)+f(k,s,t)]+f(r,s,t)\} \\
&-h(k,m,n)f(k,m,n)\{[g(r,m,n)+g(k,s,n)+g(k,m,t)] \\
&-[g(r,s,n)+g(r,m,t)+g(k,s,t)]+g(r,s,t)\} \\
&-f(k,m,n)g(k,m,n)\{[h(r,m,n)+h(k,s,n)+h(k,m,t)] \\
&-[h(r,s,n)+h(r,m,t)+h(k,s,t)]+h(r,s,t)\} \\
&= g(k,m,n)h(k,m,n)\overline{S}[f]+h(k,m,n)f(k,m,n)\overline{S}[g]+f(k,m,n)g(k,m,n)\overline{S}[h].
\end{aligned} \tag{4.6.19}$$

Summing both sides of (4.6.19), first with respect to t from 1 to c, then with respect to s from 1 to b and finally with respect to r from 1 to a and rewriting, we have

$$\overline{P}\left(f,g,h;\overline{J},\overline{L};\overline{e}\right)(k,m,n) = \frac{1}{3\overline{e}}\overline{Q}\left(f,g,h;\overline{S}\right)(k,m,n). \tag{4.6.20}$$

From (4.6.20) and using the properties of modulus, we get the required inequality in (4.6.11). The proof is complete.

Remark 4.6.3. Taking $h(k,m,n) = 1$ and hence $\overline{S}[h] = 0$ in Theorem 4.6.2 and by simple computation, it is easy to see that the inequality (4.6.11) reduces to

$$\begin{aligned}
&\left| f(k,m,n)g(k,m,n) - \frac{1}{2\overline{e}}\left[g(k,m,n)\left\{ \overline{J}[f] - \overline{L}[f] + \sum_{r=1}^{a}\sum_{s=1}^{b}\sum_{t=1}^{c} f(r,s,t) \right\} \right.\right. \\
&\left.\left. + f(k,m,n)\left\{ \overline{J}[g] - \overline{L}[g] + \sum_{r=1}^{a}\sum_{s=1}^{b}\sum_{t=1}^{c} g(r,s,t) \right\} \right] \right| \\
&\leqslant \frac{1}{2\overline{e}}\left[|g(k,m,n)| \sum_{r=1}^{a}\sum_{s=1}^{b}\sum_{t=1}^{c} |\overline{S}[f]| + |f(k,m,n)| \sum_{r=1}^{a}\sum_{s=1}^{b}\sum_{t=1}^{c} |\overline{S}[g]| \right],
\end{aligned} \tag{4.6.21}$$

for all $(k,m,n) \in E$. Further, by taking $g(k,m,n) = 1$ and hence $\overline{S}[g] = 0$ in (4.6.21) and by simple computation, we get

$$\left| f(k,m,n) - \frac{1}{\overline{e}}\left[\overline{J}[f] - \overline{L}[f] + \sum_{r=1}^{a}\sum_{s=1}^{b}\sum_{t=1}^{c} f(r,s,t) \right] \right| \leqslant \frac{1}{\overline{e}} \sum_{r=1}^{a}\sum_{s=1}^{b}\sum_{t=1}^{c} |\overline{S}[f]|, \tag{4.6.22}$$

for all $(k,m,n) \in E$.

Remark 4.6.4. Summing both sides of (4.6.20) first with respect to n from 1 to c, then with respect to m from 1 to b and finally with respect to k from 1 to a and rewriting, we have

$$\overline{M}\left(f,g,h;\overline{J},\overline{L};\overline{e}\right) = \frac{1}{3\overline{e}^2}\sum_{k=1}^{a}\sum_{m=1}^{b}\sum_{n=1}^{c}\overline{Q}(f,g,h;\overline{S})(k,m,n), \tag{4.6.23}$$

where

$$\overline{M}\left(f,g,h;\overline{J},\overline{L};\overline{e}\right) = \frac{1}{\overline{e}}\sum_{k=1}^{a}\sum_{m=1}^{b}\sum_{n=1}^{c} f(k,m,n)g(k,m,n)h(k,m,n)$$

$$-\frac{1}{3\overline{e}^2}\sum_{k=1}^{a}\sum_{m=1}^{b}\sum_{n=1}^{c}[g(k,m,n)h(k,m,n)(\overline{J}[f]-\overline{L}[f])+h(k,m,n)f(k,m,n)(\overline{J}[g]-\overline{L}[g])$$

$$+f(k,m,n)g(k,m,n)(\overline{J}[h]-\overline{L}[h])]$$

$$-\frac{1}{3}\left[\left(\frac{1}{\overline{e}}\sum_{k=1}^{a}\sum_{m=1}^{b}\sum_{n=1}^{c} g(k,m,n)h(k,m,n)\right)\left(\frac{1}{\overline{e}}\sum_{k=1}^{a}\sum_{m=1}^{b}\sum_{n=1}^{c} f(k,m,n)\right)\right.$$

$$+\left(\frac{1}{\overline{e}}\sum_{k=1}^{a}\sum_{m=1}^{b}\sum_{n=1}^{c} h(k,m,n)f(k,m,n)\right)\left(\frac{1}{\overline{e}}\sum_{k=1}^{a}\sum_{m=1}^{b}\sum_{n=1}^{c} g(k,m,n)\right)$$

$$\left.+\left(\frac{1}{\overline{e}}\sum_{k=1}^{a}\sum_{m=1}^{b}\sum_{n=1}^{c} f(k,m,n)g(k,m,n)\right)\left(\frac{1}{\overline{e}}\sum_{k=1}^{a}\sum_{m=1}^{b}\sum_{n=1}^{c} h(k,m,n)\right)\right].$$

From (4.6.23) and using the properties of modulus, we get the following Grüss-type discrete inequality

$$\left|\overline{M}(f,g,h;\overline{J},\overline{L};\overline{e})\right| \leqslant \frac{1}{3\overline{e}^2}\sum_{k=1}^{a}\sum_{m=1}^{b}\sum_{n=1}^{c}\overline{Q}(|f|,|g|,|h|;|\overline{S}|)(k,m,n). \tag{4.6.24}$$

The discrete Ostrowski-type inequalities in the following theorems, involving functions of many independent variables are investigated in [95].

Theorem 4.6.3. Let $f,\ g : Q \to \mathbb{R}$ be functions such that $\Delta_i f(x)$, $\Delta_i g(x)$ for $i = 1,\dots,n$ are bounded on Q, where Q is as defined in section 2.6. Then for all $x,\ y \in Q$,

$$\left|f(x)g(x) - \frac{1}{2M}\left[g(x)\sum_{y} f(y) + f(x)\sum_{y} g(y)\right]\right|$$

$$\leqslant \frac{1}{2M}\sum_{i=1}^{n}[|g(x)|\|\Delta_i f\|_\infty + |f(x)|\|\Delta_i g\|_\infty]H_i(x), \tag{4.6.25}$$

where, $M = \prod_{i=1}^{n} a_i$ and $H_i(x) = \sum_y |x_i - y_i|$.

Proof. For $x=(x_1,\dots,x_n)$, $y=(y_1,\dots,y_n)$ in Q, it is easy to observe that the following identities hold:

$$f(x)-f(y)=\sum_{i=1}^{n}\left\{\sum_{t_i=y_i}^{x_i-1}\Delta_i f(y_1,\dots,y_{i-1},t_i,x_{i+1},\dots,x_n)\right\}, \tag{4.6.26}$$

$$g(x)-g(y)=\sum_{i=1}^{n}\left\{\sum_{t_i=y_i}^{x_i-1}\Delta_i g(y_1,\dots,y_{i-1},t_i,x_{i+1},\dots,x_n)\right\}. \tag{4.6.27}$$

Multiplying both sides of (4.6.26) and (4.6.27) by $g(x)$ and $f(x)$ respectively and adding, we get

$$2f(x)g(x)-g(x)f(y)-f(x)g(y)=g(x)\sum_{i=1}^{n}\left\{\sum_{t_i=y_i}^{x_i-1}\Delta_i f(y_1,\dots,y_{i-1},t_i,x_{i+1},\dots,x_n)\right\}$$
$$+f(x)\sum_{i=1}^{n}\left\{\sum_{t_i=y_i}^{x_i-1}\Delta_i g(y_1,\dots,y_{i-1},t_i,x_{i+1},\dots,x_n)\right\}. \tag{4.6.28}$$

Summing both sides of (4.6.28) with respect to y over Q, using the fact that $M>0$ and rewriting, we have

$$f(x)g(x)-\frac{1}{2M}g(x)\sum_y f(y)-\frac{1}{2M}f(x)\sum_y g(y)$$
$$=\frac{1}{2M}\left[g(x)\sum_y\left[\sum_{i=1}^{n}\left\{\sum_{t_i=y_i}^{x_i-1}\Delta_i f(y_1,\dots,y_{i-1},t_i,x_{i+1},\dots,x_n)\right\}\right]\right.$$
$$\left.+f(x)\sum_y\left[\sum_{i=1}^{n}\left\{\sum_{t_i=y_i}^{x_i-1}\Delta_i g(y_1,\dots,y_{i-1},t_i,x_{i+1},\dots,x_n)\right\}\right]\right]. \tag{4.6.29}$$

From (4.6.29) and using the properties of modulus, we have

$$\left|f(x)g(x)-\frac{1}{2M}\left[g(x)\sum_y f(y)+f(x)\sum_y g(y)\right]\right|$$
$$\leqslant\frac{1}{2M}\left[|g(x)|\sum_y\left[\sum_{i=1}^{n}\left\{\left|\sum_{t_i=y_i}^{x_i-1}\Delta_i f(y_1,\dots,y_{i-1},t_i,x_{i+1},\dots,x_n)\right|\right\}\right]\right.$$
$$\left.+|f(x)|\sum_y\left[\sum_{i=1}^{n}\left\{\left|\sum_{t_i=y_i}^{x_i-1}\Delta_i g(y_1,\dots,y_{i-1},t_i,x_{i+1},\dots,x_n)\right|\right\}\right]\right]$$
$$\leqslant\frac{1}{2M}\left[|g(x)|\sum_y\left[\sum_{i=1}^{n}\left\{\|\Delta_i f\|_\infty\left|\sum_{t_i=y_i}^{x_i-1}1\right|\right\}\right]+|f(x)|\sum_y\left[\sum_{i=1}^{n}\left\{\|\Delta_i g\|_\infty\left|\sum_{t_i=y_i}^{x_i-1}1\right|\right\}\right]\right]$$
$$=\frac{1}{2M}\sum_{i=1}^{n}\left[|g(x)|\|\Delta_i f\|_\infty+|f(x)|\|\Delta_i g\|_\infty\right]\left(\sum_y|x_i-y_i|\right)$$
$$=\frac{1}{2M}\sum_{i=1}^{n}\left[|g(x)|\|\Delta_i f\|_\infty+|f(x)|\|\Delta_i g\|_\infty\right]H_i(x).$$

The proof of the inequality (4.6.25) is complete.

Remark 4.6.5. By taking $g(x) = 1$ and hence $\Delta_i g(x) = 0$ in Theorem 4.6.3 and by simple calculation, it is easy to see that the inequality (4.6.25) reduces to

$$\left| f(x) - \frac{1}{M}\sum_{y} f(y) \right| \leqslant \frac{1}{M}\sum_{i=1}^{n} \|\Delta_i f\|_\infty H_i(x), \tag{4.6.30}$$

for all $x,\ y \in Q$. The inequality (4.6.30) can be considered as the discrete version of the inequality established by Milovanović [76, Theorem 2].

Theorem 4.6.4. Let $f,\ g,\ \Delta_i f,\ \Delta_i g$ be as in Theorem 4.6.3. Then for every $x,\ y \in Q$,

$$\left| f(x)g(x) - \frac{1}{M}\left[g(x)\sum_{y} f(y) + f(x)\sum_{y} g(y) \right] + \frac{1}{M}\sum_{y} f(y)g(y) \right|$$

$$\leqslant \frac{1}{M}\sum_{y}\left[\sum_{i=1}^{n} \|\Delta_i f\|_\infty |x_i - y_i| \right]\left[\sum_{i=1}^{n} \|\Delta_i g\|_\infty |x_i - y_i| \right], \tag{4.6.31}$$

and

$$\left| f(x)g(x) - \frac{1}{M}\left[g(x)\sum_{y} f(y) + f(x)\sum_{y} g(y) \right] + \frac{1}{M^2}\left(\sum_{y} f(y)\right)\left(\sum_{y} g(y)\right) \right|$$

$$\leqslant \frac{1}{M^2}\left[\sum_{i=1}^{n} \|\Delta_i f\|_\infty H_i(x) \right]\left[\sum_{i=1}^{n} \|\Delta_i g\|_\infty H_i(x) \right], \tag{4.6.32}$$

where M and $H_i(x)$ are as defined in Theorem 4.6.3.

Proof. From the hypotheses, the identities (4.6.26) and (4.6.27) hold. Multiplying the left hand sides and right hand sides of (4.6.26) and (4.6.27), we get

$$f(x)g(x) - g(x)f(y) - f(x)g(y) + f(y)g(y)$$

$$= \left[\sum_{i=1}^{n} \left\{ \sum_{t_i=y_i}^{x_i-1} \Delta_i f(y_1,\ldots,y_{i-1},t_i,x_{i+1},\ldots,x_n) \right\} \right]$$

$$\times \left[\sum_{i=1}^{n} \left\{ \sum_{t_i=y_i}^{x_i-1} \Delta_i g(y_1,\ldots,y_{i-1},t_i,x_{i+1},\ldots,x_n) \right\} \right]. \tag{4.6.33}$$

Summing both sides of (4.6.33) with respect to y over Q and rewriting, we have

$$f(x)g(x) - \frac{1}{M}\left[g(x)\sum_{y} f(y) + f(x)\sum_{y} g(y) \right] + \frac{1}{M}\sum_{y} f(y)g(y)$$

$$= \frac{1}{M}\sum_{y}\left[\sum_{i=1}^{n} \left\{ \sum_{t_i=y_i}^{x_i-1} \Delta_i f(y_1,\ldots,y_{i-1},t_i,x_{i+1},\ldots,x_n) \right\} \right]$$

$$\times\left[\sum_{i=1}^{n}\left\{\sum_{t_i=y_i}^{x_i-1}\Delta_i g(y_1,\ldots,y_{i-1},t_i,x_{i+1},\ldots,x_n)\right\}\right]. \tag{4.6.34}$$

From (4.6.34) and using the properties of modulus, we have

$$\left|f(x)g(x)-\frac{1}{M}\left[g(x)\sum_y f(y)+f(x)\sum_y g(y)\right]+\frac{1}{M}\sum_y f(y)g(y)\right|$$
$$\leqslant\frac{1}{M}\sum_y\left[\sum_{i=1}^{n}\left|\left\{\sum_{t_i=y_i}^{x_i-1}|\Delta_i f(y_1,\ldots,y_{i-1},t_i,x_{i+1},\ldots,x_n)|\right\}\right|\right]$$
$$\times\left[\sum_{i=1}^{n}\left|\left\{\sum_{t_i=y_i}^{x_i-1}|\Delta_i g(y_1,\ldots,y_{i-1},t_i,x_{i+1},\ldots,x_n)|\right\}\right|\right]$$
$$\leqslant\frac{1}{M}\sum_y\left[\sum_{i=1}^{n}\|\Delta_i f\|_\infty|x_i-y_i|\right]\left[\sum_{i=1}^{n}\|\Delta_i g\|_\infty|x_i-y_i|\right],$$

which is the required inequality in (4.6.31).

Summing both sides of (4.6.26) and (4.6.27) with respect to y over Q and rewriting, we get

$$f(x)-\frac{1}{M}\sum_y f(y)=\frac{1}{M}\sum_y\left[\sum_{i=1}^{n}\left\{\sum_{t_i=y_i}^{x_i-1}\Delta_i f(y_1,\ldots,y_{i-1},t_i,x_{i+1},\ldots,x_n)\right\}\right], \tag{4.6.35}$$

and

$$g(x)-\frac{1}{M}\sum_y g(y)=\frac{1}{M}\sum_y\left[\sum_{i=1}^{n}\left\{\sum_{t_i=y_i}^{x_i-1}\Delta_i g(y_1,\ldots,y_{i-1},t_i,x_{i+1},\ldots,x_n)\right\}\right], \tag{4.6.36}$$

respectively. Multiplying the left hand sides and right hand sides of (4.6.35) and (4.6.36), we get

$$f(x)g(x)-\frac{1}{M}\left[g(x)\sum_y f(y)+f(x)\sum_y g(y)\right]+\frac{1}{M^2}\left(\sum_y f(y)\right)\left(\sum_y g(y)\right)$$
$$=\frac{1}{M^2}\left(\sum_y\left[\sum_{i=1}^{n}\left\{\sum_{t_i=y_i}^{x_i-1}\Delta_i f(y_1,\ldots,y_{i-1},t_i,x_{i+1},\ldots,x_n)\right\}\right]\right)$$
$$\times\left(\sum_y\left[\sum_{i=1}^{n}\left\{\sum_{t_i=y_i}^{x_i-1}\Delta_i g(y_1,\ldots,y_{i-1},t_i,x_{i+1},\ldots,x_n)\right\}\right]\right) \tag{4.6.37}$$

From (4.6.37) and using the properties of modulus, we have

$$\left|f(x)g(x)-\frac{1}{M}\left[g(x)\sum_y f(y)+f(x)\sum_y g(y)\right]+\frac{1}{M^2}\left(\sum_y f(y)\right)\left(\sum_y g(y)\right)\right|$$
$$\leqslant\frac{1}{M^2}\left(\sum_y\left[\sum_{i=1}^{n}\left|\left\{\sum_{t_i=y_i}^{x_i-1}|\Delta_i f(y_1,\ldots,y_{i-1},t_i,x_{i+1},\ldots,x_n)|\right\}\right|\right]\right)$$
$$\times\left(\sum_y\left[\sum_{i=1}^{n}\left|\left\{\sum_{t_i=y_i}^{x_i-1}|\Delta_i g(y_1,\ldots,y_{i-1},t_i,x_{i+1},\ldots,x_n)|\right\}\right|\right]\right)$$
$$\leqslant\frac{1}{M^2}\left(\sum_{i=1}^{n}\|\Delta_i f\|_\infty H_i(x)\right)\left(\sum_{i=1}^{n}\|\Delta_i g\|_\infty H_i(x)\right).$$

This is the desired inequality in (4.6.32) and the proof is complete.

4.7 Applications

In this section, we present applications of some of the inequalities given in earlier sections and hope that they will encourage to open up new vistas for future research.

4.7.1 *Applications for cubature formulae*

In [8], Barnett and Dragomir used the inequality given in Theorem 4.2.1 to obtain a general cubature formula. In the following Theorem we present a result given in [8].

Consider the arbitrary divisions $I_n = a = x_0 < x_1 < \cdots < x_{n-1} < x_n = b$ and $J_m = c = y_0 < y_1 < \cdots < y_{m-1} < y_m = d$ and $\xi_i \in [x_i, x_{i+1}]$ $(i = 0, \ldots, n-1)$, $\eta_j \in [y_j, y_{j+1}]$ $(j = 0, \ldots, m-1)$ $(n, m \in \mathbb{N})$ be intermediate points. Consider the sum

$$C(f, I_n, J_m, \xi, \eta) = \sum_{i=0}^{n-1}\sum_{j=0}^{m-1} h_i \int_{y_j}^{y_{j+1}} f(\xi_i, t)dt$$

$$+\sum_{i=0}^{n-1}\sum_{j=0}^{m-1} l_j \int_{x_i}^{x_{i+1}} f(s, \eta_j)ds - \sum_{i=0}^{n-1}\sum_{j=0}^{m-1} h_i l_j f(\xi_i, \eta_j), \tag{4.7.1}$$

for which we assume that the involved integrals can more easily be computed than the original double integral

$$D = \int_a^b \int_c^d f(s,t)dtds,$$

and $h_i = x_{i+1} - x_i$ $(i = 0, \ldots, n-1)$, $l_j = y_{j+1} - y_j$ $(j = 0, \ldots, m-1)$, $\xi = (\xi_0, \ldots, \xi_{n-1})$, $\eta = (\eta_0, \ldots, \eta_{m-1})$.

With this assumption, we state the following cubature formula given in [8].

Theorem 4.7.1. Let $f : \Delta \to \mathbb{R}$ be as in Theorem 4.2.1 and I_n, J_m, ξ_i, η_j, ξ, η be as above. Then we have the cubature formula

$$\int_a^b \int_c^d f(s,t)dtds = C(f, I_n, J_m, \xi, \eta) + R(f, I_n, J_m, \xi, \eta), \tag{4.7.2}$$

where $C(f, I_n, J_m, \xi, \eta)$ is given by (4.7.1) and the remainder term $R(f, I_n, J_m, \xi, \eta)$ satisfies the estimation

$$|R(f, I_n, J_m, \xi, \eta)|$$

$$\leqslant \|D_2D_1 f\|_\infty \sum_{i=0}^{n-1}\sum_{j=0}^{m-1} \left[\frac{1}{4}h_i^2 + \left(\xi_i - \frac{x_i + x_{i+1}}{2}\right)^2\right] \left[\frac{1}{4}l_j^2 + \left(\eta_j - \frac{y_j + y_{j+1}}{2}\right)^2\right]$$

$$\leqslant \frac{1}{4}\|D_2D_1 f\|_\infty \sum_{i=0}^{n-1} h_i^2 \sum_{j=0}^{m-1} l_j^2. \tag{4.7.3}$$

Proof. Apply Theorem 4.2.1 on $[x_i,x_{i+1}]\times[y_j,y_{j+1}]$ $(i=0,\ldots,n-1;\ j=0,\ldots,m-1)$, to get

$$\left|\int_{x_i}^{x_{i+1}}\int_{y_j}^{y_{j+1}} f(s,t)dtds-\left[l_j\int_{x_i}^{x_{i+1}} f(s,\eta_j)ds+h_i\int_{y_j}^{y_{j+1}} f(\xi_i,t)dt-h_il_jf(\xi_i,\eta_j)\right]\right|$$

$$\leqslant\left[\frac{1}{4}h_i^2+\left(\xi_i-\frac{x_i+x_{i+1}}{2}\right)^2\right]\left[\frac{1}{4}l_j^2+\left(\eta_j-\frac{y_j+y_{j+1}}{2}\right)^2\right]\|D_2D_1f\|_\infty, \tag{4.7.4}$$

for all $i=0,\ldots,n-1;\ j=0,\ldots,m-1$.

Summing both sides of (4.7.4) over i from 0 to $n-1$ and over j from 0 to $m-1$ and using the generalized triangle inequality we deduce the first inequality in (4.7.3).

For the second part, we observe that

$$\left|\xi_i-\frac{x_i+x_{i+1}}{2}\right|\leqslant\frac{1}{2}h_i,\left|\eta_j-\frac{y_j+y_{j+1}}{2}\right|\leqslant\frac{1}{2}l_j,$$

for all $i,\ j$ as above.

Now, define the sum

$$C_M(f,I_n,J_m)=\sum_{i=0}^{n-1}\sum_{j=0}^{m-1}h_i\int_{y_j}^{y_{j+1}} f\left(\frac{x_i+x_{i+1}}{2},t\right)dt$$

$$+\sum_{i=0}^{n-1}\sum_{j=0}^{m-1}l_j\int_{x_i}^{x_{i+1}} f\left(s,\frac{y_j+y_{j+1}}{2}\right)ds-\sum_{i=0}^{n-1}\sum_{j=0}^{m-1}h_il_jf\left(\frac{x_i+x_{i+1}}{2},\frac{y_j+y_{j+1}}{2}\right). \tag{4.7.5}$$

The following corollary contains the best cubature formula, we can get from (4.7.1).

Corollary 4.7.1. Let $f,\ I_n,\ J_m$ be as in Theorem 4.7.1, then we have

$$\int_a^b\int_c^d f(s,t)dtds=C_M(f,I_n,J_m)+R(f,I_n,J_m), \tag{4.7.6}$$

where $C_M(f,I_n,J_m)$ is the midpoint formula given by (4.7.5) and the remainder $R(f,I_n,J_m)$ satisfies the estimation

$$|R(f,I_n,J_m)|\leqslant\frac{1}{16}\|D_2D_1f\|_\infty\sum_{i=0}^{n-1}h_i^2\sum_{j=0}^{m-1}l_j^2. \tag{4.7.7}$$

4.7.2 *More applications for cubature formulae*

Below, we shall present an application of one case of Theorem 4.2.2, namely, when all the derivatives $D_1f,\ D_2f,\ D_2D_1f$ of a function $f:\Delta\to\mathbb{R}$ are bounded. That is,

$$\left|f(x,y)-\frac{1}{(b-a)(d-c)}\int_a^b\int_c^d f(t,s)dsdt\right|$$

$$\leqslant \frac{1}{b-a}\left[\frac{1}{4}(b-a)^2+\left(x-\frac{a+b}{2}\right)^2\right]\|D_1 f\|_\infty$$

$$+\frac{1}{d-c}\left[\frac{1}{4}(d-c)^2+\left(y-\frac{c+d}{2}\right)^2\right]\|D_2 f\|_\infty$$

$$+\frac{1}{(b-a)(d-c)}\left[\frac{1}{4}(b-a)^2+\left(x-\frac{a+b}{2}\right)^2\right]$$

$$\times\left[\frac{1}{4}(d-c)^2+\left(y-\frac{c+d}{2}\right)^2\right]\|D_2 D_1 f\|_\infty, \tag{4.7.8}$$

for all $(x,y)\in\Delta$.

Using (4.7.8) we have the following Theorem given by Dragomir, Cerone, Barnett and Roumeliotis in [37].

Theorem 4.7.2. Let $I_n, J_m, \xi_i, \eta_j, h_i, l_j, \xi, \eta$ be as given in subsection 4.7.1 and $f:\Delta\to\mathbb{R}$ be as in Theorem 4.2.2. Then we have

$$\int_a^b\int_c^d f(t,s)\,ds\,dt = R(f,I_n,J_m,\xi,\eta)+W(f,I_n,J_m,\xi,\eta), \tag{4.7.9}$$

where $R(f,I_n,J_m,\xi,\eta)$ is the Riemann sum defined by

$$R(f,I_n,J_m,\xi,\eta)=\sum_{i=0}^{n-1}\sum_{j=0}^{m-1} h_i l_j f(\xi_i,\eta_j), \tag{4.7.10}$$

and the remainder through the approximation $W(f,I_n,J_m,\xi,\eta)$ satisfies the estimation

$$|W(f,I_n,J_m,\xi,\eta)|\leqslant (d-c)\|D_1 f\|_\infty\sum_{i=0}^{n-1}\left[\frac{1}{4}h_i^2+\left(\xi_i-\frac{x_i+x_{i+1}}{2}\right)^2\right]$$

$$+(b-a)\|D_2 f\|_\infty\sum_{j=0}^{m-1}\left[\frac{1}{4}l_j^2+\left(\eta_j-\frac{y_j+y_{j+1}}{2}\right)^2\right]$$

$$+\|D_2 D_1 f\|_\infty\sum_{i=0}^{n-1}\left[\frac{1}{4}h_i^2+\left(\xi_i-\frac{x_i+x_{i+1}}{2}\right)^2\right]\sum_{j=0}^{m-1}\left[\frac{1}{4}l_j^2+\left(\eta_j-\frac{y_j+y_{j+1}}{2}\right)^2\right]$$

$$\leqslant \frac{1}{2}(d-c)\|D_1 f\|_\infty\sum_{i=0}^{n-1}h_i^2+\frac{1}{2}(b-a)\|D_2 f\|_\infty\sum_{j=0}^{m-1}l_j^2+\frac{1}{4}\|D_2 D_1 f\|_\infty\sum_{i=0}^{n-1}h_i^2\sum_{j=0}^{m-1}l_j^2$$

$$\leqslant \frac{1}{2}(d-c)(b-a)\left[\nu(h)\|D_1 f\|_\infty+\nu(l)\|D_2 f\|_\infty+\frac{1}{2}\|D_2 D_1 f\|_\infty\nu(h)\nu(l)\right], \tag{4.7.11}$$

where $\nu(h)=\max\{h_i,\ i=0,\dots,n-1\}$ and $\nu(l)=\max\{l_j,\ j=0,\dots,m-1\}$.

Proof. Apply (4.7.8) on $[x_i, x_{i+1}] \times [y_j, y_{j+1}]$ to obtain

$$\left| \int_{x_i}^{x_{i+1}} \int_{y_j}^{y_{j+1}} f(t,s) ds dt - h_i l_j f(\xi_i, \eta_j) \right|$$

$$\leqslant \left[\frac{1}{4} h_i^2 + \left(\xi_i - \frac{x_i + x_{i+1}}{2} \right)^2 \right] l_j \|D_1 f\|_\infty + \left[\frac{1}{4} l_j^2 + \left(\eta_j - \frac{y_j + y_{j+1}}{2} \right)^2 \right] h_i \|D_2 f\|_\infty$$

$$+ \left[\frac{1}{4} h_i^2 + \left(\xi_i - \frac{x_i + x_{i+1}}{2} \right)^2 \right] \left[\frac{1}{4} l_j^2 + \left(\eta_j - \frac{y_j + y_{j+1}}{2} \right)^2 \right] \|D_2 D_1 f\|_\infty, \qquad (4.7.12)$$

for all $i = 0, \ldots, n-1$; $j = 0, \ldots, m-1$.

Summing both sides of (4.7.12) over i from 0 to $n-1$ and over j from 0 to $m-1$, we get the desired estimation (4.7.11).

4.7.3 *Application to numerical integration*

In [64], Hanna, Dragomir and Cerone employed the inequality given in Theorem 4.2.3 to approximate two dimensional integrals for n-time differentiable mappings via the application of function evaluations of one dimensional integrals at the boundary and interior points.

The following application in numerical integration is found in [64].

Theorem 4.7.3. Let $f : \Delta \to \mathbb{R}$ be as in Theorem 4.2.3. In addition, let I_ν and J_μ be arbitrary divisions of $[a,b]$ and $[c,d]$ respectively, that is,

$$I_\nu : a = \xi_0 < \xi_1 < \cdots < \xi_\nu = b,$$

with $x_i \in (\xi_i, \xi_{i+1})$ for $i = 0, \ldots, \nu - 1$, and

$$J_\mu : c = \tau_0 < \tau_1 < \cdots < \tau_\mu = d,$$

with $y_j \in (\tau_j, \tau_{j+1})$ for $j = 0, \ldots, \mu - 1$, then we have the cubature formula

$$\int_a^b \int_c^d f(t,s) ds dt = \sum_{k=0}^{n-1} \sum_{l=0}^{m-1} \sum_{i=0}^{\nu-1} \sum_{j=0}^{\mu-1} X_k^{(I)}(x_i) Y_l^{(j)}(y_j) \frac{\partial^{i+j} f(x_i, y_j)}{\partial x^i \partial y^j}$$

$$+(-1)^m \sum_{k=0}^{n-1} \sum_{i=0}^{\nu-1} \sum_{j=0}^{\mu-1} X_k^{(I)}(x_i) \int_{\tau_j}^{\tau_{j+1}} S_m^{(j)}(y_j, s) \frac{\partial^{k+m} f(x_i, s)}{\partial x^k \partial s^m} ds$$

$$+(-1)^n \sum_{l=0}^{m-1} \sum_{i=0}^{\nu-1} \sum_{j=0}^{\mu-1} Y_l^{(j)}(y_j) \int_{\xi_i}^{\xi_{i+1}} K_n^{(I)}(x_i, t) \frac{\partial^{n+l} f(t, y_j)}{\partial t^n \partial y^l} dt + R(f, I_\nu, J_\mu, x, y), \qquad (4.7.13)$$

where the remainder term satisfies the estimation

$$|R(f,I_\nu,J_\mu,cx,y)|$$

$$\leqslant \begin{cases} \dfrac{\left\|\dfrac{\partial^{n+m}f}{\partial t^n\partial s^m}\right\|_\infty}{(n+1)!(m+1)!}\sum_{i=0}^{\nu-1}\left[(x_i-\xi_i)^{n+1}+(\xi_{i+1}-x_i)^{n+1}\right] \\ \qquad\times\sum_{j=0}^{\mu-1}\left[(y_j-\tau_j)^{m+1}+(\tau_{j+1}-y_j)^{m+1}\right], \\ \qquad\qquad \text{if } \dfrac{\partial^{n+m}f}{\partial t^n\partial s^m}\in L_\infty(\Delta); \\ \dfrac{\left\|\dfrac{\partial^{n+m}f}{\partial t^n\partial s^m}\right\|_p}{n!m!}\sum_{i=0}^{\nu-1}\left[\dfrac{(x_i-\xi_i)^{nq+1}+(\xi_{i+1}-x_i)^{nq+1}}{nq+1}\right]^{\frac{1}{q}} \\ \qquad\times\sum_{j=0}^{\mu-1}\left[\dfrac{(y_j-\tau_j)^{mq+1}+(\tau_{j+1}-y_j)^{mq+1}}{mq+1}\right]^{\frac{1}{q}}, \\ \qquad\qquad \text{if } \dfrac{\partial^{n+m}f}{\partial t^n\partial s^m}\in L_p(\Delta);\ \dfrac{1}{p}+\dfrac{1}{q}=1,\ p>1; \\ \dfrac{\left\|\dfrac{\partial^{n+m}f}{\partial t^n\partial s^m}\right\|_1}{4(n!m!)}\sum_{i=0}^{\nu-1}\left[(x_i-\xi_i)^n+(\xi_{i+1}-x_i)^n+\left|(x_i-\xi_i)^n-(\xi_{i+1}-x_i)^n\right|\right] \\ \qquad\times\sum_{j=0}^{\mu-1}\left[(y_j-\tau_j)^m+(\tau_{j+1}-y_j)^m+\left|(y_j-\tau_j)^m-(\tau_{j+1}-y_j)^m\right|\right], \\ \qquad\qquad \text{if } \dfrac{\partial^{n+m}f}{\partial t^n\partial s^m}\in L_1(\Delta); \end{cases}$$

where $X_k^{(I)}$ $(k=0,\dots,n-1;\ i=0,\dots,\nu-1)$, $Y_l^{(J)}$ $(l=0,\dots,m-1;\ j=0,\dots,\mu-1)$ and $K_n^{(I)}$ $(i=0,\dots,\nu-1)$, $S_m^{(J)}$ $(j=0,\dots,\mu-1)$ are defined by

$$X_k^{(I)}(x_i)=\frac{(\xi_{i+1}-x_i)^{k+1}+(-1)^k(x_i-\xi_i)^{k+1}}{(k+1)!},$$

$$Y_l^{(J)}(y_j)=\frac{(\tau_{j+1}-y_j)^{l+1}+(-1)^l(y_j-\tau_j)^{l+1}}{(l+1)!},$$

and

$$K_n^{(I)}(x_i,t)=\begin{cases}\dfrac{(t-\xi_i)^n}{n!}, & t\in[\xi_i,x_i],\\ \dfrac{(t-\xi_{i+1})^n}{n!}, & t\in(x_i,\xi_{i+1}],\end{cases}$$

$$S_m^{(J)}(y_j,s)=\begin{cases}\dfrac{(s-\tau_j)^m}{m!}, & s\in[\tau_j,y_j],\\ \dfrac{(s-\tau_{j+1})^m}{m!}, & s\in(y_j,\tau_{j+1}].\end{cases}$$

The proof is obvious by Theorem 4.2.3 applied on $[\xi_i,\xi_{i+1}]\times[\tau_j,\tau_{j+1}]$, $i=0,\dots,\nu-1$; $j=0,\dots,\mu-1$, and we omit the details.

4.8 Miscellaneous inequalities

4.8.1 *Dragomir, Barnett and Cerone [29]*

Let $f : [a,b] \times [c,d] \to \mathbb{R}$ be a continuous mapping on $[a,b] \times [c,d]$ and $f''_{x,y} = \frac{\partial^2 f}{\partial x \partial y}$ exists on $(a,b) \times (c,d)$ and is in $L_p((a,b) \times (c,d))$, i.e.,

$$\left\|f''_{s,t}\right\|_p = \left(\int_a^b \int_c^d \left|\frac{\partial^2 f(x,y)}{\partial x \partial y}\right|^p dxdy\right)^{\frac{1}{p}} < \infty, p > 1,$$

then we have the inequality

$$\left|\int_a^b \int_c^d f(s,t)dsdt - \left[(b-a)\int_c^d f(x,t)dt \right.\right.$$

$$\left.\left. +(d-c)\int_a^b f(s,y)ds - (d-c)(b-a)f(x,y)\right]\right|$$

$$\leqslant \left[\frac{(x-a)^{q+1}+(b-x)^{q+1}}{q+1}\right]^{\frac{1}{q}} \left[\frac{(y-c)^{q+1}+(d-y)^{q+1}}{q+1}\right]^{\frac{1}{q}} \|f''_{s,t}\|_p,$$

for all $(x,y) \in [a,b] \times [c,d]$, where $\frac{1}{p} + \frac{1}{q} = 1$.

4.8.2 *Hanna and Roumeliotis [63]*

Let $f : [a_1,b_1] \times [a_2,b_2] \to \mathbb{R}$ be such that all its partial derivatives upto order 2 exist and be continuous, i.e., $\frac{\partial^i f}{\partial t_1^j \partial t_2^k} < \infty$, $i = 1,2$; $j = 0,\ldots,i$; $k = i-j$. Furthermore, let $w : (a_1,b_1) \times (a_2,b_2) \to (0,\infty)$ be integrable i.e., $\int_{a_1}^{b_1}\int_{a_2}^{b_2} w(t_1,t_2)dt_2dt_1 < \infty$. Then for all $(x_1,x_2) \in [a_1,b_1] \times [a_2,b_2]$ the following integral inequality holds

$$\left|\int_{a_1}^{b_1}\int_{a_2}^{b_2} w(t_1,t_2)f(t_1,t_2)dt_2dt_1 - f(x_1,x_2)\int_{a_1}^{b_1}\int_{a_2}^{b_2} w(t_1,t_2)dt_2dt_1\right.$$

$$+\frac{\partial f(x_1,x_2)}{\partial t_1}\int_{a_1}^{b_1}\int_{a_2}^{b_2} w(t_1,t_2)(x_1-t_1)dt_2dt_1$$

$$\left.+\frac{\partial f(x_1,x_2)}{\partial t_2}\int_{a_1}^{b_1}\int_{a_2}^{b_2} w(t_1,t_2)(x_2-t_2)dt_2dt_1\right|$$

$$\leqslant \frac{1}{2}\left\|\frac{\partial^2 f}{\partial t_1^2}\right\|_\infty \int_{a_1}^{b_1}\int_{a_2}^{b_2} w(t_1,t_2)(x_1-t_1)^2 dt_2dt_1$$

$$+\frac{1}{2}\left\|\frac{\partial^2 f}{\partial t_2^2}\right\|_\infty \int_{a_1}^{b_1}\int_{a_2}^{b_2} w(t_1,t_2)(x_2-t_2)^2 dt_2dt_1$$

$$+\left\|\frac{\partial^2 f}{\partial t_1 \partial t_2}\right\|_\infty \int_{a_1}^{b_1}\int_{a_2}^{b_2} w(t_1,t_2)|x_1-t_1||x_2-t_2|dt_2dt_1.$$

4.8.3 *Hanna and Dragomir [65]*

Let $f : [a,b] \times [c,d] \to \mathbb{R}$ be a function so that the partial derivatives $D_1 f(x,y)$, $D_2 f(x,y)$ and $D_1 D_2 f(x,y)$ exist and are continuous on $[a,b] \times [c,d]$. If $|D_1 f(x,y)|$ is convex over first direction, $|D_2 f(x,y)|$ is convex over the second direction and $|D_1 D_2 f(x,y)|$ is convex in both directions, then we have the inequality

$$\left| \frac{1}{(b-a)(d-c)} \int_a^b \int_c^d f(t,s)dsdt - f(x_0,y_0) \right|$$

$$\leqslant \frac{[\|D_1 f(x_0,\cdot)\|_\infty + \|D_1 f\|_\infty]}{2(b-a)} \left[\frac{1}{4}(b-a)^2 + \left(x_0 - \frac{a+b}{2} \right)^2 \right]$$

$$+ \frac{[\|D_2 f(\cdot,y_0)\|_\infty + \|D_2 f\|_\infty]}{2(d-c)} \left[\frac{1}{4}(d-c)^2 + \left(y_0 - \frac{c+d}{2} \right)^2 \right]$$

$$+ \frac{[\|D_1 D_2 f(x_0,\cdot)\|_\infty + |D_1 D_2 f(x_0,y_0)| + \|D_1 D_2 f(\cdot,y_0)\|_\infty + \|D_1 D_2 f\|_\infty]}{4(b-a)(d-c)}$$

$$\times \left[\frac{1}{4}(b-a)^2 + \left(x_0 - \frac{a+b}{2} \right)^2 \right] \left[\frac{1}{4}(d-c)^2 + \left(y_0 - \frac{c+d}{2} \right)^2 \right],$$

for all $(x_0,y_0) \in [a,b] \times [c,d]$, where

$$\|D_1 f(x_0,\cdot)\|_\infty = \sup_{s \in [c,d]} |D_1 f(x_0,s)| < \infty,$$

$$\|D_2 f(\cdot,y_0)\|_\infty = \sup_{t \in [a,b]} |D_2 f(t,y_0)| < \infty,$$

$$\|D_1 f\|_\infty = \sup_{(t,s) \in [a,b] \times [c,d]} |D_1 f(t,s)| < \infty,$$

$$\|D_1 D_2 f(x_0,\cdot)\|_\infty = \sup_{s \in [c,d]} |D_1 D_2 f(x_0,s)| < \infty,$$

$$\|D_1 D_2 f(\cdot,y_0)\|_\infty = \sup_{t \in [a,b]} |D_1 D_2 f(t,y_0)| < \infty,$$

and

$$\|D_1 D_2 f\|_\infty = \sup_{(t,s) \in [a,b] \times [c,d]} |D_1 D_2 f(t,s)| < \infty.$$

4.8.4 *Hanna, Dragomir and Cerone [62]*

Let I, J be two closed intervals and $f: I\times J\to\mathbb{R}$ be a mapping so that the partial derivatives $\frac{\partial^{i+m+1}f(a,\cdot)}{\partial x^i\partial y^{m+1}}$ $(i=0,\ldots,n)$, $\frac{\partial^{j+n+1}f(\cdot,b)}{\partial x^{n+1}\partial y^j}$ $(j=0,\ldots,m)$ and $\frac{\partial^{n+m+2}f(\cdot,\cdot)}{\partial x^{n+1}\partial y^{m+1}}$ exist on the intervals J,I and $I\times J$ respectively, where $a\in I$ and $b\in J$ are given. Let $x\in I$ and $y\in J$ and assume that $\frac{\partial^{i+m+1}f(a,\cdot)}{\partial x^i\partial y^{m+1}}$ are continuous on $[b,y]$ $(i=0,\ldots,n)$, $\frac{\partial^{j+n+1}f(\cdot,b)}{\partial x^{n+1}\partial y^j}$ are continuous on $[a,x]$ $(j=0,\ldots,m)$ and $\frac{\partial^{n+m+2}f(\cdot,\cdot)}{\partial x^{n+1}\partial y^{m+1}}$ is continuous on $[a,x]\times[b,y]$. Then we have the inequality

$$\Bigg| f(x,y)-\sum_{i=0}^{n}\sum_{j=0}^{m}\frac{(x-a)^i}{i!}\frac{(y-b)^j}{j!}\frac{\partial^{i+j}f(a,b)}{\partial x^i\partial y^j}$$

$$-\frac{1}{m!}\sum_{i=0}^{n}\frac{(x-a)^i}{i!}\int_b^y(y-s)^m\frac{\partial^{i+m+1}f(a,s)}{\partial x^i\partial y^{m+1}}ds-\frac{1}{n!}\sum_{j=0}^{m}\frac{(y-b)^j}{j!}\int_a^x(x-t)^n\frac{\partial^{j+n+1}f(t,b)}{\partial x^{n+1}\partial y^j}dt\Bigg|$$

$$\leqslant\begin{cases}\dfrac{1}{(n+1)!(m+1)!}(x-a)^{n+1}(y-b)^{m+1}\left\|\dfrac{\partial^{n+m+2}f}{\partial x^{n+1}\partial y^{m+1}}\right\|_{\infty;[a,x]\times[b,y]},\\ \qquad\text{if } \dfrac{\partial^{n+m+2}f}{\partial x^{n+1}\partial y^{m+1}}\in L_\infty[[a,x]\times[b,y]];\\ \dfrac{1}{n!(nq+1)^{\frac{1}{q}}m!(mq+1)^{\frac{1}{q}}}(x-a)^{n+\frac{1}{q}}(y-b)^{m+\frac{1}{q}}\left\|\dfrac{\partial^{n+m+2}f}{\partial x^{n+1}\partial y^{m+1}}\right\|_{p;[a,x]\times[b,y]},\\ \qquad\text{if } \dfrac{\partial^{n+m+2}f}{\partial x^{n+1}\partial y^{m+1}}\in L_p[[a,x]\times[b,y]];\quad p>1,\ \dfrac{1}{p}+\dfrac{1}{q}=1,\\ \dfrac{1}{n!m!}(x-a)^n(y-b)^m\left\|\dfrac{\partial^{n+m+2}f}{\partial x^{n+1}\partial y^{m+1}}\right\|_{1;[a,x]\times[b,y]},\\ \qquad\text{if } \dfrac{\partial^{n+m+2}f}{\partial x^{n+1}\partial y^{m+1}}\in L_1[[a,x]\times[b,y]];\end{cases}$$

where $\|\ \|_{p;[a,x]\times[b,y]}$ is the usual p norm $(p\in[1,\infty])$ on the region $[a,x]\times[b,y]$.

4.8.5 *Milovanović [76]*

Let $f:\mathbb{R}^m\to\mathbb{R}$ be a differentiable function defined on $D=\{(x_1,\ldots,x_m): a_i\leqslant x_i\leqslant b_i\}$ for $(i=1,\ldots,m)$ and let $\left|\frac{\partial f}{\partial x_i}\right|\leqslant M_i$ $(M_i>0,\ i=1,\ldots,m)$ in D. Furthermore, let the function $x\to p(x)$ be integrable and $p(x)>0$ for every $x\in D$. Then for every $x\in D$, we have the inequality

$$\left|f(x)-\frac{\int_D p(y)f(y)dy}{\int_D p(y)dy}\right|\leqslant\frac{\sum_{i=1}^{m}M_i\int_D p(y)|x_i-y_i|dy}{\int_D p(y)dy}.$$

4.8.6 *Dragomir, Barnett and Cerone [44]*

Let the function f and the set B be as in Theorem 4.4.2. Let $w : B \to \mathbb{R}$ be a nonnegative and integrable function and $\int_B w(y)dy > 0$. Then we have the inequality

$$\left| f(x) - \frac{\int_B w(y)f(y)dy}{\int_B w(y)dy} \right| \leqslant \frac{\sum_{i=1}^{n} L_i \int_B w(y)|x_i - y_i|^{r_i} dy}{\int_B w(y)dy},$$

for all $x \in B$.

4.8.7 *Anastassiou [2]*

Let $f \in C^1\left(\prod_{i=1}^{k}[a_i,b_i]\right)$, where $a_i < b_i$; $a_i,\ b_i \in \mathbb{R}$, $i = 1,\ldots,k$, and let $\vec{x}_0 = (x_{01},\ldots,x_{0k}) \in \prod_{i=1}^{k}[a_i,b_i]$ be fixed. Then

$$\left| \frac{1}{\prod\limits_{t=1}^{k}(b_i - a_i)} \int_{a_1}^{b_1} \cdots \int_{a_k}^{b_k} f(z_1,\ldots,z_k)dz_1\cdots dz_k - f(\vec{x}_0) \right|$$

$$\leqslant \sum_{i=1}^{k} \frac{(x_{0i} - a_i)^2 + (b_i - x_{0i})^2}{2(b_i - a_i)} \left\| \frac{\partial f}{\partial z_i} \right\|_\infty .$$

The inequality is sharp, namely the optimal function is

$$f^\circ(z_1,\ldots,z_k) = \sum_{i=1}^{k} |z_i - x_{0i}|^{\alpha_i}, \quad \alpha_i > 1.$$

4.8.8 *Anastassiou [2]*

Let Q be a compact and convex subset of $\mathbb{R}^k$, $k \geqslant 1$. Let $f \in C^{n+1}(Q)$, $n \in \mathbb{N}$ and $\vec{x}_0 = (x_{01},\ldots,x_{0k}) \in Q$ be fixed such that all partial derivatives $f_\alpha = \frac{\partial^\alpha f}{\partial z^\alpha}$, where $\alpha = (\alpha_1,\ldots,\alpha_k)$, $\alpha_i \in Z^+$, $i = 1,\ldots,k$, $|\alpha| = \sum_{i=1}^{k}\alpha_i = j$, $j = 1,\ldots,n$ fulfill $f_\alpha(\vec{x}_0) = 0$. Then

$$\left| \frac{1}{\mathrm{Vol}(Q)} \int_Q f(\vec{z})d\vec{z} - f(\vec{x}_0) \right| \leqslant \frac{D_{n+1}(f)}{(n+1)!\mathrm{Vol}(Q)} \int_Q \left(\|\vec{z} - \vec{x}_0\|_l\right)^{n+1} d\vec{z},$$

where

$$D_{n+1}(f) = \max_{\alpha:|\alpha|=n+1} \|f_\alpha\|_\infty,$$

and

$$\|\vec{z} - \vec{x}_0\|_l = \sum_{i=l}^{k} |z_i - x_{0i}|.$$

4.8.9 *Pachpatte [92]*

Let D, $\overline{D}$ f, g, $\frac{\partial f}{\partial x_i}$, $\frac{\partial g}{\partial x_i}$ be as in Theorem 4.5.1. Let $w : D \to \mathbb{R}$ be a nonnegative and integrable function and $\int_D w(y)dy > 0$. Then we have the inequality

$$\left| f(x)g(x) - \left[\frac{g(x)\int_D w(y)f(y)dy + f(x)\int_D w(y)g(y)dy}{2\int_D w(y)dy} \right] \right|$$

$$\leqslant \frac{\int_D w(y) \sum_{i=1}^{n} \left[|g(x)| \left\| \frac{\partial f}{\partial x_i} \right\|_\infty + |f(x)| \left\| \frac{\partial g}{\partial x_i} \right\|_\infty \right] |x_i - y_i| dy}{2\int_D w(y)dy},$$

for all $x \in \overline{D}$.

4.8.10 *Pachpatte [86]*

Let $A = \{1,\ldots,k+1\}$, $B = \{1,\ldots,m+1\}$ $(k,m \in \mathbb{N})$ and $E = A \times B$. Let $f : E \to \mathbb{R}$ be a function such that $\Delta_1 f(x,y)$, $\Delta_2 f(x,y)$, $\Delta_2\Delta_1 f(x,y)$ exist on E. Then

$$\left| \sum_{s=1}^{k}\sum_{t=1}^{m} f(s,t) - \frac{1}{2}\left\{ m\sum_{s=1}^{k}[f(s,1)+f(s,m+1)] + k\sum_{t=1}^{m}[f(1,t)+f(k+1,t)] \right\} \right.$$

$$\left. +\frac{1}{4}km\,[f(1,1)+f(1,m+1)+f(k+1,1)+f(k+1,m+1)] \right| \leqslant \frac{1}{4}km\sum_{s=1}^{k}\sum_{t=1}^{m} |\Delta_2\Delta_1 f(s,t)|.$$

4.8.11 *Pachpatte[83]*

Let $A = \{1,\ldots,k+1\}$, $B = \{1,\ldots,m+1\}$, $C = \{1,\ldots,n+1\}$ $(k,\ m,\ n \in \mathbb{N})$ and $G = A \times B \times C$. Let $f : G \to \mathbb{R}$ be a function such that $\Delta_1 f(x,y,z)$, $\Delta_2 f(x,y,z)$, $\Delta_3\Delta_2\Delta_1 f(x,y,z)$ exist on G. Then

$$\left| \sum_{u=1}^{k}\sum_{v=1}^{m}\sum_{w=1}^{n} f(u,v,w) - \frac{1}{8}[f(1,1,1)+f(k+1,m+1,n+1)] \right.$$

$$+\frac{1}{4}mn\sum_{u=1}^{k}[f(u,1,1)+f(u,1,n+1)+f(u,m+1,1)+f(u,m+1,n+1)]$$

$$+\frac{1}{4}kn\sum_{v=1}^{m}[f(k+1,v,n+1)+f(k+1,v,1)+f(1,v,n+1)+f(1,v,1)]$$

$$+\frac{1}{4}km\sum_{w=1}^{n}[f(k+1,m+1,w)+f(k+1,1,w)+f(1,m+1,w)+f(1,1,w)]$$

$$-\frac{1}{2}k\sum_{v=1}^{m}\sum_{w=1}^{n}[f(1,v,w)+f(k+1,v,w)]-\frac{1}{2}m\sum_{u=1}^{k}\sum_{w=1}^{n}[f(u,1,w)+f(u,m+1,w)]$$

$$\left.-\frac{1}{2}n\sum_{u=1}^{k}\sum_{v=1}^{m}[f(u,v,1)+f(u,v,n+1)]\right| \leqslant \frac{1}{8}kmn\sum_{u=1}^{k}\sum_{v=1}^{m}\sum_{w=1}^{n}|\Delta_3\Delta_2\Delta_1 f(u,v,w)|.$$

4.8.12 *Pachpatte [87]*

Let $A_1 = \{1,\ldots,a_1+1\},\ldots, A_n = \{1,\ldots,a_n+1\}$ $(a_1,\ldots,a_n \in \mathbb{N})$ and $H = A_1 \times \cdots \times A_n$. Let $u : H \to \mathbb{R}$ be a function such that $\Delta_i u(x)$ $(i = 1,\ldots,n)$ exist on H. Then

$$\left|\sum_H u(x) - \frac{1}{2n}\left\{a_1\sum_{x_2=1}^{a_2}\cdots\sum_{x_n=1}^{a_n}[u(1,x_2,\ldots,x_n)+u(a_1+1,x_2,\ldots,x_n)] + \cdots +\right.\right.$$

$$\left.\left.a_n\sum_{x_1=1}^{a_1}\cdots\sum_{x_{n-1}=1}^{a_{n-1}}[u(x_1,\ldots,x_{n-1},1)+u(x_1,\ldots,x_{n-1},a_n+1)]\right\}\right| \leqslant \frac{1}{2n}\left(\sum_{i=1}^{n}a_i\sum_H|\Delta_i u(x)|\right),$$

where for a suitable function $w(x)$ defined on H the notation

$$\sum_H w(x) = \sum_{x_1=1}^{a_1}\cdots\sum_{x_n=1}^{a_n} w(x_1,\ldots,x_n),$$

is used to simplify the presentation.

4.8.13 *Pachpatte [95]*

Let $f(x)$, $\Delta_i f(x)$ and Q be as in Theorem 4.6.3. Let $w(y)$ be a real-valued nonnegative function defined on Q and $\sum_y w(y) > 0$. Then for $x,\ y \in Q$,

$$\left|f(x) - \frac{\sum_y w(y)f(y)}{\sum_y w(y)}\right| \leqslant \frac{\sum_y w(y)\sum_{i=1}^{n}\|\Delta_i f\|_\infty|x_i-y_i|}{\sum_y w(y)},$$

where for a suitable function u defined on Q, the notation $\sum_y u(y)$ is defined earlier in Section 2.6.

4.8.14 *Pachpatte [95]*

Let $f(x), g(x)$, $\Delta_i f(x)$, $\Delta_i g(x)$ and Q be as in Theorem 4.6.3. Let $w(y)$ be a real-valued nonnegative function defined on Q and $\sum_y w(y) > 0$. Then for $x,\ y \in Q$,

$$\left|f(x)g(x) - \frac{g(x)\sum_y w(y)f(y) + f(x)\sum_y w(y)g(y)}{2\sum_y w(y)}\right|$$

$$\leqslant \frac{\sum\limits_{y} w(y) \sum\limits_{i=1}^{n} [|g(x)| \|\Delta_i f\|_\infty + |f(x)| \|\Delta_i g\|_\infty] |x_i - y_i|}{2 \sum\limits_{y} w(y)},$$

where for a suitable function u defined on Q, the notation $\sum_y u(y)$ is defined earlier in Section 2.6.

4.9 Notes

A number of authors have written about multidimensional generalizations, extensions and variants of the Ostrowsk's inequality given in (7). The inequality in Theorem 4.2.1 is due to Barnett and Dragomir [8], which may be regarded as a generalization of the classical result due to Ostrowski given in (7) for double integrals. The inequalities in Theorems 4.2.2 and 4.2.3 are taken from Dragomir, Cerone, Barnett and Roumeliotis [37] and Hanna, Dragomir and Cerone [64] respectively. Theorems 4.2.4 and 4.2.5 contains the Ostrowski type inequalities for double integrals established by Pachpatte in [115] and [125] respectively. Theorem 4.3.1 deals with the Ostrowski type inequality involving functions of three independent variables and is taken from Pachpatte [83]. Lemma 4.3.1 and Theorem 4.3.2 are due to Sofo [148], while Theorems 4.3.3 and 4.3.4 are due to Pachpatte [125] and [134] respectively.

Theorem 4.4.1 contains the multivariate version of the Ostrowski's inequality (7), first proved by Milovanović in [76]. The results in Theorem 4.4.2 and Corollaries 4.4.1–4.4.3 deal with the Ostrowski type inequalities for multivariate mappings of the r-Hölder type and are taken from Dragomir, Barnett and Cerone [44] and Theorem 4.4.3 is taken from Pachpatte [87]. Theorems 4.5.1 and 4.5.2 are taken from Pachpatte [92], while Theorem 4.5.3 is new. Section 4.6 deals with the discrete inequalities of Ostrowski type involving functions of many independent variables and are taken from Pachpatte [95,125,134]. Section 4.7 is devoted to the applications of some of the inequalities given by Barnett and Dragomir [8], Dragomir, Cerone, Barnett and Roumeliotis [37] and Hanna, Dragomir and Cerone [64]. Section 4.8 deals with some miscellaneous multivariate inequalities related to the Ostrowski's inequality in (7).

Chapter 5

Inequalities via convex functions

5.1 Introduction

The lasting influence of the fundamental inequalities in (8) and (9) due to Hadamard and Jensen in the development of various branches of mathematics is enormous. In view of the usefulness of these inequalities in analysis and their applications, the study of the inequalities of the type (8) and (9) has been focus of great attention by many researchers, interested both in theory and applications. In the last two decades, a large number of papers related to these inequalities have appeared, which deal with various generalizations, numerous variants and applications. This chapter deals with some fundamental results related to these inequalities, recently investigated in the literature by various investigators. For earlier results, as well as additional references, see [45,108]. Some applications are also given to illustrate the usefulness of certain inequalities.

5.2 Integral inequalities involving convex functions

In this section, we offer some new integral inequalities analogues to that of Hadamard's inequalities (8), involving the product of two convex functions, investigated by Pachpatte in [85,97].

The following Lemma proved in [143], see also [108, pp. 61-62], which deals with simple chaterizations of convex functions is crucial in our discussion.

Lemma 5.2.1. Let $[a,b] \subset \mathbb{R}$ $(a<b)$, $t \in [0,1]$ and a function $f : [a,b] \to \mathbb{R}$. The following statements are equivalent:

(i) f is convex on $[a,b]$ and

(ii) for all $x,\ y \in [a,b]$, the function $p : [0,1] \to \mathbb{R}$ defined by $p(t) = f(tx+(1-t)y)$ is convex on $[0,1]$.

The following Theorem deals with the Hadamard type inequalities proved in [97].

Theorem 5.2.1. Let f and g be real-valued, nonnegative and convex functions on $[a,b] \subset \mathbb{R}$ $(a<b)$. Then

$$\frac{1}{b-a}\int_a^b f(x)g(x)dx \leqslant \frac{1}{3}M(a,b)+\frac{1}{6}N(a,b), \tag{5.2.1}$$

$$2f\left(\frac{a+b}{2}\right)g\left(\frac{a+b}{2}\right) \leqslant \frac{1}{b-a}\int_a^b f(x)g(x)dx+\frac{1}{6}M(a,b)+\frac{1}{3}N(a,b), \tag{5.2.2}$$

where $M(a,b)=f(a)g(a)+f(b)g(b)$, $N(a,b)=f(a)g(b)+f(b)g(a)$.

Proof. Since f and g are convex on $[a,b]$, then for $t\in[0,1]$, we have

$$f(ta+(1-t)b) \leqslant tf(a)+(1-t)f(b), \tag{5.2.3}$$

$$g(ta+(1-t)b) \leqslant tg(a)+(1-t)g(b). \tag{5.2.4}$$

From (5.2.3) and (5.2.4), we obtain

$$f(ta+(1-t)b)g(ta+(1-t)b)$$

$$\leqslant t^2f(a)g(a)+(1-t)^2f(b)g(b)+t(1-t)[f(a)g(b)+f(b)g(a)]. \tag{5.2.5}$$

By Lemma 5.2.1, $f(ta+(1-t)b)$ and $g(ta+(1-t)b)$ are convex on $[0,1]$, they are integrable on $[0,1]$ and consequently $f(ta+(1-t)b)g(ta+(1-t)b)$ is also integrable on $[0,1]$. Similarly, since f and g are convex on $[a,b]$, they are integrable on $[a,b]$ and hence fg is also integrable on $[a,b]$. Integrating both sides of (5.2.5) over $[0,1]$, we get

$$\int_0^1 f(ta+(1-t)b)g(ta+(1-t)b)dt \leqslant \frac{1}{3}M(a,b)+\frac{1}{6}N(a,b). \tag{5.2.6}$$

By substituting $ta+(1-t)b=x$, it is easy to observe that

$$\int_0^1 f(ta+(1-t)b)g(ta+(1-t)b)dt = \frac{1}{b-a}\int_a^b f(x)g(x)dx. \tag{5.2.7}$$

Using (5.2.7) in (5.2.6), we get the the desired inequality in (5.2.1).

Since f and g are convex on $[a,b]$, then for $t\in[0,1]$, we observe that

$$f\left(\frac{a+b}{2}\right)g\left(\frac{a+b}{2}\right)$$

$$=f\left(\frac{ta+(1-t)b}{2}+\frac{(1-t)a+tb}{2}\right)g\left(\frac{ta+(1-t)b}{2}+\frac{(1-t)a+tb}{2}\right)$$

$$\leqslant \frac{1}{4}[f(ta+(1-t)b)+f((1-t)a+tb)]\,[g(ta+(1-t)b)+g((1-t)a+tb)]$$

$$\leqslant \frac{1}{4}[f(ta+(1-t)b)g(ta+(1-t)b)+f((1-t)a+tb)g((1-t)a+tb)]$$

$$+\frac{1}{4}[(tf(a)+(1-t)f(b))((1-t)g(a)+tg(b))$$

$$+((1-t)f(a)+tf(b))(tg(a)+(1-t)g(b))]$$

$$=\frac{1}{4}[f(ta+(1-t)b)g(ta+(1-t)b)+f((1-t)a+tb)g((1-t)a+tb)]$$

$$+\frac{1}{4}\left[2t(1-t)(f(a)g(a)+f(b)g(b))+\left(t^2+(1-t)^2\right)(f(a)g(b)+f(b)g(a))\right]. \quad (5.2.8)$$

Again as explained in the proof of the inequality (5.2.1) given above we integrate both sides of (5.2.8) over $[0,1]$ and obtain

$$f\left(\frac{a+b}{2}\right)g\left(\frac{a+b}{2}\right)$$

$$\leqslant \frac{1}{4}\int_0^1 [f(ta+(1-t)b)g(ta+(1-t)b)+f((1-t)a+tb)g((1-t)a+tb)]\,dt$$

$$+\frac{1}{12}M(a,b)+\frac{1}{6}N(a,b). \quad (5.2.9)$$

From (5.2.9), it is easy to observe that

$$f\left(\frac{a+b}{2}\right)g\left(\frac{a+b}{2}\right)\leqslant \frac{1}{2}\int_0^1 f(ta+(1-t)b)g(ta+(1-t)b)\,dt$$

$$+\frac{1}{12}M(a,b)+\frac{1}{6}N(a,b). \quad (5.2.10)$$

Now, multiplying both sides of (5.2.10) by 2 and using (5.2.7), we get the required inequality in (5.2.2). The proof is complete.

Remark 5.2.1. If we choose $a=0$ and $b=1$ and the convex function $f(x)=cx$ and $g(x)=d(1-x)$, where c, d are positive constants, then it is easy to observe that the inequalities obtained in (5.2.1) and (5.2.2) are sharp in the sense that the equalities in (5.2.1) and (5.2.2) hold.

In the following theorem, we give a slight variant of the corresponding Theorem 2 proved in [97].

Theorem 5.2.2. Let f and g be real-valued, nonnegative and convex functions on $[a,b] \subset \mathbb{R}$ $(a<b)$. Then

$$\frac{3}{2(b-a)^2}\int_a^b\int_a^b\int_0^1 f(tx+(1-t)y)g(tx+(1-t)y)dtdydx$$

$$\leqslant \frac{1}{b-a}\int_a^b f(x)g(x)dx + \frac{1}{8}[M(a,b)+N(a,b)], \tag{5.2.11}$$

$$\frac{3}{b-a}\int_a^b\int_0^1 f\left(tx+(1-t)\left(\frac{a+b}{2}\right)\right)g\left(tx+(1-t)\left(\frac{a+b}{2}\right)\right)dtdx$$

$$\leqslant \frac{1}{b-a}\int_a^b f(x)g(x)dx + \frac{1}{2}[M(a,b)+N(a,b)], \tag{5.2.12}$$

where $t \in [0,1]$ and $M(a,b), N(a,b)$ are as in Theorem 5.2.1.

Proof. Since f and g are convex on $[a,b]$, then for $x,\ y \in [a,b]$ and $t \in [0,1]$, we have

$$f(tx+(1-t)y) \leqslant tf(x)+(1-t)f(y), \tag{5.2.13}$$

$$g(tx+(1-t)y) \leqslant tg(x)+(1-t)g(y). \tag{5.2.14}$$

From (5.2.13) and (5.2.14), we obtain

$$f(tx+(1-t)y)g(tx+(1-t)y)$$

$$\leqslant t^2 f(x)g(x)+(1-t)^2 f(y)g(y)+t(1-t)\left[f(x)g(y)+f(y)g(x)\right]. \tag{5.2.15}$$

Integrating (5.2.15) over $[0,1]$, we obtain

$$\int_0^1 f(tx+(1-t)y)g(tx+(1-t)y)dt$$

$$\leqslant \frac{1}{3}[f(x)g(x)+f(y)g(y)]+\frac{1}{6}[f(x)g(y)+f(y)g(x)]. \tag{5.2.16}$$

Integrating (5.2.16) over $[a,b]\times[a,b]$, we obtain

$$\int_a^b\int_a^b\int_0^1 f(tx+(1-t)y)g(tx+(1-t)y)dtdydx$$

$$\leqslant \frac{1}{3}(b-a)\left[\int_a^b f(x)g(x)dx+\int_a^b f(y)g(y)dy\right]$$

$$+\frac{1}{6}\left[\left(\int_a^b f(x)dx\right)\left(\int_a^b g(y)dy\right)+\left(\int_a^b f(y)dy\right)\left(\int_a^b g(x)dx\right)\right]. \tag{5.2.17}$$

By using the right half of the Hadamard's inequality given in (8) on the right hand side of (5.2.17), we have

$$\int_a^b\int_a^b\int_0^1 f(tx+(1-t)y)g(tx+(1-t)y)dtdydx$$

$$\leqslant \frac{2}{3}(b-a)\int_a^b f(x)g(x)dx+\frac{1}{12}(b-a)^2[M(a,b)+N(a,b)]. \tag{5.2.18}$$

Now, dividing both sides of (5.2.18) by $\frac{2}{3}(b-a)^2$, we get the desired inequality in (5.2.11).

Since f and g are convex on $[a,b]$, we have

$$f\left(tx+(1-t)\left(\frac{a+b}{2}\right)\right)\leqslant tf(x)+(1-t)f\left(\frac{a+b}{2}\right), \tag{5.2.19}$$

$$g\left(tx+(1-t)\left(\frac{a+b}{2}\right)\right)\leqslant tg(x)+(1-t)g\left(\frac{a+b}{2}\right), \tag{5.2.20}$$

for $x\in[a,b]$ and $t\in[0,1]$. From (5.2.19) and (5.2.20), we have

$$\begin{aligned} &f\left(tx+(1-t)\left(\frac{a+b}{2}\right)\right)g\left(tx+(1-t)\left(\frac{a+b}{2}\right)\right)\\ &\leqslant t^2f(x)g(x)+(1-t)^2f\left(\frac{a+b}{2}\right)g\left(\frac{a+b}{2}\right)\\ &+t(1-t)\left[f(x)g\left(\frac{a+b}{2}\right)+f\left(\frac{a+b}{2}\right)g(x)\right]. \end{aligned} \tag{5.2.21}$$

Integrating (5.2.21) over $[0,1]$, we obtain

$$\begin{aligned} &\int_0^1 f\left(tx+(1-t)\left(\frac{a+b}{2}\right)\right)g\left(tx+(1-t)\left(\frac{a+b}{2}\right)\right)dt\\ &\leqslant \frac{1}{3}\left[f(x)g(x)+f\left(\frac{a+b}{2}\right)g\left(\frac{a+b}{2}\right)\right]\\ &+\frac{1}{6}\left[f(x)g\left(\frac{a+b}{2}\right)+f\left(\frac{a+b}{2}\right)g(x)\right]. \end{aligned} \tag{5.2.22}$$

Now, integrating (5.2.22) over $[a,b]$ and using the right half of the Hadamard's inequality given in (8) and the convexity of the functions f, g; we observe that

$$\begin{aligned} &\int_a^b\int_0^1 f\left(tx+(1-t)\left(\frac{a+b}{2}\right)\right)g\left(tx+(1-t)\left(\frac{a+b}{2}\right)\right)dtdx\\ &\leqslant \frac{1}{3}\int_a^b f(x)g(x)dx+\frac{1}{3}(b-a)f\left(\frac{a+b}{2}\right)g\left(\frac{a+b}{2}\right)\\ &+\frac{1}{6}\left[g\left(\frac{a+b}{2}\right)\int_a^b f(x)dx+f\left(\frac{a+b}{2}\right)\int_a^b g(x)dx\right]\\ &\leqslant \frac{1}{3}\int_a^b f(x)g(x)dx+\frac{1}{12}(b-a)(f(a)+f(b))(g(a)+g(b))\\ &+\frac{1}{6}\left[\left(\frac{g(a)+g(b)}{2}\right)(b-a)\left(\frac{f(a)+f(b)}{2}\right)+\left(\frac{f(a)+f(b)}{2}\right)(b-a)\left(\frac{g(a)+g(b)}{2}\right)\right]\\ &=\frac{1}{3}\int_a^b f(x)g(x)dx+\frac{1}{12}(b-a)[M(a,b)+N(a,b)]\\ &+\frac{1}{12}(b-a)[M(a,b)+N(a,b)]. \end{aligned} \tag{5.2.23}$$

Now, multiplying both sides of (5.2.23) by $\frac{3}{b-a}$, we get the required inequality in (5.2.12). The proof is complete.

Remark 5.2.2. It should be noted that, in [70], Klaričić Bakula and Pečarić have noticed some errors while calculating the bounds on inequalities in Theorem 2 given in [97]. In fact, in [97, Theorem 2], $\frac{1}{8}\left[\frac{M(a,b)+N(a,b)}{(b-a)^2}\right]$ stands in place of the term $\frac{1}{8}[M(a,b)+N(a,b)]$ in (5.2.11) and $\frac{1}{4}\left(\frac{1+b-a}{B-1}\right)[M(a,b)+N(a,b)]$ stands in place of the term $\frac{1}{2}[M(a,b)+N(a,b)]$ in (5.2.12), see also [20].

In [24], Dragomir, Pečarić and Persson have proved certain Hadamard-type inequalities for the following classes of functions.

Let I be an interval of R and $a,\ b \in I$ with $a < b$. In [60], Godunova and Levin introduced the following class of functions.

A map $f : I \to \mathbb{R}$ is said to belongs to the class $Q(I)$, if it is nonnegative and for all $x,\ y \in I$ and $\lambda \in (0,1)$, satisfies the inequality

$$f(\lambda x+(1-\lambda)y) \leqslant \frac{f(x)}{\lambda}+\frac{f(y)}{1-\lambda}. \tag{5.2.24}$$

In [60] it is noted that all nonnegative monotone and nonnegative convex functions belong to this class. In [24], Dragomir, Pečarić and Persson restricted the above class of functions and introduced the following class of functions.

A map $f : I \to \mathbb{R}$ is said to belongs to the class $P(I)$ if it is nonnegative and for all $x,\ y \in I$ and $\lambda \in (0,1)$, satisfies the following inequality

$$f(\lambda x+(1-\lambda)y) \leqslant f(x)+f(y). \tag{5.2.25}$$

Obviously, $Q(I) \supset P(I)$ and as noted in [60], $P(I)$ also contain all monotone, convex and quasi-convex functions.

The following two Theorems contains the Hadamard-type inequalities recently established in [85], involving the product of two functions belonging to the above classes of functions.

Theorem 5.2.3. Let $f,g \in Q(I)$ and $f,g \in L_1[a,b]$. Then

$$f\left(\frac{a+b}{2}\right)g\left(\frac{a+b}{2}\right) \leqslant \frac{240}{b-a}\int_a^b L(x)f(x)g(x)dx+40\left[M(a,b)+2N(a,b)\right], \tag{5.2.26}$$

$$\frac{1}{b-a}\int_a^b L(x)f(x)g(x)dx \leqslant \frac{1}{3}M(a,b)+\frac{1}{6}N(a,b), \tag{5.2.27}$$

where

$$L(x) = \left[\frac{(b-x)(x-a)}{(b-a)^2}\right], \quad x \in I, \tag{5.2.28}$$

and $M(a,b) = f(a)g(a)+f(b)g(b)$, $N(a,b) = f(a)g(b)+f(b)g(a)$.

Proof. Since $f,\ g \in Q(I)$, we have for $x,\ y \in I$ (with $\lambda = \frac{1}{2}$ in (5.2.24))

$$f\left(\frac{X+Y}{2}\right) \leqslant 2[f(x)+f(y)], \tag{5.2.29}$$

$$g\left(\frac{X+Y}{2}\right) \leqslant 2[g(x)+g(y)]. \tag{5.2.30}$$

Substituting $x = ta+(1-t)b$, $y = (1-t)a+tb$; $t \in (0,1)$ in (5.2.9) and (5.2.30), we have

$$f\left(\frac{a+b}{2}\right) \leqslant 2[f(ta+(1-t)b)+f((1-t)a+tb)], \tag{5.2.31}$$

$$g\left(\frac{a+b}{2}\right) \leqslant 2[g(ta+(1-t)b)+g((1-t)a+tb)]. \tag{5.2.32}$$

From (5.2.31) and (5.2.32), we observe that

$$\begin{aligned} f\left(\frac{a+b}{2}\right)g\left(\frac{a+b}{2}\right) &\leqslant 4[f(ta+(1-t)b)+f((1-t)a+tb)] \\ &\quad \times [g(ta+(1-t)b)+g((1-t)a+tb)] \\ &= 4[f(ta+(1-t)b)g(ta+(1-t)b)+f((1-t)a+tb)g((1-t)a+tb)] \\ &\quad +4[f(ta+(1-t)b)g((1-t)a+tb)+f((1-t)a+tb)g(ta+(1-t)b)] \\ &\leqslant 4H_1(t)+4H_2(t)+4\left[\left(\frac{f(a)}{t}+\frac{f(b)}{1-t}\right)\left(\frac{g(a)}{1-t}+\frac{g(b)}{t}\right)\right. \\ &\quad \left.+\left(\frac{f(a)}{1-t}+\frac{f(b)}{t}\right)\left(\frac{g(a)}{t}+\frac{g(b)}{1-t}\right)\right], \end{aligned} \tag{5.2.33}$$

where

$$H_1(t) = f(ta+(1-t)b)g(ta+(1-t)b), \tag{5.2.34}$$

$$H_2(t) = f((1-t)a+tb)g((1-t)a+tb). \tag{5.2.35}$$

From (5.2.33), it is easy to observe that

$$\begin{aligned} &t^2(1-t)^2 f\left(\frac{a+b}{2}\right)g\left(\frac{a+b}{2}\right) \\ &\leqslant 4t^2(1-t)^2\left[H_1(t)+H_2(t)\right]+4\left[2t(1-t)M(a,b)+\left(t^2+(1-t)^2\right)N(a,b)\right]. \end{aligned} \tag{5.2.36}$$

Integrating both sides of (5.2.36) with respect to t from 0 to 1, we have

$$\begin{aligned} \frac{1}{30} f\left(\frac{a+b}{2}\right)g\left(\frac{a+b}{2}\right) &\leqslant 4\int_0^1 t^2(1-t)^2 H_1(t)dt \\ &\quad +4\int_0^1 t^2(1-t)^2 H_2(t)dt+\frac{4}{3}M(a,b)+\frac{8}{3}N(a,b). \end{aligned} \tag{5.2.37}$$

It is easy to observe that

$$\int_0^1 t^2(1-t)^2 H_1(t)dt = \frac{1}{b-a}\int_a^b L(x)f(x)g(x)dx, \tag{5.2.38}$$

$$\int_0^1 t^2(1-t)^2 H_2(t)dt = \frac{1}{b-a}\int_a^b L(x)f(x)g(x)dx. \tag{5.2.39}$$

Using (5.2.38) and (5.2.39) in (5.2.37) and rewriting, we get the desired inequality in (5.2.26).

Since $f,\ g \in Q(I)$, we have for $a,\ b \in I$ and $t \in (0,1)$,

$$t(1-t)f(ta+(1-t)b) \leqslant (1-t)f(a)+tf(b), \tag{5.2.40}$$

$$t(1-t)g(ta+(1-t)b) \leqslant (1-t)g(a)+tg(b). \tag{5.2.41}$$

From (5.2.40) and (5.2.41), we observe that

$$t^2(1-t)^2 f(ta+(1-t)b)g(ta+(1-t)b) \leqslant [(1-t)f(a)+tf(b)][(1-t)g(a)+tg(b)]$$

$$= (1-t)^2 f(a)g(a)+t^2 f(b)g(b)+t(1-t)[f(a)g(b)+f(b)g(a)]. \tag{5.2.42}$$

Using (5.2.34) in (5.2.42) and integrating with respect to t from 0 to 1, we have

$$\int_0^1 t^2(1-t)^2 H_1(t)dt \leqslant \frac{1}{3}M(a,b)+\frac{1}{6}N(a,b). \tag{5.2.43}$$

Using (5.2.38) in (5.2.43), we get the required inequality in (5.2.27). The proof is complete.

Theorem 5.2.4. Let $f,\ g \in P(I)$ and $f,g \in L_1[a,b]$. Then

$$f\left(\frac{a+b}{2}\right)g\left(\frac{a+b}{2}\right) \leqslant \frac{2}{b-a}\int_a^b f(x)g(x)dx+2[M(a,b)+N(a,b)], \tag{5.2.44}$$

$$\frac{1}{b-a}\int_a^b f(x)g(x)dx \leqslant M(a,b)+N(a,b), \tag{5.2.45}$$

where $M(a,b)$ and $N(a,b)$ are as defined in Theorem 5.2.3.

Proof. Since $f,\ g \in P(I)$, from (5.2.25) with $x = ta+(1-t)b,\ y = (1-t)a+tb;\ t \in (0,1)$ and $\lambda = \frac{1}{2}$, we have

$$f\left(\frac{a+b}{2}\right) \leqslant f(ta+(1-t)b)+f((1-t)a+tb), \tag{5.2.46}$$

$$g\left(\frac{a+b}{2}\right) \leqslant g(ta+(1-t)b)+g((1-t)a+tb). \tag{5.2.47}$$

From (5.2.46) and (5.2.47), we have

$$f\left(\frac{a+b}{2}\right)g\left(\frac{a+b}{2}\right) \leqslant [f(ta+(1-t)b)+f((1-t)a+tb)]$$

$$\times [g(ta+(1-t)b)+g((1-t)a+tb)]$$

$$= H_1(t)+H_2(t)+f(ta+(1-t)b)g((1-t)a+tb)+f((1-t)a+tb)g(ta+(1-t)b)$$

$$\leqslant H_1(t)+H_2(t)+2[f(a)+f(b)][g(a)+g(b)], \tag{5.2.48}$$

where $H_1(t)$ and $H_2(t)$ are defined by (5.2.34) and (5.2.35). Integrating both sides of (5.2.48) with respect to t from 0 to 1, we obtain

$$f\left(\frac{a+b}{2}\right)g\left(\frac{a+b}{2}\right) \leqslant \int_0^1 H_1(t)dt + \int_0^1 H_2(t)dt + 2\left[M(a,b)+N(a,b)\right]. \tag{5.2.49}$$

It is easy to observe that

$$\int_0^1 H_1(t)dt = \int_0^1 H_2(t)dt = \frac{1}{b-a}\int_a^b f(x)g(x)dx. \tag{5.2.50}$$

Using (5.2.50) in (5.2.49), we get the required inequality in (5.2.44).

Since $f,g \in P(I)$, we have

$$f(ta+(1-t)b) \leqslant f(a)+f(b), \tag{5.2.51}$$

$$g(ta+(1-t)b) \leqslant g(a)+g(b). \tag{5.2.52}$$

From (5.2.51) and (5.2.52), we observe that

$$f(ta+(1-t)b)g(ta+(1-t)b) \leqslant [f(a)+f(b)][g(a)+g(b)]. \tag{5.2.53}$$

Using (5.2.34) in (5.2.53) and integrating with respect to t from 0 to 1 we get

$$\int_0^1 H_1(t)dt \leqslant M(a,b)+N(a,b). \tag{5.2.54}$$

Using (5.2.50) in (5.2.54), we get the desired inequality in (5.2.45). The proof is complete.

5.3 Further integral inequalities involving convex functions

The present section is devoted to some integral inequalities involving convex functions investigated by Pachpatte in [84].

Let $f,\ g : [a,b] \subset \mathbb{R} \to \mathbb{R}$ $(a<b)$ be convex mappings. For $x,\ y \in [a,b]$, we shall define the mappings $F(x,y)(t),\ G(x,y)(t) : [0,1] \to \mathbb{R}$ given by (see [23])

$$F(x,y)(t) = \frac{1}{2}\left[f(tx+(1-t)y)+f((1-t)x+ty)\right], \tag{5.3.1}$$

$$G(x,y)(t) = \frac{1}{2}\left[g(tx+(1-t)y)+g((1-t)x+ty)\right]. \tag{5.3.2}$$

In [23], Dragomir and Ionescu established some interesting properties of such mappings. In particular in [23], it is shown that $F(x,y)(t),\ G(x,y)(t)$ are convex on $[0,1]$. In [143],

Pečarić and Dragomir proved that the following statements are equivalent for mappings $f,\ g : [a,b] \to \mathbb{R}$:

(i) $f,\ g$ are convex on $[a,b]$;

(ii) for all $x,\ y \in [a,b]$ the mappings $f_0,\ g_0 : [0,1] \to \mathbb{R}$ defined by $f_0(t) = f(tx+(1-t)y)$ or $f((1-t)x+ty)$, $g_0(t) = g(tx+(1-t)y)$ or $g((1-t)x+ty)$ are convex on $[0,1]$.

From these properties, it is easy to observe that if f_0 and g_0 are convex on $[0,1]$, then they are integrable on $[0,1]$ and hence f_0g_0 is also integrable on $[0,1]$. Similarly, if f and g are convex on $[a,b]$, they are integrable on $[a,b]$ and hence fg is also integrable on $[a,b]$. Consequently, it is easy to see that if f and g are convex on $[a,b]$, then $F = F(x,y)$ and $G = G(x,y)$ are convex and hence $Fg,\ Gf,\ Ff,\ Gg$ are also integrable on $[a,b]$. We shall use these facts in our discussion without further mention.

The following Theorem deals with the integral inequalities involving product of two functions, recently established in [84].

Theorem 5.3.1. Let f and g be real-valued, nonnegative and convex functions on $[a,b]$ and the mappings $F(x,y)(t)$ and $G(x,y)(t)$ be defined by (5.3.1) and (5.3.2). Then for all $t \in [0,1]$ we have

$$\frac{1}{(b-a)^2}\int_a^b (b-y)f(y)g(y)dy$$

$$\leqslant \frac{2}{5}\frac{1}{(b-a)^2}\int_a^b \left(\int_a^y [F(x,y)(t)g(x)+G(x,y)(t)f(x)]dx\right)dy + \frac{1}{10}f(a)g(a), \tag{5.3.3}$$

$$\frac{1}{(b-a)^2}\int_a^b (y-a)f(y)g(y)dy$$

$$\leqslant \frac{2}{5}\frac{1}{(b-a)^2}\int_a^b \left(\int_y^b [F(x,y)(t)g(x)+G(x,y)(t)f(x)]dx\right)dy + \frac{1}{10}f(b)g(b), \tag{5.3.4}$$

$$\frac{1}{b-a}\int_a^b f(y)g(y)dy$$

$$\leqslant \frac{2}{5}\frac{1}{(b-a)^2}\int_a^b \int_a^b [F(x,y)(t)g(x)+G(x,y)(t)f(x)]dxdy + \frac{1}{10}[f(a)g(a)+f(b)g(b)]. \tag{5.3.5}$$

Proof. The assumptions that f and g are nonnegative and convex imply that, we may assume that $f,\ g \in C^1$ and that we have the following estimates

$$f(tx+(1-t)y) \geqslant f(x)+(1-t)(y-x)f'(x), \tag{5.3.6}$$

$$f((1-t)x+ty) \geqslant f(x)+t(y-x)f'(x), \tag{5.3.7}$$

$$g(tx+(1-t)y) \geqslant g(x)+(1-t)(y-x)g'(x), \tag{5.3.8}$$

$$g((1-t)x+ty) \geqslant g(x)+t(y-x)g'(x), \tag{5.3.9}$$

for $x,\ y \in [a,b]$ and $t \in [0,1]$. From (5.3.6), (5.3.7), (5.3.1) and (5.3.8), (5.3.9), (5.3.2), it is easy to observe that

$$F(x,y)(t) \geqslant f(x)+\frac{1}{2}(y-x)f'(x), \tag{5.3.10}$$

$$G(x,y)(t) \geqslant g(x)+\frac{1}{2}(y-x)g'(x), \tag{5.3.11}$$

for $x,\ y \in [a,b]$ and $t \in [0,1]$. Multiplying (5.3.10) by $g(x)$ and (5.3.11) by $f(x)$ and then adding, we obtain

$$F(x,y)(t)g(x)+G(x,y)(t)f(x) \geqslant 2f(x)g(x)+\frac{1}{2}(y-x)\frac{d}{dx}(f(x)g(x)). \tag{5.3.12}$$

Integrating the inequality (5.3.12) over x from a to y, we have

$$\int_a^y [F(x,y)(t)g(x)+G(x,y)(t)f(x)]dx \geqslant \frac{5}{2}\int_a^y f(x)g(x)dx - \frac{1}{2}(y-a)f(a)g(a). \tag{5.3.13}$$

Further, integrating both sides of (5.3.13) with respect to y from a to b, we get

$$\int_a^b \left(\int_a^y [F(x,y)(t)g(x)+G(x,y)(t)f(x)]dx\right)dy$$

$$\geqslant \frac{5}{2}\int_a^b (b-y)f(y)g(y)dy - \frac{1}{4}(b-a)^2 f(a)g(a). \tag{5.3.14}$$

Multiplying both sides of (5.3.14) by $\frac{2}{5}\frac{1}{(b-a)^2}$ and rewriting, we get the required inequality in (5.3.3).

Similarly, by first integrating (5.3.12) over x from y to b and then integrating the resulting inequality over y from a to b, and rewriting we get the required inequality in (5.3.4). The inequality (5.3.5) is obtained by adding the inequalities (5.3.3) and (5.3.4). The proof is complete.

The slight variants of the inequalities given in Theorem 5.3.1, also established in [84], are embodied in the following theorem.

Theorem 5.3.2. Let f and g be real-valued, nonnegative and convex functions on $[a,b]$ and the mappings $F(x,y)(t)$ and $G(x,y)(t)$ be defined by (5.3.1) and (5.3.2). Then for all $t\in[0,1]$, we have

$$\begin{aligned}&\frac{1}{(b-a)^2}\int_a^b (b-y)\left[f^2(y)+g^2(y)\right]dy\\ &\leqslant \frac{4}{5}\frac{1}{(b-a)^2}\int_a^b\left(\int_a^y [F(x,y)(t)f(x)+G(x,y)(t)g(x)]\,dx\right)dy\\ &\quad+\frac{1}{10}\left[f^2(a)+g^2(a)\right],\end{aligned} \tag{5.3.15}$$

$$\begin{aligned}&\frac{1}{(b-a)^2}\int_a^b (y-a)\left[f^2(y)+g^2(y)\right]dy\\ &\leqslant \frac{4}{5}\frac{1}{(b-a)^2}\int_a^b\left(\int_y^b [F(x,y)(t)f(x)+G(x,y)(t)g(x)]\,dx\right)dy\\ &\quad+\frac{1}{10}\left[f^2(b)+g^2(b)\right],\end{aligned} \tag{5.3.16}$$

$$\begin{aligned}\frac{1}{b-a}\int_a^b\left[f^2(y)+g^2(y)\right]dy &\leqslant \frac{4}{5}\frac{1}{(b-a)^2}\int_a^b\int_a^b [F(x,y)(t)f(x)+G(x,y)(t)g(x)]\,dxdy\\ &\quad+\frac{1}{10}\left[f^2(a)+g^2(a)+f^2(b)+g^2(b)\right].\end{aligned} \tag{5.3.17}$$

Proof. As in the proof of Theorem 5.3.1, from the assumptions we have the estimates (5.3.10) and (5.3.11). Multiplying (5.3.10) by $f(x)$ and (5.3.11) by $g(x)$ and then adding, we obtain

$$\begin{aligned}&F(x,y)(t)f(x)+G(x,y)(t)g(x)\\ &\geqslant f^2(x)+g^2(x)+\frac{1}{2}(y-x)\left[f(x)f'(x)+g(x)g'(x)\right].\end{aligned} \tag{5.3.18}$$

Integrating (5.3.18) over x from a to y, we have

$$\begin{aligned}&\int_a^y [F(x,y)(t)f(x)+G(x,y)(t)g(x)]\,dx\\ &\geqslant \frac{5}{4}\int_a^y \left[f^2(x)+g^2(x)\right]dx-\frac{1}{4}(y-a)\left[f^2(a)+g^2(a)\right].\end{aligned} \tag{5.3.19}$$

Further, integrating both sides of (5.3.19) with respect to y from a to b, we have

$$\begin{aligned}&\int_a^b\left(\int_a^y [F(x,y)(t)f(x)+G(x,y)(t)g(x)]dx\right)dy\\ &\geqslant \frac{5}{4}\int_a^b (b-y)\left[f^2(y)+g^2(y)\right]dy-\frac{1}{8}(b-a)^2\left[f^2(a)+g^2(a)\right].\end{aligned} \tag{5.3.20}$$

Multiplying both sides of (5.3.20) by $\frac{4}{5}\frac{1}{(b-a)^2}$ and rewriting, we get the required inequality in (5.3.15).

The proofs of the inequalities (5.3.16) and (5.3.17) follows by the same arguments as in the proof of Theorem 5.3.1 with suitable modifications. We omit the details.

Below, we shall give the inequalities proved in [84], similar to those of given in Theorem 5.3.1, involving only one convex function.

Theorem 5.3.3. Let f be a real-valued, nonnegative and convex function on $[a,b]$. Then for all $t \in [0,1]$ we have

$$\frac{1}{(b-a)^2}\int_a^b (b-y)f(y)dy$$

$$\leqslant \frac{2}{3}\frac{1}{(b-a)^2}\int_a^b \left(\int_a^y \left(\int_0^1 f(tx+(1-t)y)dt\right)dx\right)dy + \frac{1}{6}f(a), \tag{5.3.21}$$

$$\frac{1}{(b-a)^2}\int_a^b (y-a)f(y)dy$$

$$\leqslant \frac{2}{3}\frac{1}{(b-a)^2}\int_a^b \left(\int_y^b \left(\int_0^1 f(tx+(1-t)y)dt\right)dx\right)dy + \frac{1}{6}f(b), \tag{5.3.22}$$

$$\frac{1}{b-a}\int_a^b f(y)dy$$

$$\leqslant \frac{2}{3}\frac{1}{(b-a)^2}\int_a^b \left(\int_a^b \left(\int_0^1 f(tx+(1-t)y)dt\right)dx\right)dy + \frac{1}{6}[f(a)+f(b)]. \tag{5.3.23}$$

Proof. To prove the inequality (5.3.21), as in the proof of Theorem 5.3.1, from the assumptions we have the estimate (5.3.6). Integrating both sides of (5.3.6) over t from 0 to 1, we have

$$\int_0^1 f(tx+(1-t)y)dt \geqslant f(x) + \frac{1}{2}(y-x)f'(x). \tag{5.3.24}$$

Now, first integrating both sides of (5.3.24) over x from a to y and after that integrating the resulting inequality over y from a to b, we get the required inequality in (5.3.21).
Similarly, by first integrating both sides of (5.3.24) over x from y to b and then integrating the resulting inequality over y from a to b, we get the inequality in (5.3.22). By adding the inequalities (5.3.21) and (5.3.22) we get the inequality (5.3.23). The proof is complete.

5.4 Integral inequalities involving log-convex functions

In this section we present some new integral inequalities involving log-convex functions, recently investigated by Pachpatte in [104,110,136].
Let I be an interval of R and $a,\ b \in I$ with $a < b$. A function $f : I \to (0,\infty)$ is said to be log-convex function, if for all $x,\ y \in I$ and $t \in [0,1]$ one has the inequality (see [45,46,108])

$$f(tx+(1-t)y) \leqslant [f(x)]^t[f(y)]^{1-t}.$$

In [30], Dragomir and Mond proved that the following inequalities hold for log-convex functions

$$f\left(\frac{a+b}{2}\right) \leqslant \exp\left[\frac{1}{b-a}\int_a^b \log[f(x)]dx\right] \leqslant \frac{1}{b-a}\int_a^b G(f(x), f(a+b-x))dx$$

$$\leqslant \frac{1}{b-a}\int_a^b f(x)dx \leqslant L(f(a), f(b)) \leqslant \frac{f(a)+f(b)}{2},$$

where, $G(p,q) = \sqrt{pq}$ is the Geometric mean and

$$L(p,q) = \frac{p-q}{\log p - \log q} \quad (p \neq q),$$

is the Logarithmic mean of the positive real numbers p,q (for $p = q$, we put $L(p,p) = p$). For the further refinements of Hadamard's inequalities in (8) for log-convex functions, see [45,46].

The first two Theorems deals with the Hadamard-type integral inequalities established in [104], involving two log-convex functions.

Theorem 5.4.1. Let $f,\ g : I \to (0,\infty)$ be log-convex functions on I and $a,\ b \in I$ with $a < b$. Then the following inequality holds

$$\frac{4}{b-a}\int_a^b f(x)g(x)dx \leqslant [f(a)+f(b)]L(f(a),f(b)) + [g(a)+g(b)]L(g(a),g(b)), \tag{5.4.1}$$

where $L(\cdot,\cdot)$ is a Logarithmic mean of positive real numbers.

Proof. Since $f,\ g$ are log-convex functions, we have

$$f(ta+(1-t)b) \leqslant [f(a)]^t[f(b)]^{1-t}, \tag{5.4.2}$$

$$g(ta+(1-t)b) \leqslant [g(a)]^t[g(b)]^{1-t}, \tag{5.4.3}$$

for all $t \in [0,1]$. It is easy to observe that

$$\int_a^b f(x)g(x)dx = (b-a)\int_0^1 f(ta+(1-t)b)g(ta+(1-t)b)dt. \tag{5.4.4}$$

Using the elementary inequality $cd \leqslant \frac{1}{2}[c^2+d^2]$ $c,\ d \geqslant 0$ reals, (5.4.2), (5.4.3) on the right side of (5.4.4) and by making the change of variable, we have

$$\int_a^b f(x)g(x)dx \leqslant \frac{1}{2}(b-a)\int_0^1 [\{f(ta+(1-t)b)\}^2 + \{g(ta+(1-t)b)\}^2]\,dt$$

$$\leqslant \frac{1}{2}(b-a)\int_0^1 \left[\{[f(a)]^t[f(b)]^{1-t}\}^2 + \{[g(a)]^t[g(b)]^{1-t}\}^2\right] dt$$

$$= \frac{1}{2}(b-a)\left\{f^2(b)\int_0^1 \left[\frac{f(a)}{f(b)}\right]^{2t} dt + g^2(b)\int_0^1 \left[\frac{g(a)}{g(b)}\right]^{2t} dt\right\}$$

$$= \frac{1}{4}(b-a)\left\{f^2(b)\int_0^2\left[\frac{f(a)}{f(b)}\right]^{\sigma}d\sigma + g^2(b)\int_0^2\left[\frac{g(a)}{g(b)}\right]^{\sigma}d\sigma\right\}$$

$$= \frac{1}{4}(b-a)\left\{f^2(b)\left[\frac{\left[\frac{f(a)}{f(b)}\right]^{\sigma}}{\log\frac{f(a)}{f(b)}}\right]_0^2 + g^2(b)\left[\frac{\left[\frac{g(a)}{g(b)}\right]^{\sigma}}{\log\frac{g(a)}{g(b)}}\right]_0^2\right\}$$

$$= \frac{1}{4}(b-a)\left\{\frac{[f(a)+f(b)][f(a)-f(b)]}{\log f(a)-\log f(b)} + \frac{[g(a)+g(b)][g(a)-g(b)]}{\log g(a)-\log g(b)}\right\}$$

$$= \frac{1}{4}(b-a)\left\{[f(a)+f(b)]L(f(a),f(b)) + [g(a)+g(b)]L(g(a),g(b))\right\}. \tag{5.4.5}$$

Rewriting (5.4.5), we get the required inequality in (5.4.1). The proof is complete.

Theorem 5.4.2. Let $f,\ g : I \to (0,\infty)$ be differentiable log-convex functions on the interval I of real numbers, $\mathring{I}$ the interior of I and $a,\ b \in \mathring{I}$ with $a < b$. Then the following inequality holds

$$\frac{2}{b-a}\int_a^b f(x)g(x)dx \geqslant \frac{1}{b-a}f\left(\frac{a+b}{2}\right)\int_a^b g(x)\exp\left[\frac{f'\left(\frac{a+b}{2}\right)}{f\left(\frac{a+b}{2}\right)}\left(x-\frac{a+b}{2}\right)\right]dx$$

$$+\frac{1}{b-a}g\left(\frac{a+b}{2}\right)\int_a^b f(x)\exp\left[\frac{g'\left(\frac{a+b}{2}\right)}{g\left(\frac{a+b}{2}\right)}\left(x-\frac{a+b}{2}\right)\right]dx. \tag{5.4.6}$$

Proof. Since $f,\ g$ are differentiable and log-convex functions on $\mathring{I}$, we have that

$$\log f(x) - \log f(y) \geqslant \frac{d}{dy}\left(\log f(y)\right)(x-y), \tag{5.4.7}$$

$$\log g(x) - \log g(y) \geqslant \frac{d}{dy}\left(\log g(y)\right)(x-y), \tag{5.4.8}$$

for all $x,\ y \in \mathring{I}$, gives that

$$\log\left(\frac{f(x)}{f(y)}\right) \geqslant \frac{f'(y)}{f(y)}(x-y), \tag{5.4.9}$$

$$\log\left(\frac{g(x)}{g(y)}\right) \geqslant \frac{g'(y)}{g(y)}(x-y), \tag{5.4.10}$$

for all $x,\ y \in \overset{\circ}{I}$. That is

$$f(x) \geqslant f(y)\exp\left[\frac{f'(y)}{f(y)}(x-y)\right], \tag{5.4.11}$$

$$g(x) \geqslant g(y)\exp\left[\frac{g'(y)}{g(y)}(x-y)\right]. \tag{5.4.12}$$

Multiplying both sides of (5.4.11) and (5.4.12) by $g(x)$ and $f(x)$ respectively and adding the resulting inequalities, we have

$$2f(x)g(x) \geqslant g(x)f(y)\exp\left[\frac{f'(y)}{f(y)}(x-y)\right] + f(x)g(y)\exp\left[\frac{g'(y)}{g(y)}(x-y)\right]. \tag{5.4.13}$$

Now, if we choose $y = \frac{a+b}{2}$, from (5.4.13), we obtain

$$2f(x)g(x) \geqslant g(x)f\left(\frac{a+b}{2}\right)\exp\left[\frac{f'\left(\frac{a+b}{2}\right)}{f\left(\frac{a+b}{2}\right)}\left(x-\frac{a+b}{2}\right)\right]$$

$$+f(x)g\left(\frac{a+b}{2}\right)\exp\left[\frac{g'\left(\frac{a+b}{2}\right)}{g\left(\frac{a+b}{2}\right)}\left(x-\frac{a+b}{2}\right)\right]. \tag{5.4.14}$$

Integrating both sides of (5.4.14) with respect to x from a to b and dividing both sides of the resulting inequality by $b-a$, we get the desired inequality in (5.4.6). The proof is complete.

In the next Theorem we present Hadamard-type integral inequalities recently proved in [136], involving three log-convex functions.

Theorem 5.4.3. Let $f,\ g,\ h : I \to (0,\infty)$ be log-convex functions on I with $a < b$. Then we have

$$\begin{aligned}&\frac{2}{b-a}\int_a^b [f(x)g(x)+g(x)h(x)+h(x)f(x)]\,dx\\ &\leqslant [f(a)+f(b)]L(f(a),f(b)) + [g(a)+g(b)]L(g(a),g(b))\\ &\quad +[h(a)+h(b)]L(h(a),h(b)),\end{aligned} \tag{5.4.15}$$

$$\begin{aligned}&\frac{4}{b-a}\int_a^b f(x)g(x)h(x)[f(x)+g(x)+h(x)]\,dx\\ &\leqslant [f(a)+f(b)]\left[f^2(a)+f^2(b)\right]L(f(a),f(b))\\ &\quad +[g(a)+g(b)]\left[g^2(a)+g^2(b)\right]L(g(a),g(b))\\ &\quad +[h(a)+h(b)]\left[h^2(a)+h^2(b)\right]L(h(a),h(b)),\end{aligned} \tag{5.4.16}$$

where $L(\cdot,\cdot)$ is a Logarithmic mean of positive real numbers.

Proof. Since $f,\ g,\ h$ are log-convex functions, we have

$$f(ta+(1-t)b) \leqslant [f(a)]^t[f(b)]^{1-t}, \tag{5.4.17}$$

$$g(ta+(1-t)b) \leqslant [g(a)]^t[g(b)]^{1-t}, \tag{5.4.18}$$

$$h(ta+(1-t)b) \leqslant [h(a)]^t[h(b)]^{1-t}, \tag{5.4.19}$$

for $t \in [0,1]$. It is easy to observe that

$$\int_a^b [f(x)g(x)+g(x)h(x)+h(x)f(x)]\,dx = (b-a)\int_0^1 [f(ta+(1-t)b)g(ta+(1-t)b)$$

$$+g(ta+(1-t)b)h(ta+(1-t)b)\ +h(ta+(1-t)b)f(ta+(1-t)b)]\,dt. \tag{5.4.20}$$

Using the elementary inequality (see [78]) $c_1c_2+c_2c_3+c_3c_1 \leqslant c_1^2+c_2^2+c_3^2$ (for $c_1,\ c_2,\ c_3$ reals) and (5.4.17), (5.4.18), (5.4.19) on the right hand side of (5.4.20) and making the change of variables, we have

$$\int_a^b [f(x)g(x)+g(x)h(x)+h(x)f(x)]dx$$

$$\leqslant (b-a)\int_0^1 [\{f(ta+(1-t)b)\}^2+\{g(ta+(1-t)b)\}^2+\{h(ta+(1-t)b)\}^2]\,dt$$

$$\leqslant (b-a)\int_0^1 \left[\left\{[f(a)]^t[f(b)]^{1-t}\right\}^2+\left\{[g(a)]^t[g(b)]^{1-t}\right\}^2+\left\{[h(a)]^t[h(b)]^{1-t}\right\}^2\right] dt$$

$$= (b-a)\left[f^2(b)\int_0^1 \left[\frac{f(a)}{f(b)}\right]^{2t} dt+g^2(b)\int_0^1 \left[\frac{g(a)}{g(b)}\right]^{2t} dt+h^2(b)\int_0^1 \left[\frac{h(a)}{h(b)}\right]^{2t} dt\right]$$

$$= (b-a)\left[\frac{1}{2}f^2(b)\int_0^2 \left[\frac{f(a)}{f(b)}\right]^{\sigma} d\sigma+\frac{1}{2}g^2(b)\int_0^2 \left[\frac{g(a)}{g(b)}\right]^{\sigma} d\sigma+\frac{1}{2}h^2(b)\int_0^2 \left[\frac{h(a)}{h(b)}\right]^{\sigma} d\sigma\right]$$

$$= \frac{1}{2}(b-a)\left[f^2(b)\left\{\frac{\left[\frac{f(a)}{f(b)}\right]^{\sigma}}{\log\frac{f(a)}{f(b)}}\right\}_0^2+g^2(b)\left\{\frac{\left[\frac{g(a)}{g(b)}\right]^{\sigma}}{\log\frac{g(a)}{g(b)}}\right\}_0^2+h^2(b)\left\{\frac{\left[\frac{h(a)}{h(b)}\right]^{\sigma}}{\log\frac{h(a)}{h(b)}}\right\}_0^2\right]$$

$$= \frac{1}{2}(b-a)\left[[f(a)+f(b)]\frac{f(a)-f(b)}{\log f(a)-\log f(b)}\right.$$

$$\left.+[g(a)+g(b)]\frac{g(a)-g(b)}{\log g(a)-\log g(b)}+[h(a)+h(b)]\frac{h(a)-h(b)}{\log h(a)-\log h(b)}\right]$$

$$= \frac{1}{2}(b-a)\left[[f(a)+f(b)]L(f(a),f(b))+[g(a)+g(b)]L(g(a),g(b))\right.$$

$$+[h(a)+h(b)]L(h(a),h(b))\big]. \tag{5.4.21}$$

The desired inequality in (5.4.15) follows by rewriting (5.4.21).

It is easy to observe that

$$\begin{aligned}
&\int_a^b f(x)g(x)h(x)[f(x)+g(x)+h(x)]dx \\
&= (b-a)\int_0^1 f(ta+(1-t)b)g(ta+(1-t)b)h(ta+(1-t)b) \\
&\times [f(ta+(1-t)b)+g(ta+(1-t)b)+h(ta+(1-t)b)]\,dt.
\end{aligned} \tag{5.4.22}$$

for $t \in [0,1]$. Using the elementary inequalities (see [78]),

$$c_1c_2c_3[c_1+c_2+c_3] \leqslant \frac{1}{3}(c_1c_2+c_2c_3+c_3c_1)^2,$$

$$c_1c_2+c_2c_3+c_3c_1 \leqslant c_1^2+c_2^2+c_3^2,$$

and

$$(c_1+c_2+c_3)^2 \leqslant 3(c_1^2+c_2^2+c_3^2)$$

(for c_1, c_2, c_3 reals), from (5.4.22), we observe that

$$\begin{aligned}
&\int_a^b f(x)g(x)h(x)[f(x)+g(x)+h(x)]dx \\
&\leqslant (b-a)\int_0^1 \left[\left\{[f(a)]^t[f(b)]^{1-t}\right\}^4+\left\{[g(a)]^t[g(b)]^{1-t}\right\}^4+\left\{[h(a)]^t[h(b)]^{1-t}\right\}^4\right]dt.
\end{aligned}$$

The rest of the proof of (5.4.16) can be completed by following the proof of inequality (5.4.15) with suitable modifications. Here we omit the further details.

The following Theorems contains the Hadamard-type integral inequalities established in [110], involving several log-convex functions.

Theorem 5.4.4. Let $f_i : I \to (0,\infty)$ $(i=1,\dots,n)$ be differentiable log-convex functions on the interval I of real numbers and a, $b \in \mathring{I}$ with $a<b$ ($\mathring{I}$ the interior of I). Then the following inequalities hold

$$\begin{aligned}
&\frac{\dfrac{1}{b-a}\displaystyle\int_a^b \prod_{i=1}^n f_i(x)dx}{\displaystyle\prod_{i=1}^n f_i\left(\frac{a+b}{2}\right)} \\
&\geqslant L\left(\exp\left[\sum_{i=1}^n \frac{f_i'\left(\frac{a+b}{2}\right)}{f_i\left(\frac{a+b}{2}\right)}\left(\frac{b-a}{2}\right)\right],\ \exp\left[-\sum_{i=1}^n \frac{f_i'\left(\frac{a+b}{2}\right)}{f_i\left(\frac{a+b}{2}\right)}\left(\frac{b-a}{2}\right)\right]\right) \\
&\geqslant 1,
\end{aligned} \tag{5.4.23}$$

where $L(\cdot,\cdot)$ is a Logarithmic mean of positive real numbers.

Proof. Since f_i $(i=1,\ldots,n)$ are differentiable and log-convex functions on $\mathring{I}$, we have that

$$\log f_i(x) - \log f_i(y) \geqslant \frac{d}{dy}(\log f_i(y))(x-y),$$

i.e.,

$$\log f_i(x) - \log f_i(y) \geqslant \frac{f_i'(y)}{f_i(y)}(x-y), \tag{5.4.24}$$

for all $x,\ y \in \mathring{I}$. Writing (5.4.24) for $i=1,\ldots,n$; adding the resulting inequalities and using the properties of log, it is easy to observe that

$$\log\left[\frac{\prod_{i=1}^{n} f_i(x)}{\prod_{i=1}^{n} f_i(y)}\right] \geqslant \sum_{i=1}^{n} \frac{f_i'(y)}{f_i(y)}(x-y), \tag{5.4.25}$$

for all $x,\ y \in \mathring{I}$. From (5.4.25), we have

$$\frac{\prod_{i=1}^{n} f_i(x)}{\prod_{i=1}^{n} f_i(y)} \geqslant \exp\left[\sum_{i=1}^{n} \frac{f_i'(y)}{f_i(y)}(x-y)\right], \tag{5.4.26}$$

for all $x,\ y \in \mathring{I}$. By taking $y=\frac{a+b}{2}$ in (5.4.26), we get

$$\frac{\prod_{i=1}^{n} f_i(x)}{\prod_{i=1}^{n} f_i\left(\frac{a+b}{2}\right)} \geqslant \exp\left[\sum_{i=1}^{n} \frac{f_i'\left(\frac{a+b}{2}\right)}{f_i\left(\frac{a+b}{2}\right)}\left(x-\frac{a+b}{2}\right)\right]. \tag{5.4.27}$$

Integrating (5.4.27) over x on $[a,b]$ and using Jensen's integral inequality for $\exp(\cdot)$ functions, we have

$$\frac{\frac{1}{b-a}\int_a^b \prod_{i=1}^{n} f_i(x)dx}{\prod_{i=1}^{n} f_i\left(\frac{a+b}{2}\right)} \geqslant \frac{1}{b-a}\int_a^b \exp\left[\sum_{i=1}^{n} \frac{f_i'\left(\frac{a+b}{2}\right)}{f_i\left(\frac{a+b}{2}\right)}\left(x-\frac{a+b}{2}\right)\right]dx$$

$$\geqslant \exp\left[\frac{1}{b-a}\int_a^b \sum_{i=1}^{n} \frac{f_i'\left(\frac{a+b}{2}\right)}{f_i\left(\frac{a+b}{2}\right)}\left(x-\frac{a+b}{2}\right)dx\right] = 1. \tag{5.4.28}$$

Now, as for $\alpha \neq 0$, we have that

$$\frac{1}{b-a}\int_a^b \exp(\alpha x)dx = \frac{\exp(\alpha b)-\exp(\alpha a)}{\alpha(b-a)} = L(\exp[\alpha b],\exp[\alpha a]),$$

where $L(\cdot,\cdot)$ is the usual Logarithmic mean, then

$$\frac{1}{b-a}\int_a^b \exp\left[\alpha\left(x-\frac{a+b}{2}\right)\right]dx = \frac{\exp\left[\alpha\left(\frac{b-a}{2}\right)\right]-\exp\left[-\alpha\left(\frac{b-a}{2}\right)\right]}{\alpha\left[\left(\frac{b-a}{2}\right)-\left(-\left(\frac{b-a}{2}\right)\right)\right]}$$

$$= L\left(\exp\left[\alpha\left(\frac{b-a}{2}\right)\right], \exp\left[-\alpha\left(\frac{b-a}{2}\right)\right]\right).$$

Using the above equality for $\alpha = \sum_{i=1}^n \frac{f_i'\left(\frac{a+b}{2}\right)}{f_i\left(\frac{a+b}{2}\right)}$, the inequality (5.4.28) gives the desired inequality in (5.4.23).

Remark 5.4.1. By taking $n = 1$ and $f_1 = f$ in Theorem 5.4.4, we get

$$\frac{\frac{1}{b-a}\int_a^b f(x)dx}{f\left(\frac{a+b}{2}\right)} \geqslant L\left(\exp\left[\frac{f'\left(\frac{a+b}{2}\right)}{f\left(\frac{a+b}{2}\right)}\left(\frac{b-a}{2}\right)\right], \exp\left[-\frac{f'\left(\frac{a+b}{2}\right)}{f\left(\frac{a+b}{2}\right)}\left(\frac{b-a}{2}\right)\right]\right)$$

$$\geqslant 1, \tag{5.4.29}$$

which is the inequality given by Dragomir in [46, Theorem 2.1].

Theorem 5.4.5. Let f_i be as in Theorem 5.4.4. Then the following inequalities hold

$$\frac{\int_a^b \left(\prod_{i=1}^n f_i(y)\right)\left[\sum_{i=1}^n \frac{f_i'(y)}{f_i(y)}\left(\frac{a+b}{2}-y\right)\right]dy}{\int_a^b \left(\prod_{i=1}^n f_i(y)\right)dy}$$

$$\leqslant \log\left[\frac{\int_a^b \left(\prod_{i=1}^n f_i(y)\right)\exp\left[\sum_{i=1}^n \frac{f_i'(y)}{f_i(y)}\left(\frac{a+b}{2}-y\right)\right]dy}{\int_a^b \left(\prod_{i=1}^n f_i(y)\right)dy}\right]$$

$$\leqslant \log\left[\frac{\prod_{i=1}^n f_i\left(\frac{a+b}{2}\right)}{\frac{1}{b-a}\int_a^b \left(\prod_{i=1}^n f_i(y)\right)dy}\right]. \tag{5.4.30}$$

Proof. By following the proof of Theorem 5.4.4, the inequality (5.4.26) holds. Taking $x = \frac{a+b}{2}$ in the inequality (5.4.26), we have

$$\prod_{i=1}^{n} f_i\left(\frac{a+b}{2}\right) \geqslant \left(\prod_{i=1}^{n} f_i(y)\right) \exp\left[\sum_{i=1}^{n} \frac{f_i'(y)}{f_i(y)}\left(\frac{a+b}{2} - y\right)\right], \tag{5.4.31}$$

for all $y \in [a,b]$. Integrating (5.4.31) over y and using Jensen's integral inequality for $\exp(\cdot)$ functions, we have

$$(b-a)\prod_{i=1}^{n} f_i\left(\frac{a+b}{2}\right) \geqslant \int_a^b \left(\prod_{i=1}^{n} f_i(y)\right) \exp\left[\sum_{i=1}^{n} \frac{f_i'(y)}{f_i(y)}\left(\frac{a+b}{2} - y\right)\right] dy$$

$$\geqslant \int_a^b \left(\prod_{i=1}^{n} f_i(y)\right) dy \times \exp\left[\frac{\int_a^b \left(\prod_{i=1}^{n} f_i(y)\right) \left[\sum_{i=1}^{n} \frac{f_i'(y)}{f_i(y)}\left(\frac{a+b}{2} - y\right)\right] dy}{\int_a^b \left(\prod_{i=1}^{n} f_i(y)\right) d}\right]. \tag{5.4.32}$$

From (5.4.32), we have

$$\exp\left[\frac{\int_a^b \left(\prod_{i=1}^{n} f_i(y)\right) \left[\sum_{i=1}^{n} \frac{f_i'(y)}{f_i(y)}\left(\frac{a+b}{2} - y\right)\right] dy}{\int_a^b \left(\prod_{i=1}^{n} f_i(y)\right) dy}\right]$$

$$\leqslant \frac{\int_a^b \left(\prod_{i=1}^{n} f_i(y)\right) \exp\left[\sum_{i=1}^{n} \frac{f_i'(y)}{f_i(y)}\left(\frac{a+b}{2} - y\right)\right] dy}{\int_a^b \left(\prod_{i=1}^{n} f_i(y)\right) dy} \leqslant \frac{(b-a)\left(\prod_{i=1}^{n} f_i\left(\frac{a+b}{2}\right)\right)}{\int_a^b \left(\prod_{i=1}^{n} f_i(y)\right) dy},$$

which is equivalent to

$$\frac{\int_a^b \left(\prod_{i=1}^{n} f_i(y)\right) \left[\sum_{i=1}^{n} \frac{f_i'(y)}{f_i(y)}\left(\frac{a+b}{2} - y\right)\right] dy}{\int_a^b \left(\prod_{i=1}^{n} f_i(y)\right) dy}$$

$$\leqslant \log\left[\frac{\int_a^b \left(\prod_{i=1}^{n} f_i(y)\right) \exp\left[\sum_{i=1}^{n} \frac{f_i'(y)}{f_i(y)}\left(\frac{a+b}{2} - y\right)\right] dy}{\int_a^b \left(\prod_{i=1}^{n} f_i(y)\right) dy}\right] \leqslant \log\left[\frac{\prod_{i=1}^{n} f_i\left(\frac{a+b}{2}\right)}{\frac{1}{b-a}\int_a^b \left(\prod_{i=1}^{n} f_i(y)\right) dy}\right].$$

This is the desired inequality in (5.4.30) and the proof is complete.

Remark 5.4.2. By taking $n = 1$ and $f_1 = f$ in Theorem 5.4.5, from (5.4.32), we obtain

$$
\begin{aligned}
(b-a)f\left(\frac{a+b}{2}\right) &\geqslant \int_a^b f(y)\exp\left[\frac{f'(y)}{f(y)}\left(\frac{a+b}{2}-y\right)\right]dy \\
&\geqslant \int_a^b f(y)dy \times \exp\left(\frac{\displaystyle\int_a^b f(y)\left[\frac{f'(y)}{f(y)}\left(\frac{a+b}{2}-y\right)\right]dy}{\displaystyle\int_a^b f(y)dy}\right) \\
&= \int_a^b f(y)dy \times \exp\left(\frac{\displaystyle\int_a^b f'(y)\left(\frac{a+b}{2}-y\right)dy}{\displaystyle\int_a^b f(y)dy}\right). \qquad (5.4.33)
\end{aligned}
$$

A simple integration by parts gives

$$
\int_a^b f'(y)\left(\frac{a+b}{2}-y\right)dy = \int_a^b f(y)dy - \frac{f(a)+f(b)}{2}(b-a). \qquad (5.4.34)
$$

Using (5.4.34) in (5.4.33), it is easy to observe that

$$
\begin{aligned}
\exp\left(1-\frac{\frac{f(a)+f(b)}{2}(b-a)}{\displaystyle\int_a^b f(y)dy}\right) &\leqslant \frac{\displaystyle\int_a^b f(y)\exp\left[\frac{f'(y)}{f(y)}\left(\frac{a+b}{2}-y\right)\right]dy}{\displaystyle\int_a^b f(y)dy} \\
&\leqslant \frac{(b-a)f\left(\frac{a+b}{2}\right)}{\displaystyle\int_a^b f(y)dy},
\end{aligned}
$$

which is equivalent to

$$
\begin{aligned}
1-\frac{\frac{f(a)+f(b)}{2}(b-a)}{\displaystyle\int_a^b f(y)dy} &\leqslant \log\left[\frac{\displaystyle\int_a^b f(y)\exp\left[\frac{f'(y)}{f(y)}\left(\frac{a+b}{2}-y\right)\right]dy}{\displaystyle\int_a^b f(y)dy}\right] \\
&\leqslant \log\left[\frac{f\left(\frac{a+b}{2}\right)}{\frac{1}{b-a}\displaystyle\int_a^b f(y)dy}\right],
\end{aligned}
$$

from which we get

$$
\begin{aligned}
\frac{\frac{f(a)+f(b)}{2}}{\frac{1}{b-a}\displaystyle\int_a^b f(y)dy} &\geqslant 1+\log\left[\frac{\displaystyle\int_a^b f(x)dx}{\displaystyle\int_a^b f(x)\exp\left[\frac{f'(x)}{f(x)}\left(\frac{a+b}{2}-x\right)\right]dx}\right] \\
&\geqslant 1+\log\left[\frac{\frac{1}{b-a}\displaystyle\int_a^b f(x)dx}{f\left(\frac{a+b}{2}\right)}\right] \geqslant 1. \qquad (5.4.35)
\end{aligned}
$$

We note that, the inequality (5.4.35) is given by Dragomir in [46, Theorem 2.3].

5.5 Discrete inequalities involving log-convex functions

In this section, we give some results on refinements and converse for discrete Jensen's inequality established by Dragomir and Mond in [40], involving log-convex functions.

The following refinement of discrete Jensen's inequality for log-convex functions is given in [40].

Theorem 5.5.1. Let $f : I \to (0,\infty)$ be a log-convex function on the interval I of real numbers and $x_i \in \overset{\circ}{I}$, the interior of I, $p_i \geqslant 0$ $(i=1,\dots,n)$ with $P_n > 0$, where $P_k = \sum_{i=1}^{k} p_i$, $(k=1,\dots,n)$. Then the following inequalities hold

$$\frac{\frac{1}{P_n}\sum_{i=1}^{n} p_i f(x_i)}{f\left(\frac{1}{P_n}\sum_{i=1}^{n} p_i x_i\right)} \geqslant \frac{1}{P_n}\sum_{j=1}^{n} p_j \exp\left(\frac{f'_+\left(\frac{1}{P_n}\sum_{i=1}^{n} p_i x_i\right)}{f\left(\frac{1}{P_n}\sum_{i=1}^{n} p_i x_i\right)}\left(x_j - \frac{1}{P_n}\sum_{i=1}^{n} p_i x_i\right)\right) \geqslant 1, \quad (5.5.1)$$

where f'_+ is the right derivative of f on $\overset{\circ}{I}$.

Proof. As f is log-convex on I, it follows that f is convex on I and thus [144] the right derivative f'_+ of f exists on $\overset{\circ}{I}$. Since $logf$ is convex on I, we have

$$\log f(x) - \log f(y) \geqslant \frac{d^+(\log f)}{dt}(y)(x-y),$$

for all $x,\ y \in \overset{\circ}{I}$, which gives

$$\log\left[\frac{f(x)}{f(y)}\right] \geqslant \frac{f'_+(y)}{f(y)}(x-y),$$

for all $x,\ y \in \overset{\circ}{I}$, i.e.,

$$f(x) \geqslant f(y)\exp\left[\frac{f'_+(y)}{f(y)}(x-y)\right], \quad (5.5.2)$$

for all $x,\ y \in \overset{\circ}{I}$. Now, choose $x = x_j$ $(j=1,\dots,n)$ and $y = \frac{1}{P_n}\sum_{i=1}^{n} p_i x_i$ in (5.5.2). We obtain

$$f(x_j) \geqslant f\left(\frac{1}{P_n}\sum_{i=1}^{n} p_i x_i\right)\exp\left[\frac{f'_+\left(\frac{1}{P_n}\sum_{i=1}^{n} p_i x_i\right)}{f\left(\frac{1}{P_n}\sum_{i=1}^{n} p_i x_i\right)}\left(x_j - \frac{1}{P_n}\sum_{i=1}^{n} p_i x_i\right)\right], \quad (5.5.3)$$

for all $j=1,\dots,n$. If we multiply the inequality (5.5.3) by p_j and sum over $j=1,\dots,n$, we derive

$$\frac{1}{P_n}\sum_{i=1}^{n} p_i f(x_i) \geqslant f\left(\frac{1}{P_n}\sum_{i=1}^{n} p_i x_i\right)\frac{1}{P_n}\sum_{j=1}^{n} p_j \exp\left[\frac{f'_+\left(\frac{1}{P_n}\sum_{i=1}^{n} p_i x_i\right)}{f\left(\frac{1}{P_n}\sum_{i=1}^{n} p_i x_i\right)}\left(x_j - \frac{1}{P_n}\sum_{i=1}^{n} p_i x_i\right)\right],$$

and the first inequality in (5.5.1) is proved.

To prove the second inequality in (5.5.1), we use the following Jensen's discrete inequality for $\exp(\cdot)$ functions, i.e.,

$$\frac{1}{P_n}\sum_{j=1}^{n} p_j \exp(y_j) \geqslant \exp\left(\frac{1}{P_n}\sum_{j=1}^{n} p_j y_j\right), \tag{5.5.4}$$

where, $y_j \in \mathbb{R}$ $(j = 1,\dots,n)$ and p_j are as above. If we choose

$$y_j = \frac{f'_+\left(\frac{1}{P_n}\sum_{i=1}^{n} p_i x_i\right)}{f\left(\frac{1}{P_n}\sum_{i=1}^{n} p_i x_i\right)}\left(x_j - \frac{1}{P_n}\sum_{i=1}^{n} p_i x_i\right), \tag{5.5.5}$$

then we deduce

$$\frac{1}{P_n}\sum_{j=1}^{n} p_j y_j = \frac{1}{P_n}\sum_{j=1}^{n} p_j \frac{f'_+\left(\frac{1}{P_n}\sum_{i=1}^{n} p_i x_i\right)}{f\left(\frac{1}{P_n}\sum_{i=1}^{n} p_i x_i\right)}\left(x_j - \frac{1}{P_n}\sum_{i=1}^{n} p_i x_i\right) = 0, \tag{5.5.6}$$

and the second inequality in (5.5.1) follows by using (5.5.5) and (5.5.6) in (5.5.4).

The following corollary holds.

Corollary 5.5.1. Let $f : I \to \mathbb{R}$ be a convex mapping on the interval I of real numbers and $x_i \in \mathring{I}$, $p_i \geqslant 0$ $(i = 1,\dots,n)$ with $P_n > 0$, where $P_k = \sum_{i=1}^{k} p_i$ $(k = 1,\dots,n)$. Then we have the inequality

$$\frac{\frac{1}{P_n}\sum_{i=1}^{n} p_i \exp[f(x_i)]}{\exp\left[f\left(\frac{1}{P_n}\sum_{i=1}^{n} p_i x_i\right)\right]} \geqslant \frac{1}{P_n}\sum_{j=1}^{n} p_j \exp\left[f'_+\left(\frac{1}{P_n}\sum_{i=1}^{n} p_i x_i\right)\left(x_j - \frac{1}{P_n}\sum_{i=1}^{n} p_i x_i\right)\right] \geqslant 1. \tag{5.5.7}$$

Proof. Define the mapping $g : I \to (0,\infty)$ by $g(x) = \exp[f(x)]$. It is clear that, the mapping g is log-convex on I. Now, applying Theorem 5.5.1 for the log-convex mapping g, we easily deduce the inequality (5.5.7).

The next Theorem deals with the converse of Jensen's discrete inequality established in [40], for log-convex mappings.

Theorem 5.5.2. Let $f : I \to (0,\infty)$ be a log-convex function on the interval I of real numbers and $x_i \in \mathring{I}$, $p_i \geqslant 0$ $(i = 1,\dots,n)$ with $P_n > 0$, where $P_k = \sum_{i=1}^{k} p_i$ $(k = 1,\dots,n)$. Then the following inequalities hold

$$1 \geqslant \frac{f\left(\frac{1}{P_n}\sum_{i=1}^{n} p_i x_i\right)}{\frac{1}{P_n}\sum_{i=1}^{n} p_i f(x_i)} \geqslant \exp\left[\frac{\frac{1}{P_n}\sum_{i=1}^{n} p_i x_i \frac{1}{P_n}\sum_{i=1}^{n} p_i f'_+(x_i) - \frac{1}{P_n}\sum_{i=1}^{n} p_i x_i f'_+(x_i)}{\frac{1}{P_n}\sum_{i=1}^{n} p_i f(x_i)}\right] > 0, \tag{5.5.8}$$

where f'_+ is the right derivative of f on $\mathring{I}$.

Proof. By following the proof of Theorem 5.5.1, we have

$$f(x) \geqslant f(y)\exp\left[\frac{f'_+(y)}{f(y)}(x-y)\right], \tag{5.5.9}$$

for all $x,\ y \in \overset{\circ}{I}$. Now, if we choose in the inequality (5.5.9), $x = \frac{1}{P_n}\sum_{i=1}^n p_i x_i$ and $y = x_j$ $(j = 1,\dots,n)$, we get that

$$f\left(\frac{1}{P_n}\sum_{i=1}^n p_i x_i\right) \geqslant f(x_j)\exp\left[\frac{f'_+(x_j)}{f(x_j)}\left(\frac{1}{P_n}\sum_{i=1}^n p_i x_i - x_j\right)\right], \tag{5.5.10}$$

for all $j = 1,\dots,n$. If we multiply the inequality (5.5.10) by $p_j \geqslant 0$ and sum over $j = 1,\dots,n$, we derive that

$$f\left(\frac{1}{P_n}\sum_{i=1}^n p_i x_i\right) \geqslant \frac{1}{P_n}\sum_{j=1}^n p_j f(x_j)\exp\left[\frac{f'_+(x_j)}{f(x_j)}\left(\frac{1}{P_n}\sum_{i=1}^n p_i x_i - x_j\right)\right]. \tag{5.5.11}$$

Since the mapping $\exp(\cdot)$ is convex, we can use Jensen's discrete inequality

$$\frac{1}{Q_n}\sum_{j=1}^n q_j \exp(y_j) \geqslant \exp\left[\frac{1}{Q_n}\sum_{j=1}^n q_j y_j\right], \tag{5.5.12}$$

where $q_j \geqslant 0$ with $Q_n = \sum_{j=1}^n q_j > 0$ and $y_j \in \mathbb{R}$ $(j = 1,\dots,n)$. If we choose in (5.5.12), $q_j = p_j f(x_j) \geqslant 0$ and $y_j = \frac{f'_+(x_j)}{f(x_j)}\left(\frac{1}{P_n}\sum_{i=1}^n p_i x_i - x_j\right)$ for $j = 1,\dots,n$, then we get

$$\begin{aligned}
&\frac{1}{\sum\limits_{j=1}^n p_j f(x_j)}\sum_{j=1}^n p_j f(x_j)\exp\left[\frac{f'_+(x_j)}{f(x_j)}\left(\frac{1}{P_n}\sum_{i=1}^n p_i x_i - x_j\right)\right] \\
&\geqslant \exp\left[\frac{1}{\sum\limits_{j=1}^n p_j f(x_j)}\sum_{j=1}^n p_j f(x_j)\frac{f'_+(x_j)}{f(x_j)}\left(\frac{1}{P_n}\sum_{i=1}^n p_i x_i - x_j\right)\right] \\
&= \exp\left[\frac{\frac{1}{P_n}\sum\limits_{i=1}^n p_i x_i \sum\limits_{i=1}^n p_i f'_+(x_i) - \sum\limits_{i=1}^n p_i x_i f'_+(x_i)}{\sum\limits_{i=1}^n p_i f(x_i)}\right] \\
&= \exp\left[\frac{\frac{1}{P_n}\sum\limits_{i=1}^n p_i x_i \frac{1}{P_n}\sum\limits_{i=1}^n p_i f'_+(x_i) - \frac{1}{P_n}\sum\limits_{i=1}^n p_i x_i f'_+(x_i)}{\frac{1}{P_n}\sum\limits_{i=1}^n p_i f(x_i)}\right].
\end{aligned} \tag{5.5.13}$$

From (5.5.13), we get

$$\begin{aligned}
&\frac{1}{P_n}\sum_{j=1}^n p_j f(x_j)\exp\left[\frac{f'_+(x_j)}{f(x_j)}\left(\frac{1}{P_n}\sum_{i=1}^n p_i x_i - x_j\right)\right] \\
&\geqslant \frac{1}{P_n}\sum_{i=1}^n p_i f(x_i)\exp\left[\frac{\frac{1}{P_n}\sum\limits_{i=1}^n p_i x_i \frac{1}{P_n}\sum\limits_{i=1}^n p_i f'_+(x_i) - \frac{1}{P_n}\sum\limits_{i=1}^n p_i x_i f'_+(x_i)}{\frac{1}{P_n}\sum\limits_{i=1}^n p_i f(x_i)}\right].
\end{aligned} \tag{5.5.14}$$

Using (5.5.11) and (5.5.14), we derive the desired result (5.5.8) and the Theorem is proved.

The following corollary also holds.

Corollary 5.5.2. Let $f : I \to \mathbb{R}$ be a convex mapping on the interval I of real numbers and $x_i \in \mathring{I}$, $p_i \geqslant 0$ $(i = 1, \ldots, n)$ with $P_n > 0$, where $P_k = \sum_{i=1}^{k} p_i$ $(k = 1, \ldots, n)$. Then we have the inequality

$$1 \geqslant \frac{\exp f\left(\frac{1}{P_n}\sum_{i=1}^{n} p_i x_i\right)}{\frac{1}{P_n}\sum_{i=1}^{n} p_i \exp f(x_i)}$$

$$\geqslant \exp\left[\frac{\frac{1}{P_n}\sum_{i=1}^{n} p_i x_i \frac{1}{P_n}\sum_{i=1}^{n} p_i f'_+(x_i)\exp f(x_i) - \frac{1}{P_n}\sum_{i=1}^{n} p_i x_i f'_+(x_i)\exp f(x_i)}{\frac{1}{P_n}\sum_{i=1}^{n} p_i \exp f(x_i)}\right]$$

$$> 0. \tag{5.5.15}$$

Proof. Define the mapping $g : I \to (0,\infty)$ by $g(x) = \exp[f(x)]$. It is clear that the mapping g is log-convex on I. Now applying Theorem 5.5.2 for the log-convex mapping g, we easily deduce inequality (5.5.15).

If $f : I \to (0,\infty)$ is a log-convex function on I, an interval of real numbers, then the following refinement of the discrete Jensen's inequality (9) holds:

$$f\left(\frac{1}{P_n}\sum_{i=1}^{n} p_i x_i\right) \leqslant \left[\prod_{i=1}^{n}[f(x_i)]^{p_i}\right]^{\frac{1}{P_n}} \leqslant \frac{1}{P_n}\sum_{i=1}^{n} p_i f(x_i), \tag{5.5.16}$$

where, $x_i \in I$, $p_i \geqslant 0$ $(i = 1, \ldots, n)$ and $P_n = \sum_{i=1}^{n} p_i > 0$.

Indeed the first inequality in (5.5.16) follows by Jensen's discrete inequality applied to the convex map $\log(f(x))$ and the second is the classical arithmetic-geometric mean inequality. In the following theorem, we present an inequality proved in [40], related to the first inequality in (5.5.16).

Theorem 5.5.3. Let $f : I \to (0,\infty)$ be a log-convex mapping on the interval I of real numbers and $x_i \in \mathring{I}$, $p_i \geqslant 0$ $(i = 1, \ldots, n)$ with $P_n > 0$, where $P_k = \sum_{i=1}^{k} p_i$ $(k = 1, \ldots, n)$. Then one has the inequality

$$1 \leqslant \frac{\left[\prod_{i=1}^{n}[f(x_i)]^{p_i}\right]^{\frac{1}{P_n}}}{f\left(\frac{1}{P_n}\sum_{i=1}^{n} p_i x_i\right)} \leqslant \exp\left[\frac{1}{P_n}\sum_{i=1}^{n} p_i \frac{f'_+(x_i)}{f(x_i)} x_i - \frac{1}{P_n}\sum_{i=1}^{n} p_i x_i \frac{1}{P_n}\sum_{i=1}^{n} p_i \frac{f'_+(x_i)}{f(x_i)}\right], \tag{5.5.17}$$

where f'_+ is the right derivative of f on $\mathring{I}$.

Proof. By the convexity of $log f$ we can write that

$$\log f(x) - \log f(y) \geqslant \frac{f'_+(y)}{f(y)}(x-y), \tag{5.5.18}$$

for all $x,\ y \in \overset{\circ}{I}$. If we choose in the inequality (5.5.18), $x = \frac{1}{P_n}\sum_{i=1}^{n} p_i x_i$ and $y = x_j$ ($j = 1,\ldots,n$), then we get

$$\log f\left(\frac{1}{P_n}\sum_{i=1}^{n} p_i x_i\right) - \log f(x_j) \geqslant \frac{f'_+(x_j)}{f(x_j)}\left(\frac{1}{P_n}\sum_{i=1}^{n} p_i x_i - x_j\right), \tag{5.5.19}$$

for all $j = 1,\ldots,n$. If we multiply the inequality (5.5.19) by $p_j \geqslant 0$ and sum over $j = 1,\ldots,n$, we obtain

$$\log f\left(\frac{1}{P_n}\sum_{i=1}^{n} p_i x_i\right) - \frac{1}{P_n}\sum_{j=1}^{n} p_j \log f(x_j) \geqslant \frac{1}{P_n}\sum_{j=1}^{n} p_j \frac{f'_+(x_j)}{f(x_j)}\left(\frac{1}{P_n}\sum_{i=1}^{n} p_i x_i - x_j\right)$$

$$= \frac{1}{P_n}\sum_{i=1}^{n} p_i x_i \frac{1}{P_n}\sum_{j=1}^{n} p_j \frac{f'_+(x_j)}{f(x_j)} - \frac{1}{P_n}\sum_{j=1}^{n} p_j \frac{f'_+(x_j)}{f(x_j)} x_j. \tag{5.5.20}$$

From (5.5.20), we get

$$\log\left[\prod_{i=1}^{n}[f(x_i)]^{p_i}\right]^{\frac{1}{P_n}} - \log f\left(\frac{1}{P_n}\sum_{i=1}^{n} p_i x_i\right)$$

$$\leqslant \frac{1}{P_n}\sum_{j=1}^{n} p_j \frac{f'_+(x_j)}{f(x_j)} x_j - \frac{1}{P_n}\sum_{i=1}^{n} p_i x_i \frac{1}{P_n}\sum_{i=1}^{n} p_i \frac{f'_+(x_i)}{f(x_i)},$$

from which we get inequality (5.5.17).

The following corollary holds.

Corollary 5.5.3. Let $f : I \to \mathbb{R}$ be a convex mapping on the interval I of real numbers and $x_i \in \overset{\circ}{I}$, $p_i \geqslant 0$ ($i = 1,\ldots,n$) with $P_n > 0$, where $P_k = \sum_{i=1}^{k} p_i$ ($k = 1,\ldots,n$). Then

$$0 \leqslant \frac{1}{P_n}\sum_{i=1}^{n} p_i f(x_i) - f\left(\frac{1}{P_n}\sum_{i=1}^{n} p_i x_i\right)$$

$$\leqslant \frac{1}{P_n}\sum_{i=1}^{n} p_i f'_+(x_i) x_i - \frac{1}{P_n}\sum_{i=1}^{n} p_i x_i \frac{1}{P_n}\sum_{i=1}^{n} p_i f'_+(x_i). \tag{5.5.21}$$

Proof. Define the mapping $g : I \to (0,\infty)$ by $g(x) = \exp f(x)$. Then g is log-convex on I. If we apply inequality (5.5.17) for the mapping g we obtain

$$1 \leqslant \frac{\exp\left[\frac{1}{P_n}\sum_{i=1}^{n} p_i f(x_i)\right]}{\exp\left[f\left(\frac{1}{P_n}\sum_{i=1}^{n} p_i f(x_i)\right)\right]} \leqslant \exp\left[\frac{1}{P_n}\sum_{i=1}^{n} p_i f'_+(x_i) x_i - \frac{1}{P_n}\sum_{i=1}^{n} p_i x_i \frac{1}{P_n}\sum_{i=1}^{n} p_i f'_+(x_i)\right],$$

from which we get inequality (5.5.21).

5.6 Discrete inequalities for differentiable convex functions

This section deals with some converses of the discrete Jensen's inequality for differentiable convex mappings, recently investigated in [11,25,38,41].

First we shall present the following version of the discrete Jensen's inequality given by Dragomir in [41].

Theorem 5.6.1. Let $f : I \subset \mathbb{R} \to \mathbb{R}$ be a convex function on the interval I, $x \in \overset{\circ}{I}$ (the interior of I), $p_i \geqslant 0$ $(i = 1, \ldots, n)$ and $\sum_{i=1}^{n} p_i = 1$. Then we have the inequality

$$0 \leqslant \sum_{i=1}^{n} p_i f(x_i) - f\left(\sum_{i=1}^{n} p_i x_i\right) \leqslant \sum_{i=1}^{n} p_i x_i l(x_i) - \sum_{i=1}^{n} p_i x_i \sum_{i=1}^{n} p_i l(x_i), \tag{5.6.1}$$

where $l(x_i) \in [f'_-(x_i), f'_+(x_i)]$ $(i = 1, \ldots, n)$ and f'_-, f'_+ are the left and right derivatives of f respectively.

Proof. By the convexity of f on I, we have that

$$f(x) - f(y) \geqslant l(y)(x - y), \tag{5.6.2}$$

for all x, $y \in \overset{\circ}{I}$ and $l(y) \in [f'_-(y), f'_+(y)]$. Choosing in (5.6.2), $x = \sum_{j=1}^{n} p_j x_j$ and $y = x_i$ $(i = 1, \ldots, n)$, we get

$$f\left(\sum_{j=1}^{n} p_j x_j\right) - f(x_i) \geqslant l(x_i)\left(\sum_{j=1}^{n} p_j x_j - x_i\right), \tag{5.6.3}$$

for all $i = 1, \ldots, n$. If we multiply the inequality (5.6.3) with $p_i \geqslant 0$ and sum over i from 1 to n, we can easily deduce (5.6.1).

The next result also proved by Dragomir in [41], deals with useful upper bounds for the right membership in inequality (5.6.1).

Theorem 5.6.2. Let f be a differentiable convex mapping on $\overset{\circ}{I}$. If m, $M \in \overset{\circ}{I}$ and $m \leqslant x_i \leqslant M$ $(i = 1, \ldots, n)$, then we have the inequality

$$0 \leqslant \sum_{i=1}^{n} p_i f(x_i) - f\left(\sum_{i=1}^{n} p_i x_i\right) \leqslant \frac{1}{4}(M - m)(f'(m) - f'(m)). \tag{5.6.4}$$

Proof. We shall use the following discrete inequality of Grüss-type

$$\left|\frac{1}{\sum_{i=1}^{n} t_i} \sum_{i=1}^{n} t_i a_i b_i - \frac{1}{\sum_{i=1}^{n} t_i} \sum_{i=1}^{n} t_i a_i \frac{1}{\sum_{i=1}^{n} t_i} \sum_{i=1}^{n} t_i b_i\right| \leqslant \frac{1}{4}(A - a)(B - b), \tag{5.6.5}$$

provided that $a \leqslant a_i \leqslant A$, $b \leqslant b_i \leqslant B$ and $t_i \geqslant 0$ $(i = 1, \ldots, n)$ with $\sum_{i=1}^{n} t_i > 0$.

Now, if we choose in (5.6.5), $t_i = p_i, a_i = x_i$ and $b_i = f'(x_i)$, and taking into account that $f'(\cdot)$ is monotonic nondecreasing on $\overset{\circ}{I}$, we can state that $a = m, A = M$, $b = f'(m), B = f'(m)$ and we have the inequality

$$\sum_{i=1}^{n} p_i x_i f'(x_i) - \sum_{i=1}^{n} p_i x_i \sum_{i=1}^{n} p_i f'(x_i) \leqslant \frac{1}{4}(M-m)(f'(m) - f'(m)).$$

Now, using (5.6.1), we deduce (5.6.4).

In the following theorem, we give the inequality obtained by Dragomir and Goh in [25] (see also [38]).

Theorem 5.6.3. Let $f : \mathbb{R}^n \to \mathbb{R}$ be a differentiable convex mapping and

$$(\nabla f)(x) = \left(\frac{\partial f(x)}{\partial x^1}, \dots, \frac{\partial f(x)}{\partial x^n} \right),$$

the vector of the partial derivatives of $x = (x^1, \dots, x^n) \in \mathbb{R}^n$. If $x_i \in \mathbb{R}^n$ $(i = 1, \dots, m)$, $p_i \geqslant 0$, $i = 1, \dots, m$, with $P_m = \sum_{i=1}^{m} p_i > 0$, then

$$0 \leqslant \frac{1}{P_m} \sum_{i=1}^{m} p_i f(x_i) - f\left(\frac{1}{P_m} \sum_{i=1}^{m} p_i x_i \right)$$

$$\leqslant \frac{1}{P_m} \sum_{i=1}^{m} p_i \langle \nabla f(x_i), x_i \rangle - \left\langle \frac{1}{P_m} \sum_{i=1}^{m} p_i \nabla f(x_i), \frac{1}{P_m} \sum_{i=1}^{m} p_i x_i \right\rangle, \tag{5.6.6}$$

where $\langle \cdot, \cdot \rangle$ is the usual inner product on $\mathbb{R}^n$.

Proof. The first inequality in (5.6.6) is just the usual discrete Jensen's inequality. As $f : \mathbb{R}^n \to \mathbb{R}$ is differentiable convex function, we have the inequality

$$f(x) - f(y) \geqslant \langle \nabla f(y), x - y \rangle, \tag{5.6.7}$$

for all x, $y \in \mathbb{R}^n$. Choose in (5.6.7), $x = \frac{1}{P_m} \sum_{i=1}^{m} p_i x_i$ and $y = x_j$ to obtain

$$f\left(\frac{1}{P_m} \sum_{i=1}^{m} p_i x_i \right) - f(x_j) \geqslant \left\langle \nabla f(x_j), \frac{1}{P_m} \sum_{i=1}^{m} p_i x_i - x_j \right\rangle, \tag{5.6.8}$$

for all $j = \{1, \dots, n\}$.

If we multiply (5.6.8) by $p_j \geqslant 0$ and sum over j from 1 to m, then we obtain

$$P_m f\left(\frac{1}{P_m} \sum_{i=1}^{m} p_i x_i \right) - \sum_{j=1}^{m} p_j f(x_j)$$

$$\geqslant \frac{1}{P_m} \left\langle \sum_{j=1}^{m} p_j \nabla f(x_j), \sum_{i=1}^{m} p_i x_i \right\rangle - \sum_{j=1}^{m} \langle \nabla f(x_j), x_j \rangle. \tag{5.6.9}$$

Dividing (5.6.9) by $P_m > 0$, we obtain (5.6.6).

In [38], Dragomir provided an upper bound for Jensen's difference

$$\frac{1}{P_m}\sum_{i=1}^{m} p_i f(x_i) - f\left(\frac{1}{P_m}\sum_{i=1}^{m} p_i x_i\right), \tag{5.6.10}$$

which, even though it is not as sharp as (5.6.6), provides a simpler way, and for applications, a better way, of estimating the Jensen's differences. His result is embodied in the following theorem.

Theorem 5.6.4. Let $f : \mathbb{R}^n \to \mathbb{R}$ be a differentiable convex mapping and $x_i \in \mathbb{R}^n$, $i = 1,\ldots,m$. Suppose that there exists the vectors ϕ, $\Phi \in \mathbb{R}^n$ such that

$$\phi \leqslant x_i \leqslant \Phi, \tag{5.6.11}$$

(the order is considered on the co-ordinates) and k, $K \in \mathbb{R}^n$ are such that

$$k \leqslant \nabla f(x_i) \leqslant K, \tag{5.6.12}$$

for all $i \in \{1,\ldots,m\}$. Then for all $p_i \geqslant 0$ $(i = 1,\ldots,m)$ with $P_m = \sum_{i=1}^{m} p_i > 0$, we have the inequality

$$0 \leqslant \frac{1}{P_m}\sum_{i=1}^{m} p_i f(x_i) - f\left(\frac{1}{P_m}\sum_{i=1}^{m} p_i x_i\right) \leqslant \frac{1}{4}\|\Phi-\phi\|\|K-k\|, \tag{5.6.13}$$

where $\|\cdot\|$ is the usual Euclidean norm on $\mathbb{R}^n$.

Proof. A simple calculation shows that

$$\frac{1}{P_m}\sum_{i=1}^{m} p_i\langle x_i, \nabla f(x_i)\rangle - \left\langle \frac{1}{P_m}\sum_{i=1}^{m} p_i x_i, \frac{1}{P_m}\sum_{i=1}^{m} p_i \nabla f(x_i)\right\rangle$$

$$= \frac{1}{2P_m^2}\sum_{i,j=1}^{m} p_i p_j \langle x_i - x_j, \nabla f(x_i) - \nabla f(x_j)\rangle. \tag{5.6.14}$$

Taking the modulus in (5.6.14), and in view of the inequality (5.6.6), we obtain, by Schwarz's inequality in inner product spaces, i.e., $|\langle a,b\rangle| \leqslant \|a\|\|b\|$, a, $b \in \mathbb{R}^n$, that

$$\frac{1}{P_m}\sum_{i=1}^{m} p_i\langle x_i, \nabla f(x_i)\rangle - \left\langle \frac{1}{P_m}\sum_{i=1}^{m} p_i x_i, \frac{1}{P_m}\sum_{i=1}^{m} p_i \nabla f(x_i)\right\rangle$$

$$\leqslant \frac{1}{2P_m^2}\sum_{i,j=1}^{m} p_i p_j |\langle x_i - x_j, \nabla f(x_i) - \nabla f(x_j)\rangle|$$

$$\leqslant \frac{1}{2P_m^2}\sum_{i,j=1}^{m} p_i p_j \|x_i - x_j\|\|\nabla f(x_i) - \nabla f(x_j)\|. \tag{5.6.15}$$

Using the Cauchy-Buniakowsky-Schwarz inequality for double sums, we can state that

$$\frac{1}{2P_m^2}\sum_{i,j=1}^{m} p_i p_j \|x_i - x_j\| \|\nabla f(x_i) - \nabla f(x_j)\|$$

$$\leqslant \left(\frac{1}{2P_m^2}\sum_{i,j=1}^{m} p_i p_j \|x_i - x_j\|^2\right)^{\frac{1}{2}} \left(\frac{1}{2P_m^2}\sum_{i,j=1}^{m} p_i p_j \left\|\nabla f(x_i) - \nabla f(x_j)\right\|^2\right)^{\frac{1}{2}}. \tag{5.6.16}$$

As a simple calculation shows that

$$\frac{1}{2P_m^2}\sum_{i,j=1}^{m} p_i p_j \|x_i - x_j\|^2 = \frac{1}{P_m}\sum_{i=1}^{m} p_i \|x_i\|^2 - \left\|\frac{1}{P_m}\sum_{i=1}^{m} p_i x_i\right\|^2,$$

and

$$\frac{1}{2P_m^2}\sum_{i,j=1}^{m} p_i p_j \left\|\nabla f(x_i) - \nabla f(x_j)\right\|^2 = \frac{1}{P_m}\sum_{i=1}^{m} p_i \|\nabla f(x_i)\|^2 - \left\|\frac{1}{P_m}\sum_{i=1}^{m} f(x_i)\right\|^2,$$

we can state, by (5.6.15) and (5.6.16), that

$$\frac{1}{P_m}\sum_{i=1}^{m} p_i \langle x_i, \nabla f(x_i)\rangle - \left\langle \frac{1}{P_m}\sum_{i=1}^{m} p_i x_i, \frac{1}{P_m}\sum_{i=1}^{m} p_i \nabla f(x_i)\right\rangle$$

$$\leqslant \left(\frac{1}{P_m}\sum_{i=1}^{m} p_i \|x_i\|^2 - \left\|\frac{1}{P_m}\sum_{i=1}^{m} p_i x_i\right\|^2\right)^{\frac{1}{2}}$$

$$\times \left(\frac{1}{P_m}\sum_{i=1}^{m} p_i \|\nabla f(x_i)\|^2 - \left\|\frac{1}{P_m}\sum_{i=1}^{m} f(x_i)\right\|^2\right)^{\frac{1}{2}}. \tag{5.6.17}$$

Now, by simple calculation, we observe that

$$\frac{1}{P_m}\sum_{i=1}^{m} p_i \|x_i\|^2 - \left\|\frac{1}{P_m}\sum_{i=1}^{m} p_i x_i\right\|^2$$

$$= \left\langle \Phi - \frac{1}{P_m}\sum_{i=1}^{m} p_i x_i, \frac{1}{P_m}\sum_{i=1}^{m} p_i x_i - \phi \right\rangle - \frac{1}{P_m}\sum_{i=1}^{m} p_i \langle \Phi - x_i, x_i - \phi\rangle. \tag{5.6.18}$$

As $\phi \leqslant x_i \leqslant \Phi$ $(i \in \{1,\dots,m\})$, then $\langle \Phi - x_i, x_i - \phi\rangle \geqslant 0$ for all $i \in \{1,\dots,n\}$ and then

$$\sum_{i=1}^{m} p_i \langle \Phi - x_i, x_i - \phi\rangle \geqslant 0,$$

and, by (5.6.18), we obtain

$$\frac{1}{P_m}\sum_{i=1}^{m} p_i \|x_i\|^2 - \left\|\frac{1}{P_m}\sum_{i=1}^{m} p_i x_i\right\|^2 \leqslant \left\langle \Phi - \frac{1}{P_m}\sum_{i=1}^{m} p_i x_i, \frac{1}{P_m}\sum_{i=1}^{m} p_i x_i - \phi \right\rangle. \tag{5.6.19}$$

It is known that if $y,\ z \in \mathbb{R}^n$, then

$$4\langle z,y\rangle \leqslant \|z+y\|^2, \tag{5.6.20}$$

with equality if and only if $z=y$. Now, if we apply (5.6.20) for the vectors $z=\Phi-\frac{1}{P_m}\sum_{i=1}^m p_i x_i$, $y=\frac{1}{P_m}\sum_{i=1}^m p_i x_i-\phi$, we deduce

$$\left\langle \Phi-\frac{1}{P_m}\sum_{i=1}^m p_i x_i, \frac{1}{P_m}\sum_{i=1}^m p_i x_i-\phi \right\rangle \leqslant \frac{1}{4}\|\Phi-\phi\|^2,$$

and then, by (5.6.18), (5.6.19), we deduce that

$$\frac{1}{P_m}\sum_{i=1}^m p_i\|x_i\|^2-\left\|\frac{1}{P_m}\sum_{i=1}^m p_i x_i\right\|^2 \leqslant \frac{1}{4}\|\Phi-\phi\|^2. \tag{5.6.21}$$

Similarly, we can state that

$$\frac{1}{P_m}\sum_{i=1}^m p_i\|\nabla f(x_i)\|^2-\left\|\frac{1}{P_m}\sum_{i=1}^m f(x_i)\right\|^2 \leqslant \frac{1}{4}\|K-k\|^2. \tag{5.6.22}$$

Finally, by (5.6.17) and (5.6.21), (5.6.22), we deduce

$$\frac{1}{P_m}\sum_{i=1}^m p_i\langle x_i,\nabla f(x_i)\rangle-\left\langle \frac{1}{P_m}\sum_{i=1}^m p_i x_i, \frac{1}{P_m}\sum_{i=1}^m p_i\nabla f(x_i) \right\rangle \leqslant \frac{1}{4}\|\Phi-\phi\|\|K-k\|, \tag{5.6.23}$$

which in view of (5.6.6) gives the desired inequality in (5.6.13). The proof is complete.

Another result which provides an upper bound for Jensen's difference, established by Budimir, Dragomir and Pečarić [11] is embodied in the following theorem.

Theorem 5.6.5. Let $f:\mathbb{R}^n\to\mathbb{R}$ be a differentiable convex mapping and $x_i\in\mathbb{R}^n$, $p_i\geqslant 0$ $(i=1,\dots,m)$ with $P_m=\sum_{i=1}^m p_i>0$. Suppose that the ∇-operator satisfies a condition of $r-H$-Hölder-type, i.e.,

$$\|\nabla f(x)-\nabla f(y)\|\leqslant H\|x-y\|^r,$$

for all $x,\ y\in\mathbb{R}^n$, where $H>0$, $r\in(0,1]$ and $\|\cdot\|$ is the Euclidean norm. Then we have the inequality

$$0\leqslant\frac{1}{P_m}\sum_{i=1}^m p_i f(x_i)-f\left(\frac{1}{P_m}\sum_{i=1}^m p_i x_i\right)\leqslant\frac{H}{2P_m^2}\sum_{i,j=1}^m p_i p_j\|x_i-x_j\|^{r+1}. \tag{5.6.24}$$

Proof. Using Korkine's identity, that is, we recall it:

$$\frac{1}{P_m}\sum_{i=1}^{m} p_i \langle y_i, x_i\rangle - \left\langle \frac{1}{P_m}\sum_{i=1}^{m} p_i y_i, \frac{1}{P_m}\sum_{i=1}^{m} p_i x_i \right\rangle = \frac{1}{2P_m^2}\sum_{i,j=1}^{m} p_i p_j \langle y_i - y_j, x_i - x_j\rangle,$$

for $x,\ y \in \mathbb{R}^n$. We may simply write that

$$\frac{1}{P_m}\sum_{i=1}^{m} p_i \langle \nabla f(x_i), x_i\rangle - \left\langle \frac{1}{P_m}\sum_{i=1}^{m} p_i \nabla f(x_i), \frac{1}{P_m}\sum_{i=1}^{m} p_i x_i \right\rangle$$

$$= \frac{1}{2P_m^2}\sum_{i,j=1}^{m} p_i p_j \langle \nabla f(x_i) - \nabla f(x_j), x_i - x_j\rangle. \tag{5.6.25}$$

Using (5.6.6) and the properties of modulus, from (5.6.25) we have

$$\begin{aligned}
0 &\leqslant \frac{1}{P_m}\sum_{i=1}^{m} p_i f(x_i) - f\left(\frac{1}{P_m}\sum_{i=1}^{m} p_i x_i\right) \\
&\leqslant \frac{1}{2P_m^2}\sum_{i,j=1}^{m} p_i p_j \left|\langle \nabla f(x_i) - \nabla f(x_j), x_i - x_j\rangle\right| \\
&\leqslant \frac{1}{2P_m^2}\sum_{i,j=1}^{m} p_i p_j \left\|\nabla f(x_i) - \nabla f(x_j)\right\| \|x_i - x_j\| \\
&\leqslant \frac{H}{2P_m^2}\sum_{i,j=1}^{m} p_i p_j \|x_i - x_j\|^{r+1},
\end{aligned}$$

and the inequality (5.6.24) is proved.

5.7 Applications

In this section we point out applications of some of the inequalities given in earlier sections. The inequalities given above are recently developed and we hope that they will be a source for future research work.

5.7.1 *Applications for special means*

In this section, first we present applications of inequalities (5.4.29) and (5.4.35) given by Dragomir in [46].

The function $f(x) = \frac{1}{x}$, $x \in (0,\infty)$ is log-convex on $(0,\infty)$. Then we have

$$\frac{1}{b-a}\int_a^b \frac{dx}{x} = L^{-1}(a,b),$$

$$f\left(\frac{a+b}{2}\right) = A^{-1}(a,b),$$

$$\frac{f'\left(\frac{a+b}{2}\right)}{f\left(\frac{a+b}{2}\right)} = -\frac{1}{A}.$$

Now, applying the inequality (5.4.29) for the function $f(x) = \frac{1}{x}$, we get the inequality

$$\frac{A(a,b)}{L(a,b)} \geqslant L\left(\exp\left(-\frac{b-a}{2A}\right), \exp\left(\frac{b-a}{2A}\right)\right) \geqslant 1, \tag{5.7.1}$$

which is a refinement of the well-known inequality

$$A(a,b) \geqslant L(a,b), \tag{5.7.2}$$

where $A(a,b)$ is the Arithmetic mean and $L(a,b)$ is the Logarithmic mean of a, b, that is $A(a,b) = \frac{a+b}{2}$ and $L(a,b) = \frac{a-b}{\log a - \log b}$, $a \neq b$ (for $a = b$, $L(a,a) = a$).

For $f(x) = \frac{1}{x}$, we also get

$$\frac{f(a)+f(b)}{2} = H^{-1}(a,b),$$

where $H(a,b) = \frac{1}{\frac{1}{a}+\frac{1}{b}}$ is the Harmonic mean of a, b. Now, using the inequality (5.4.35) we obtain another interesting inequality

$$\frac{L(a,b)}{H(a,b)} \geqslant 1 + \log\left[\frac{A(a,b)}{L(a,b)}\right] \geqslant 1, \tag{5.7.3}$$

which is a refinement of the following well known inequality

$$L(a,b) \geqslant H(a,b). \tag{5.7.4}$$

Similar inequalities may be stated for the log-convex functions $f(x) = x^x$, $x > 0$ or $f(x) = e^x + 1$, $x \in \mathbb{R}$ etc. We omit the details.

The following inequality is well known in the literature as the Arithmetic mean-Geometric mean-Harmonic mean inequality

$$A_n(p,x) \geqslant G_n(p,x) \geqslant H_n(p,x),$$

where

$$A_n(p,x) = \sum_{i=1}^n p_i x_i, \qquad \text{the weighted Arithmetic mean,}$$

$$G_n(p,x) = \prod_{i=1}^n x_i^{p_i}, \qquad \text{the weighted Geometric mean,}$$

$$H_n(p,x) = \frac{1}{\sum\limits_{i=1}^n \frac{p_i}{x_i}}, \qquad \text{the weighted Harmonic mean,}$$

and $\sum_{i=1}^n p_i = 1$ $(p_i \geqslant 0, i = 1, \ldots, n)$.

The following results are obtained by Dragomir in [41], by applying Theorem 5.6.2.

Theorem 5.7.1. Let $0 < m \leqslant x_i \leqslant M < \infty$. Then for all $p_i \geqslant 0$ $(i = 1, \ldots, n)$, with $\sum_{i=1}^{n} p_i = 1$, we have the inequality

$$1 \leqslant \frac{A_n(p,x)}{G_n(p,x)} \leqslant \exp\left[\frac{(M-m)^2}{4mM}\right]. \tag{5.7.5}$$

The proof follows by applying the inequality (5.6.4) in Theorem 5.6.2, choosing $f(x) = -\log x$, $x > 0$.

Theorem 5.7.2. Let $0 < m \leqslant y_i \leqslant M < \infty$. Then for $p_i \geqslant 0$ with $\sum_{i=1}^{n} p_i = 1$, we have the inequality

$$1 \leqslant \frac{G_n(p,y)}{H_n(p,y)} \leqslant \exp\left[\frac{(M-m)^2}{4mM}\right]. \tag{5.7.6}$$

The proof follows by Theorem 5.7.1, choosing $x_i = \frac{1}{y_i}$ $(i = 1, \ldots, n)$.
For similar results obtained by applying Theorem 5.6.2 for the mappings $f(x) = x^p$, $p > 1$, $x \geqslant 0$ and $f(x) = x \log x$, $x > 0$, we refer the interested reader to [41].

5.7.2 *Applications for some inequalities related to Ky Fan's inequality*

In 1961, Beckenbach and Bellman published the following unpublished result due to Ky Fan in their book, *Inequalities* [10].
If $x_i \in \left(0, \frac{1}{2}\right)$, $i = 1, \ldots, n$, then one has the inequality

$$\left[\frac{\prod_{i=1}^{n} x_i}{\prod_{i=1}^{n}(1-x_i)}\right]^{\frac{1}{n}} \leqslant \frac{\sum_{i=1}^{n} x_i}{\sum_{i=1}^{n}(1-x_i)}, \tag{5.7.7}$$

with equality only if $x_1 = \cdots = x_n$.
In this section, we present some inequalities related to Ky Fan's inequality given by Dragomir and Mond in [40].
Consider the mapping $f : (0,1) \to \mathbb{R}$, $f(x) = \log\left(\frac{1-x}{x}\right)^r$, $r \geqslant 0$. Then we have

$$f'(x) = -\frac{r}{x(1-x)}, \quad x \in (0,1),$$

and

$$f''(x) = \frac{r(1-2x)}{[x(1-x)]^2}, \quad x \in (0,1),$$

which shows that the map f is convex on $\left(0, \frac{1}{2}\right]$, i.e., the mapping $g : (0,1) \to \mathbb{R}$, $g(x) = \left(\frac{1-x}{x}\right)^r$ is log-convex on $\left(0, \frac{1}{2}\right]$.
In [40], applying the inequalities (5.5.1), (5.5.8) Dragomir and Mond proved the following Ky Fan-type inequalities.

Theorem 5.7.3. Let $x_i \in \left(0, \frac{1}{2}\right]$ and $p_i \geqslant 0$ with $P_n = \sum_{i=1}^{n} p_i > 0$. Then one has

$$\frac{\frac{1}{P_n}\sum\limits_{i=1}^{n} p_i \left(\frac{1-x_i}{x_i}\right)^r}{\left(\frac{\frac{1}{P_n}\sum\limits_{i=1}^{n} p_i(1-x_i)}{\frac{1}{P_n}\sum\limits_{i=1}^{n} p_i x_i}\right)^r} \geqslant \frac{1}{P_n}\sum_{j=1}^{n} p_j \left[\exp\left(\frac{\frac{1}{P_n}\sum\limits_{i=1}^{n} p_i(x_i - x_j)}{\frac{1}{P_n}\sum\limits_{i=1}^{n} p_i x_i \frac{1}{P_n}\sum\limits_{i=1}^{n} p_i(1-x_i)}\right)\right]^r \geqslant 1, \quad (5.7.8)$$

where $r > 0$.

Proof. We shall apply the inequality (5.5.1) for the log-convex map $g(x) = \left(\frac{1-x}{x}\right)^r$, $x \in \left(0, \frac{1}{2}\right]$. We now have that $\frac{g'(x)}{g(x)} = \frac{-r}{x(1-x)}$ and from the inequality (5.5.1), we have

$$\frac{\frac{1}{P_n}\sum\limits_{i=1}^{n} p_i \left(\frac{1-x_i}{x_i}\right)^r}{\left(\frac{1-\frac{1}{P_n}\sum\limits_{i=1}^{n} p_i x_i}{\frac{1}{P_n}\sum\limits_{i=1}^{n} p_i x_i}\right)^r}$$

$$\geqslant \frac{1}{P_n}\sum_{j=1}^{n} p_j \exp\left\{\left[r\frac{\frac{1}{P_n}\sum\limits_{i=1}^{n} p_i x_i}{\frac{1}{P_n}\sum\limits_{i=1}^{n} p_i x_i\left(1-\frac{1}{P_n}\sum\limits_{i=1}^{n} p_i x_i\right)}\right] - \frac{r x_j}{\frac{1}{P_n}\sum\limits_{i=1}^{n} p_i x_i\left(1-\frac{1}{P_n}\sum\limits_{i=1}^{n} p_i x_i\right)}\right\}$$

$$= \frac{1}{P_n}\sum_{j=1}^{n} p_j \left[\exp\left(\frac{\frac{1}{P_n}\sum\limits_{i=1}^{n} p_i(x_i - x_j)}{\frac{1}{P_n}\sum\limits_{i=1}^{n} p_i x_i\left(1-\frac{1}{P_n}\sum\limits_{i=1}^{n} p_i x_i\right)}\right)\right]^r,$$

and the inequality (5.7.8) is established.

Theorem 5.7.4. Let $x_i \in \left(0, \frac{1}{2}\right]$ and $p_i \geqslant 0$ with $P_n = \sum_{i=1}^{n} p_i > 0$. Then one has the inequality

$$1 \geqslant \frac{\left(\frac{\frac{1}{P_n}\sum\limits_{i=1}^{n} p_i(1-x_i)}{\frac{1}{P_n}\sum\limits_{i=1}^{n} p_i x_i}\right)^r}{\frac{1}{P_n}\sum\limits_{i=1}^{n} p_i \left(\frac{1-x_i}{x_i}\right)^r}$$

$$\geqslant \left(\exp\left[\frac{1}{P_n}\sum_{i=1}^{n} p_i \frac{(1-x_i)^{r-1}}{x_i^r} - \frac{1}{P_n}\sum_{i=1}^{n} p_i x_i \frac{1}{P_n}\sum_{i=1}^{n} p_i \frac{(1-x_i)^{r-1}}{x_i^{r+1}}\right]\right)^r$$

$$> 0, \quad (5.7.9)$$

where $r > 0$.

Proof. The result is obvious by applying the inequality (5.5.8) in Theorem 5.5.2 to the log-convex mapping $g:\left(0,\frac{1}{2}\right]\to\mathbb{R}$, $g(x)=\left(\frac{1-x}{x}\right)^r$. We omit the details.

For further inequalities related to Ky Fan's inequality, see [40] and also the papers appeared in RGMIA Research Report Collections.

5.7.3 *Applications for Shannon's entropy*

In this section, we present applications of the inequality (5.6.4) given by Dragomir in [38], for Shannon's entropy mappings.

Following [38], let X be a random variable with range $R=\{x_1,\ldots,x_n\}$ and probability distribution $p_1,\ldots,p_n$ ($p_i>0$, $i=1,\ldots,n$). Define the Shannon entropy mapping

$$H(x)=-\sum_{i=1}^{n}p_i\log p_i. \tag{5.7.10}$$

The following result is well known in the literature and concerns the maximum possible value of $H(x)$ in terms of the size of $\mathbb{R}$ (see [75]).

Lemma 5.7.1 (see [41]). Let X be defined as above. Then

$$0\leqslant H(x)\leqslant\log n. \tag{5.7.11}$$

Furthermore, $H(X)=0$ if and only if $p_i=1$ for some i and $H(X)=\log n$ if and only if $p_i=\frac{1}{n}$ for all $i\in\{1,\ldots,n\}$.

The following useful analytic inequality is proved in [38].

Lemma 5.7.2. Let $0<m\leqslant\xi_i\leqslant M<\infty$, ($p_i>0$, $i=1,\ldots,n$) with $\sum_{i=1}^{n}p_i=1$. Then

$$0\leqslant\log\left(\sum_{i=1}^{n}p_i\xi_i\right)-\sum_{i=1}^{n}p_i\log\xi_i\leqslant\frac{(M-m)^2}{4mM}. \tag{5.7.12}$$

The proof follows by Theorem 5.6.2, choosing $f(x)=-\log x$ and $x_i=\xi_i$, $i=1,\ldots,n$.

The Lemma 5.7.2 provides the following converse inequality for the Shannon entropy mapping (see [38]).

Theorem 5.7.5. Let X be as above and put $p=\min\{p_i,\ i=1,\ldots,n\}$ and $P=\max\{p_i,\ i=1,\ldots,n\}$. Then we have

$$0\leqslant\log n-H(x)\leqslant\frac{(P-p)^2}{4pP}. \tag{5.7.13}$$

Proof. Choose in Lemma 5.7.2, $\xi_i=\frac{1}{p_i}\in\left[\frac{1}{P},\frac{1}{p}\right]$ and $m=\frac{1}{P}$, $M=\frac{1}{p}$ to get the desired inequality in (5.7.13).

Another analytic inequality proved in [38], which can be applied for the entropy mapping is embodied in the following lemma.

Lemma 5.7.3. Let $0 < m \leqslant \xi_i \leqslant M < \infty$, $(p_i > 0, i = 1, \dots, n)$ with $\sum_{i=1}^{n} p_i = 1$. Then

$$0 \leqslant \sum_{i=1}^{n} p_i \xi_i \log \xi_i - \sum_{i=1}^{n} p_i \xi_i \log \left(\sum_{i=1}^{n} p_i \xi_i \right)$$
$$\leqslant \frac{1}{4}(M-m)(\log M - \log m)$$
$$\leqslant \frac{1}{4} \frac{(M-m)^2}{\sqrt{mM}}. \tag{5.7.14}$$

Proof. The first inequality in (5.7.14) follows by Theorem 5.6.2, choosing $f(x) = x \log x$, which is a convex mapping on $(0, \infty)$, and $x_i = \xi_i$, $i = 1, \dots, n$.

The second inequality in (5.7.14) follows by the celebrated inequality between the Geometric mean $G(a,b) = \sqrt{ab}$ and the Logarithmic mean

$$L(a,b) = \begin{cases} a & \text{if } b = a, \\ \dfrac{b-a}{\log b - \log a} & \text{if } b \neq a, \end{cases} \qquad a,\ b > 0,$$

which states that

$$G(a,b) \leqslant L(a,b), \quad a, b > 0,$$

i.e.,

$$\frac{\log b - \log a}{b-a} \leqslant \frac{1}{\sqrt{ab}}.$$

Choosing $b = M$, $a = m$, we obtain

$$\frac{1}{4}(M-m)(\log M - \log m) \leqslant \frac{1}{4\sqrt{Mm}}(M-m)^2.$$

The proof is complete.

The Lemma 5.7.3 provides the following converse inequality for the entropy mapping, see [38].

Theorem 5.7.6. Let X, p, P be as in Theorem 5.7.5. Then we have

$$0 \leqslant \log n - H(x) \leqslant \frac{n}{4}(P-p)(\log P - \log p) \leqslant \frac{n}{4} \frac{(P-p)^2}{\sqrt{pP}}. \tag{5.7.15}$$

Proof. Firstly, choose $p_i = \frac{1}{n}$ in (5.7.14) to get

$$0 \leqslant \frac{1}{n} \sum_{i=1}^{n} \xi_i \log \xi_i - \frac{1}{n} \sum_{i=1}^{n} \xi_i \log \left(\frac{1}{n} \sum_{i=1}^{n} \xi_i \right)$$
$$\leqslant \frac{1}{4}(M-m)(\log M - \log m)$$
$$\leqslant \frac{1}{4\sqrt{mM}}(M-m)^2. \tag{5.7.16}$$

Now, if in (5.7.16) we assume that $\xi_i = p_i \in [p, P]$, then we obtain

$$0 \leqslant \frac{1}{n} \log n - \frac{1}{n} H(x) \leqslant \frac{1}{4}(P-p)(\log P - \log p) \leqslant \frac{1}{4\sqrt{pP}}(P-p)^2,$$

from which we get (5.7.15).

5.8 Miscellaneous inequalities

5.8.1 *Klaričić Bakula and Pečarić [70]*

Let f be a nonnegative convex function on $[m_1,M_1]$, g a nonnegative convex function on $[m_2,M_2]$, $u:[a,b]\to[m_1,M_1]$ and $v:[a,b]\to[m_2,M_2]$ continuous functions, and $p:[a,b]\to\mathbb{R}$ be a positive integrable function. Then

$$\frac{1}{P}\int_a^b p(x)f(u(x))g(v(x))dx$$

$$\leqslant [f;m_1,M_1][g;m_2,M_2]\frac{1}{P}\int_a^b p(x)u(x)v(x)dx+[f;m_1,M_1][\overline{g};m_2,M_2]\frac{1}{P}\int_a^b p(x)u(x)dx$$

$$+[\overline{f};m_1,M_1][g;m_2,M_2]\frac{1}{P}\int_a^b p(x)v(x)dx+[\overline{f};m_1,M_1][\overline{g};m_2,M_2],$$

$$f\left(\frac{m_1+M_1}{2}\right)g\left(\frac{m_2+M_2}{2}\right)\leqslant\frac{1}{4P}\left[\int_a^b p(x)f(u(x))g(v(x))dx\right.$$

$$\left.+\int_a^b p(x)f(M_1+m_1-u(x))g(M_2+m_2-v(x))dx\right]$$

$$+\frac{1}{4P}\left[-2[f;m_1,M_1][g;m_2,M_2]\int_a^b p(x)u(x)v(x)dx\right.$$

$$+([\hat{g};m_2,M_2]-[\overline{g};m_2,M_2])[f;m_1,M_1]\int_a^b p(x)u(x)dx$$

$$\left.+([\hat{f};m_1,M_1]-[\overline{f};m_1,M_1])[g;m_2,M_2]\int_a^b p(x)v(x)dx\right]$$

$$+\frac{1}{4}\left([\overline{f};m_1,M_1][\hat{g};m_2,M_2]+[\hat{f};m_1,M_1][\overline{g};m_2,M_2]\right),$$

where $P=\int_a^b p(x)dx$ and for a suitable function $h:[\alpha,\beta]\subset\mathbb{R}\to\mathbb{R}$ the notations

$$[h;x,y]=\frac{h(y)-h(x)}{y-x},\quad x\neq y, \tag{5.8.1}$$

$$\overline{h}(t)=t\,h(\alpha+\beta-t), \tag{5.8.2}$$

$$\hat{h}(t)=t\,h(t), \tag{5.8.3}$$

for $x,\ y,\ t\in[\alpha,\beta]$ are used to simplify the presentation.

5.8.2 *Klaričić Bakula and Pečarić [70]*

Let f be a nonnegative convex function on $[m_1, M_2]$, g a nonnegative convex function on $[m_2, M_2]$, $u : [a,b] \to [m_1, M_1]$ and $v : [a,b] \to [m_2, M_2]$ continuous functions, and $p, q : [a,b] \to \mathbb{R}$ positive integrable functions. Then

$$\frac{1}{PQ}\int_a^b\int_a^b\int_0^1 p(x)q(y)f(tu(x)+(1-t)u(y))g(tv(x)+(1-t)v(y))dtdxdy$$
$$\leqslant \frac{1}{3PQ}\left[Q\int_a^b f(u(x))g(v(x))p(x)dx + P\int_a^b f(u(y))g(v(y))q(y)dy\right]$$
$$+\frac{1}{3PQ}\int_a^b f(u(x))p(x)dx\int_a^b g\left(v(y)\right)q(y)dy,$$
$$\frac{1}{P}\int_a^b\int_0^1 p(x)f(tu(x)+(1-t)\overline{u})g(tv(x)+(1-t)\overline{v})dtdx$$
$$\leqslant \frac{1}{3P}\int_a^b p(x)f(u(x))g(v(x))dx + \frac{1}{3}f(\overline{u})g(\overline{v})$$
$$+\frac{1}{6P}\left[g(\overline{v})\int_a^b p(x)f(u(x))dx + f(\overline{u})\int_a^b p(x)g(v(x))dx\right],$$

where $P = \int_a^b p(x)dx, Q = \int_a^b q(x)dx, \overline{u} = \frac{1}{P}\int_a^b p(x)u(x)dx, \overline{v} = \frac{1}{Q}\int_a^b q(x)v(x)dx.$

5.8.3 *Pachpatte [98]*

Let $f : [a,b] \to \mathbb{R}$ be a differentiable convex function and $F(x,y)(t)$ be as defined in (5.3.1). Then

$$\int_a^b F\left(x, \frac{a+b}{2}\right)(t)dx \leqslant \int_a^b f(x)dx,$$
$$\frac{3}{2}\frac{1}{b-a}\int_a^b f(x)dx \leqslant \frac{1}{b-a}\int_a^b F\left(x, \frac{a+b}{2}\right)(t)dx + \frac{f(a)+f(b)}{4},$$

for all $t \in [0,1]$.

5.8.4 *Pachpatte [98]*

Let $f : [a,b] \to \mathbb{R}$ be a differentiable convex function and $F(x,y)(t)$ be as defined in (5.3.1). Then

$$\frac{1}{b-a}\int_a^b\int_a^b F(x,y)(t)dxdy \leqslant \int_a^b f(x)dx,$$
$$\frac{3}{2}\frac{1}{b-a}\int_a^b f(x)dx \leqslant \frac{1}{(b-a)^2}\int_a^b\int_a^b F(x,y)(t)dxdy + \frac{f(a)+f(b)}{4},$$

for all $t \in [0,1]$.

5.8.5 *Pachpatte [98]*

Let $f : [a,b] \to \mathbb{R}$ be a convex function, $F(x,y)(t)$ be as defined in (5.3.1) and define for $t \in [0,1]$

$$G(t) = \frac{1}{b-a}\int_a^b F\left(x, \frac{a+b}{2}\right)(t)dx,$$

$$H(t) = \frac{1}{(b-a)^2}\int_a^b\int_a^b F(x,y)(t)dxdy.$$

Then

(i) G and H are convex on $[0,1]$,

(ii) $G(t) \leqslant H(t)$ for all $t \in [0,1]$.

5.8.6 *Pachpatte [136]*

Let f and g be real-valued, nonnegative and convex functions on $[a,b] \subset \mathbb{R}$ and $F(x,y)(t)$ and $G(x,y)(t)$ be as defined in (5.3.1) and (5.3.2). Then for all $t \in [0,1]$ we have

$$\frac{1}{b-a}\int_a^b F\left(x, \frac{a+b}{2}\right)(t)G\left(x, \frac{a+b}{2}\right)(t)dx$$

$$\leqslant \frac{1}{4(b-a)}\int_a^b f(x)g(x)dx + \frac{3}{16}[M(a,b)+N(a,b)],$$

$$\frac{2}{(b-a)^2}\int_a^b\int_a^b F(x,y)(t)G(x,y)(t)dxdy$$

$$\leqslant \frac{1}{b-a}\int_a^b f(x)g(x)dx + \frac{1}{4}[M(a,b)+N(a,b)],$$

where $M(a,b) = f(a)g(a)+f(b)g(b)$, $N(a,b) = f(a)g(b)+f(b)g(a)$.

5.8.7 *Pachpatte [82]*

Let f and g be real-valued, nonnegative and convex functions on $[0,1]$. Then

$$\int_0^1 (1-x)f(x)g(x)dx \leqslant \frac{1}{3}\left(\int_0^1 f(x)dx\right)\left(\int_0^1 g(x)dx\right) + \frac{1}{6}f(0)g(0),$$

$$\int_0^1 xf(x)g(x)dx \leqslant \frac{1}{3}\left(\int_0^1 f(x)dx\right)\left(\int_0^1 g(x)dx\right) + \frac{1}{6}f(1)g(1),$$

$$\int_0^1 f(x)g(x)dx \leqslant \frac{2}{3}\left(\int_0^1 f(x)dx\right)\left(\int_0^1 g(x)dx\right) + \frac{1}{6}[f(0)g(0)+f(1)g(1)].$$

5.8.8 *Pachpatte [82]*

Let f and g be real-valued, nonnegative and convex functions on $[0,1]$. Then

$$\int_0^1 (1-x)\left[f^2(x)+g^2(x)\right]dx \leqslant \frac{1}{3}\left[\left(\int_0^1 f(x)dx\right)^2+\left(\int_0^1 g(x)dx\right)^2\right]+\frac{1}{6}\left[f^2(0)+g^2(0)\right],$$

$$\int_0^1 x\left[f^2(x)+g^2(x)\right]dx \leqslant \frac{1}{3}\left[\left(\int_0^1 f(x)dx\right)^2+\left(\int_0^1 g(x)dx\right)^2\right]+\frac{1}{6}\left[f^2(1)+g^2(1)\right],$$

$$\int_0^1 \left[f^2(x)+g^2(x)\right]dx$$

$$\leqslant \frac{2}{3}\left[\left(\int_0^1 f(x)dx\right)^2+\left(\int_0^1 g(x)dx\right)^2\right]+\frac{1}{6}\left[f^2(0)+g^2(0)+f^2(1)+g^2(1)\right].$$

5.8.9 *Teng, Yang and Dragomir [149]*

Let $s:[a,b]\to\mathbb{R}$ be nonnegative, integrable and symmetric to $\frac{a+b}{2}$ and $f\in Q(I)\cap L_1[a,b]$. Then

$$f\left(\frac{a+b}{2}\right)\int_a^b s(x)dx \leqslant 4\int_a^b f(x)s(x)dx, \tag{5.8.4}$$

$$\int_a^b \frac{(b-x)(x-a)}{(b-a)^2}f(x)s(x)dx \leqslant \frac{f(a)+f(b)}{2}\int_a^b s(x)dx,$$

where $Q(I)$ is defined as in section 5.2. The constant 4 in (5.8.4) is the best possible.

5.8.10 *Tseng, Yang and Dragomir [149]*

Let $s:[a,b]\to\mathbb{R}$ be nonnegative, integrable and symmetric to $\frac{a+b}{2}$ and $f\in P(I)\cap L_1[a,b]$. Then

$$f\left(\frac{a+b}{2}\right)\int_a^b s(x)dx \leqslant 2\int_a^b f(x)s(x)dx \leqslant 2[f(a)+f(b)]\int_a^b s(x)dx, \tag{5.8.5}$$

where $P(I)$ is defined as in Section 5.2. Both inequalities in (5.8.5) are sharp.

5.8.11 *Pachpatte [107]*

Let $f,\ g:[a,b]\to\mathbb{R}$ be absolutely continuous functions on $[a,b]$. Let

$$S(f,g)=f(x)g(x)-\frac{1}{2(b-a)}\left[g(x)\int_a^b f(t)dt+f(x)\int_a^b g(t)dt\right],$$

$$A = \frac{|f'(t)|}{|f'(x)|}, \quad B = \frac{|g'(t)|}{|g'(x)|}. \tag{5.8.6}$$

(a_1) If $|f'|$, $|g'|$ are convex on $[a,b]$, then

$$|S(f,g)| \leqslant \frac{1}{4}\left[\frac{1}{4} + \left(\frac{\left(x - \frac{a+b}{2}\right)}{b-a}\right)^2\right](b-a)$$

$$\times \left\{|g(x)|\left[|f'(x)| + \|f'\|_\infty\right] + |f(x)|\left[|g'(x)| + \|g'\|_\infty\right]\right\},$$

for $x \in [a,b]$.

(a_2) If $|f'|$, $|g'|$ are log-convex on $[a,b]$, then

$$|S(f,g)| \leqslant \frac{1}{2(b-a)}\left\{|g(x)||f'(x)| \int_a^b |x-t| \left(\frac{A-1}{\log A}\right) dt\right.$$

$$\left. + |f(x)||g'(x)| \int_a^b |x-t| \left(\frac{B-1}{\log B}\right) dt\right\},$$

for $x \in [a,b]$.

5.8.12 *Pachpatte [107]*

Let $f, g : [a,b] \to \mathbb{R}$ be absolutely continuous functions on $[a,b]$. Let

$$T(f,g) = \frac{1}{b-a}\int_a^b f(x)g(x)dx - \left(\frac{1}{b-a}\int_a^b f(x)dx\right)\left(\frac{1}{b-a}\int_a^b g(x)dx\right),$$

$$E(x) = \frac{(x-a)^2 + (b-x)^2}{2}, x \in [a,b].$$

(b_1) If $|f'|$, $|g'|$ are convex on $[a,b]$, then

$$|T(f,g)| \leqslant \frac{1}{4(b-a)^3}\int_a^b \left[|f'(x)| + \|f'\|_\infty\right]\left[|g'(x)| + \|g'\|_\infty\right] E^2(x)dx,$$

(b_2) If $|f'|$, $|g'|$ are log-convex on $[a,b]$, then

$$|T(f,g)| \leqslant \frac{1}{(b-a)^3}\int_a^b \left[\left\{\int_a^b |x-t||f'(x)| \left(\frac{A-1}{\log A}\right) dt\right\}\right.$$

$$\left. \times \left\{\int_a^b |x-t||g'(x)| \left(\frac{B-1}{\log B}\right) dt\right\}\right] dx,$$

where A, B are defined by (5.8.6).

5.9 Notes

An enormous amount has been written about inequalities involvibg convex functions and their applications. The results in Theorems 5.2.1–5.2.4 deals with integral inequalities similar to that of the Hadamard's inequalities involving the product of two convex functions and are taken from Pachpatte [97] and [85]. The integral inequalities in Theorems 5.3.1–5.3.3 contains Hadamard-type inequalities involving convex functions and are established by Pachpatte in [84]. Theorems 5.4.1 and 5.4.2 contains the integral inequalities involving the product of two log-convex functions and are taken from Pachpatte [104], while Theorem 5.4.3 contains Hadamard-type integral inequalities involving three log-convex functions and taken from Pachpatte [136]. The results in Theorems 5.4.4 and 5.4.5 contains the Hadamard-type integral inequalities investigated by Pachpatte in [110].

Theorem 5.5.1 deals with a refinement of discrete Jensen's inequality for log-convex function and is taken from Dragomir and Mond [40] and Theorem 5.5.2 is the converse of Jensen's discrete inequality for log-convex functions and is also taken from Dragomir and Mond [40]. Theorem 5.5.3 contains another refinement of Jensen's discrete inequality and is taken from Dragomir and Mond [40]. Theorem 5.6.1 contains a version of the discrete Jensen's inequality and is taken from Dragomir [41] and Theorem 5.6.2 contains a useful version of Theorem 5.6.1 and is taken from Dragomir [41]. Theorems 5.6.3 and 5.6.4 are taken from Dragomir and Goh [25] and Dragomir [38], while Theorem 5.6.5 is due to Budimir, Dragomir and Pečarić [11]. Section 5.7.1 is devoted to the applications of certain inequalities for special means and taken from Dragomir [41,46]. Section 5.7.2 contains applications to obtain Ky Fan-type inequalities and are taken from Dragomir and Mond [40]. Section 5.7.3 deals with the applications for Shannon's entropy and are taken from Dragomir [38]. In Section 5.8 some useful miscellaneous inequalities investigated by various investigators are given.

Bibliography

[1] Aglić Aljinović, A. and J. Pečarić, A discrete Euler identity, J. Inequal. Pure and Appl. Math. 5 (3) Art. 58, 2004.

[2] Anastassiou, G. A., Multivariate Ostrowski type inequalities, Acta Math. Hungar. 76 (1997), 267–278.

[3] Atkinson, K. E., An introduction to numerical analysis, Wiley and Sons, second edition, 1989.

[4] Barnett, N. S. and S.S. Dragomir, An inequality of Ostrowski's type for comulative distributation functions, Kyungpook Math. J. 39 (2) (1999), 301–311.

[5] 5. Barnett, N. S. and S.S. Dragomir, Some inequalities for random variables whose probability density functions are absolutely continuous using a pre-Chebychev inequality, Kyungpook Math. J. 40 (2) (2000), 299–311.

[6] Barnett, N. S. and S.S. Dragomir, A Trapezoid type inequality for double integrals, Nonlinear Analysis 47 (4) (2001), 2321–2332.

[7] Barnett, N. S., S.S. Dragomir and A. Sofo, Better bounds for an inequality of Ostrowski type with applications, RGMIA Res. Rep. Coll. 3 (1) (2000), 113–122.

[8] Barnett, N. S. and S.S. Dragomir, An Ostrowski type inequality for double integrals and applications for cubature formulae, Soochow J. Math. 27 (1) (2001), 1–10.

[9] Barnett, N.S. and S.S. Dragomir, On the perturbed trapezoid formula, Tamkang Jour. Math. 33 (2) (2002), 119–128.

[10] Beckenbach, E.F. and R. Bellman, Inequalities, Springer-Verlag, Berlin-New York, 1965.

[11] Budimir, I., S.S. Dragomir and J. Pečarić, Further reverse results for Jensen's discrete inequality and applications in information theory, J. Inequal. Pure and Appl. Math. 2 (1) Art. 5, 2001.

[12] Buşe, C., P. Cerone, S.S. Dragomir and J. Roumeliotis, A refinement of Grüss type inequality for the Bochner integral of vecter-valued functions in Hilbert spaces and applications, J. Korean Math. Soc. 43 (2006), 911–929.

[13] Čebyšev, P.L., Sur les expressions approximatives des intégrales définies par les aures prises entre les mêmes limites, Proc. Math. Soc. Charkov 2 (1882), 93–98.

[14] Cerone, P., S.S. Dragomir and J. Roumelotis, An inequality of Ostrowski type for mappings whose second derivatives are bounded and applications, East Asian J, Math. 15 (1) (1999), 1–9.

[15] Cerone, P., S.S. Dragomir and J. Roumelotis, An inequality of Ostrowski-Grüss type for twice differentiable mappings and applications in numerical integration, Kyungpook Math. J. 39 (1999), 333–341.

[16] Cerone, P., S.S. Dragomir and J. Roumelotis, Some Ostrowski-type inequalities for n-time differentiable mappings and applications, Demonstratio Math. 32 (4) (1999), 697–712.

[17] Cerone, P. and S.S. Dragomir, Trapezoidal type rules from an inequalities point of view, in "Analytic computational method in applied mathematics" (ed.), G. Anastassiou, CRC Press, N.Y., 2000, pp. 65–134.

[18] Cerone, P. and S.S. Dragomir, Generalizations of the Grüss, Chebychev and Lupaş inequalities for integrals over different intervals, Int. J. Appl. Math. 6 (2) (2001), 117–128.

[19] Cheng, X.L., Improvement of some Ostrowski-Grüss type inequalities, Computers Math. Applic. 42 (1/2) (2001), 109–114.

[20] Cristescu, G., Improved integral inequalities for product of convex functions, J. Inequal. Pure and Appl. Math. 6 (2) Art. 35, 2005.

[21] Dedić, Lj, M. Matić and J.E. Pečarić, On some generalizations of Ostrowski inequality for Lipschitz functions and functions of bounded variation, Math. Inequal. Appl. 3 (1) (2000), 1–14.

[22] Dedić, Lj, J. Pečarić and N. Ujević, On generalizations of Ostrowski inequality and some related results, Czechoslovak Math. J. 53 (128) (2003), 173–189.

[23] Dragomir, S.S. and N. M. Ionescu, Some remarks on convex functions, Anal. Num. Theor. Approx. 21 (1992), 31–36.

[24] Dragomir, S.S., J.E. Pečarić and L.E. Persson, Some inequalities of Hadamard type, Soochow J. Math. 21 (1995), 335–341.

[25] Dragomir, S.S. and C. J. Goh, A counterpart of Jensen's discrete inequality for differentiable convex mappings and applications in information theory, Math. Comput. Modelling 24 (2) (1996), 1–11.

[26] Dragomir, S.S. and S. Wang, An inequality of Ostrowski-Grüss type and its applications to the estimation of error bounds for some special means and for some numerical quadrature rules, Computers Math. Applic. 33 (11) (1997), 15–20.

[27] Dragomir, S.S. and S. Wang, A new inequality of Ostrowski's type in L_p norm, Indian Jour. Math. 40 (3) (1998), 299–304.

[28] Dragomir, S.S. and N. S. Barnett, An Ostrowski type inequality for mappings whose second derivatives are bounded and applications, RGMIA Res. Rep. Coll. 1 (2) (1998), 69–77.

[29] Dragomir, S.S., N. S. Barnett and P. Cerone, An Ostrowski type inequality for double integrals in terms of L_p-norms and applications in numerical integration, RGMIA Res. Rep. Coll. 1 (2) (1998), 79–87.

[30] Dragomir, S.S. and B. Mond, Integral inequalities of Hadamard's type for log-convex functions, Demonstratio Math. 31 (2) (1998), 354–364.

[31] Dragomir, S.S., New estimation of the remainder in Taylor's formula using Grüss type inequalities and applications, Math. Inequal. Appl. 2 (2) (1999), 183–193.

[32] Dragomir, S.S., A generalization of Grüss inequality in inner product spaces and applications, J. Math. Anal. Appl. 237 (1999), 74–82.

[33] Dragomir, S.S., N.S. Barnett and S. Wang, An Ostrowski type inequality for a random variable whose probability density function belongs to $L_p[a,b]$, $p > 1$, Math. Ineq. Appl. 2 (4) (1999), 501–508.

[34] Dragomir, S.S. and A. McAndrew, On the Trapezoid inequality via a Grüss type result and applications, Tamkang Jour. Math. 31 (3) (2000), 193–201.

[35] Dragomir, S.S. and A. Sofo, An integral inequality for twice differentiable mappings and applications, Tamkang Jour. Math. 31 (4) (2000), 257–266.

[36] Dragomir, S.S. and S. Mabizela, Some error estimates in the Trapezoidal quadrature rule, Tamsui Oxford J. Math. Sci. 16 (2) (2000), 259–272.

[37] Dragomir, S.S., P. Cerone, N. S. Barnett and J. Roumeliotis, An inequality of the Ostrowski type for double integrals and applications for cubature formula, Tamsui Oxford J. Math. Sci. 16 (1) (2000), 1–6.

[38] Dragomir, S.S., A converse of the Jensen inequality for convex mappings of several variables and applications, RGMIA Res. Rep. Coll. 2 (7) (1999), 977–989.

[39] Dragomir, S.S., N.S. Barnett and C.E.M. Pearce, A Quasi Trapezoid inequality for double integrals and applications, ANZIAM J. 44 (2003), 355–364.

[40] Dragomir, S.S., and B. Mond, A refinement of Jensen's discrete inequality for log-convex mappings and applications, manuscript.

[41] Dragomir, S.S., A converse result for Jensen's discrete inequality via Grüss inequality and applications in information theory, Analele Univ. Ordea, Fasc. Math. 7 (1999-2000), 178–189.

[42] Dragomir, S.S., Some integral inequalities of Grüss type, Indian J. Pure and Appl. Math. 31 (2000), 379–415.

[43] Dragomir, S.S., A Grüss type inequality for sequences of vectors in inner product spaces and applications, J. Inequal. Pure and Appl. Math. 1 (2) Art. 12, 2000.

[44] Dragomir, S.S., N. S. Barnett and P. Cerone, An n-dimensional version of Ostrowski's inequality for mappings of the Hölder type, Kyungpook Math. J. 40 (1) (2000), 65–75.

[45] Dragomir, S.S. and C. E. M. Pearce, Selected topics on Hermite-Hadamard Inequalities and Applications, RGMIA Monographs, Victoria University, 2000.

[46] Dragomir, S.S., Refinements of the Hermite-Hadamard integral inequality for log-convex functions, RGMIA Res. Rep. Coll. 3 (4) (2000), 527–533.

[47] Dragomir, S.S., P. Cerone and J. Roumeliotis, A new generalization of Ostrowski integral inequality for mappings whose derivatives are bounded and applications in numerical integration and for special means, Appl. Math. Lett. 13 (1) (2000), 19–25.

[48] Dragomir, S.S., A counterpart of Bessel's inequality in inner product spaces and some Grüss type related results, RGMIA Res. Rep. Coll. 6 (2003), supplement.

[49] Dragomir, S.S., The discrete version of Ostrowski's inequality in normed linear spaces, J. Inequal. Pure and Appl. Math. 3 (1) Art. 2, 2002.

[50] Dragomir, S.S. and Th. M. Rassias (Eds.), Ostrowski type Inequalities and Applications in Numerical Integration, Kluwer Academic Publishers, Dordrecht, 2002.

[51] Dragomir, S.S. and L. Khan, Two discrete inequalities of Grüss type via Pólya-Szegö and Shisha results for real numbers, RGMIA Res. Rep. Coll. 5 (3) (2002), 515–526.

[52] Dragomir, S.S., On the Ostrowski integral inequality for Lipschitzian mappings and applications, Comput. and Math. Appl. 38 (11-12) (1999), 33–37.

[53] Dragomir, S.S., Some Grüss type inequalities in inner product spaces, J. Inequal. Pure and Appl. Math. 4 (2) Art. 42, 2003.

[54] Dragomir, S.S. and I. Gomm, Some integral and discrete versions of the Grüss inequality for real and complex functions and sequences, Tamsui Oxford J. Math. Sci. 19 (1) (2003), 67–77.

[55] Dragomir, S.S., On Bessel and Grüss inequalities for orthonormal families in inner product spaces, RGMIA Res. Rep. Coll. 6 (2003), supplement.

[56] Dragomir, S.S., J.E. Pečarić and B. Tepeš, Pre-Grüss type inequalities in inner product spaces, RGMIA Res. Rep. Coll. 6 (3) (2003), 591–601.

[57] Dragomir, S.S., An inequality of Ostrowski type via Pompeiu's mean value theorem, JIPAM. J. Ineq. Math. Pure Appl. Math. 6 (3), Art. 83, 2005.

[58] Engels, H., Numerical Quadrature and Cubature, Academic Press, New York, 1980.

[59] Fink, A.M., Bounds on the deviation of a function from its averages, Chechoslovak Math. J. 42 (1992), 289–310.

[60] Godunova, E.K. and V.I. Levin, Nervenstka Dlja Funkcii Širokogo Klassa, Moskva, 1985, 138–142.

[61] Grüss, G., Über das Maximum des absoluten Betrages von

$$\frac{1}{b-a}\int_a^b f(x)g(x)dx - \frac{1}{(b-a)^2}\int_a^b f(x)dx\int_a^b g(x)dx,$$

Math. Z. 39 (1935), 215–226.

[62] Hanna, G., S.S. Dragomir and P. Cerone, A Taylor like formula for mappings of two variables defined on a rectangle in the plane, Tamusi Oxford J. Math. Sci. 18 (1) (2002), 1–16.

[63] Hanna, G. and Roumeliotis, Weighted integral inequalities in two dimensions, RGMIA Res. Rep. Coll. 4 (3) (2001), 487–510.

[64] Hanna, G., S.S. Dragomir and P. Cerone, A general Ostrowski type inequality for double integrals, Tamsui Oxford J. Math. Sci. 18 (1) (2002), 1–16.

[65] Hanna, G. and S.S. Dragomir, Some Ostrowski type inequalities for double integrals of functions whose partial derivatives satisfy certain convexity properties, RGMIA Res. Rep. Coll. 5 (1) (2002), 193–205.

[66] Hanna, G., S.S. Dragomir and J. Roumeliotis, An approximation for the finite Fourier transform of two independent variables, RGMIA Res. Rep. Coll. 6 (1) (2003), 31–41.

[67] Hanna, G., S.S. Dragomir and J. Roumeliotis, Error estimates on approximating the finite Fourier transform of complex-valued functions via pre-Grüss inequality, RGMIA Res. Rep. Coll. 7 (2) (2004), 229–237.

[68] Hardy, G.H., J.E. Littlewood and G. Polya, Inequalities, Cambridge University Press, 1934.

[69] Heinig, H.P. and L. Maligranda, Chebysev inequality in function spaces, Real Analysis Exchange 17 (1991-92), 211–247.

[70] Klaričić Bakula, M. and J. Pečarić, Note on some Hadamard-type inequalities, J. Inequal. Pure and Appl. Math. 5 (3) Art. 74, 2004.

[71] Maligranda, L., J.E. Pečarić and L.E. Persson, On some inequalities of the Grüss-Barnes and Borell type, J. Math. Anal. Appl. 187 (1994), 306–323.

[72] Matić, M., J.E. Pečarić and N. Ujević, On new estimation of the remainder in generalized Taylor's formula, Math. Inequal. Appl. 2 (3) (1999), 343–361.

[73] Matić, M., J.E. Pečarić and N. Ujević, Improvement and further generalization of inequalities of Ostrowski-Grüss type, Computers Math. Appl. 39 (2000), 161–175.

[74] Matić, M., J.E. Pečarić and N. Ujević, Weighted version of multivariate Ostrowski type inequalities, Rocky Mountain Jour. Math. 31 (2) (2001), 511–538.

[75] Mceliece, R.J., The Theory of Information and Coding, Addition Wesley Publ. Comp., Reading, 1977.

[76] Milovanović, G.V., On some integral inequalities, Univ. Beograd Publ. Elek. Fak. Ser. Mat. Fiz. No 496-No 541 (1975), 119–124.

[77] Milovanović, G.V. and J.E. Pečarić, On generalization of the inequality of A. Ostrowski and some related applications, Univ. Beograd Publ. Elek. Fak. Ser. Mat. Fiz. No 544-No 576 (1976), 155–158.

[78] Mitrinović, D.S., Analytic Inequalities, Springer-Verlag, Berlin-New York, 1970.

[79] Mitrinović, D.S., J.E. Pečarić and A.M. Fink, Classical and New Inequalities in Analysis, Kluwer Academic Publishers, Dordrecht, 1993.

[80] Mitrinović, D.S., J.E. Pečarić and A.M. Fink, Inequalities for functions and their integrals and derivatives, Kluwer Academic Publishers, Dordrecht, 1994.

[81] Ostrowski, A.M., Über die Absolutabweichung einer differentiebaren Funktion van ihrem Integralmittewert, Comment Math. Helv. 10 (1938), 226–227.

[82] Pachpatte, B.G., A note on some inequalities analogous to Grüss inequality, Octogon Math. Magazine 5 (2) (1997), 62–66

[83] Pachpatte, B.G., On an inequality of Ostrowski type in three independent variables, J. Math. Anal. Appl. 249 (2000), 583–591.

[84] Pachpatte, B.G., On some integral inequalities involving convex functions, RGMIA Res. Rep. Coll. 3 (3) (2000), 487–492.
[85] Pachpatte, B.G., A note on Hadamard like inequalities, Octogon Math. Magazin 8 (1) (2000), 37–42.
[86] Pachpatte, B.G., On a new Ostrowski type inequality in two independent variables, Tamkang J. Math. 32 (1) (2001), 45–49.
[87] Pachpatte, B.G., An inequality of Ostrowski type in n independent variables, Facta Univ. (Niš) Ser. Math. Inform. 16 (2001), 21–24.
[88] Pachpatte, B.G., A note on Ostrowski type inequalities, Demonstratio Math. 35 (2002), 27–30.
[89] Pachpatte, B.G., On Grüss type inequalities for double integrals, J. Math. Anal. Appl. 267 (2002), 454–459.
[90] Pachpatte, B.G., On Grüss type integral inequalities, J. Inequal. Pure and Appl. Math. 3 (1) Art. 11, 2002.
[91] Pachpatte, B.G., On multidimensional Grüss type inequalities, J. Inequal. Pure and Appl. Math. 3 (2) Art. 27, 2002.
[92] Pachpatte, B.G., On multivariate Ostrowski type inequalities, J. Inequal. Pure and Appl. Math. 3 (4) Art. 58, 2002.
[93] Pachpatte, B.G., New integral inequalities for differentiable functions, Tamkang J. Math. 34 (3) (2003), 249–253.
[94] Pachpatte, B.G., New weighted multivariate Grüss type inequalities, J. Inequal. Pure and Appl. Math. 4 (5) Art. 108, 2003.
[95] Pachpatte, B.G., On multidimensional Ostrowski and Grüss type finite difference inequalities, J. Inequal. Pure and Appl. Math. 4 (1) Art. 7, 2003.
[96] Pachpatte, B.G., On Trapezoid and Grüss like integral inequalities, Tamkang J. Math. 34 (4) (2003), 365–369.
[97] Pachpatte, B.G., On some inequalities for convex functions, RGMIA Res. Rep. Coll. 6 (2003), supplement.
[98] Pachpatte, B.G., A note on variants of certain inequalities for convex functions, Tamkang J. Math. 34 (2) (2003), 163–167.
[99] Pachpatte, B.G., A note on Ostrowski and Grüss type discrete inequalities, Tamkang J. Math. 35 (1) (2004), 61–65.
[100] Pachpatte, B.G., New inequalities of Ostrowski type for twice differentiable mappings, Tamkang J. Math. 35 (3) (2004), 219–226.
[101] Pachpatte, B.G., On a new generalization of Ostrowski's inequality, J. Inequal. Pure and Appl. Math. 5 (2) Art. 36, 2004.
[102] Pachpatte, B.G., New inequalities of Ostrowski and Trapezoid type for n-time differentiable functions, Bull. Korean Math. Soc. 41 (4) (2004), 633–639.
[103] Pachpatte, B.G., On Grüss type discrete inequalities, Math. Inequal. Appl. 7 (1) (2004), 13–17.
[104] Pachpatte, B.G., A note on integral inequalities involving two log-convex functions, Math. Inequal. Appl. 7 (4) (2004), 511–515.
[105] Pachpatte, B.G., New Ostrowski and Grüss type inequalities, An. Sti. Univ. Al. I. Cuza Iaşi Math. LI (2005), 377–386.
[106] Pachpatte, B.G., On Grüss like integral inequalities via Pompeiu's mean value theorem, J. Inequal. Pure and Appl. Math. 6 (3) Art. 82, 2005.
[107] Pachpatte, B.G., On Ostrowski-Grüss-Čebyšev type inequalities for functions whose modulus of derivatives are convex, J. Inequal. Pure and Appl. Math. 6 (4) Art. 128, 2005.
[108] Pachpatte, B.G., Mathematical Inequalities, North-Holland Mathematical Library, Vol. 67, Elsevier Science B. V., Amsterdam, 2005.
[109] Pachpatte, B.G., A note on Ostrowski like inequalities, J. Inequal. Pure and Appl. Math. 6 (4) Art. 114, 2005.

[110] Pachpatte, B.G., A note on Hadamard type integral inequalities involving several log-convex functions, Tamkang J. Math. 36 (1) (2005), 43–47.

[111] Pachpatte, B.G., On Čebyšev-Grüss type inequalities via Pečarić's extension of Montgomery identity, J. Inequal. Pure and Appl. Math. 7 (1) Art. 11, 2006.

[112] Pachpatte, B.G., New Čebyšev type inequalities via trapezoidal like rules, J. Inequal. Pure and Appl. Math. 7 (1) Art. 31, 2006.

[113] Pachpatte, B.G., On Čebyšev type inequalities involving functions whose derivatives belong to L_p spaces, J. Inequal. Pure and Appl. Math. 7 (2) Art. 58, 2006.

[114] Pachpatte, B.G., A note on integral inequalities involving product of two functions, J. Inequal. Pure and Appl. Math. 7 (2) Art. 78, 2006.

[115] Pachpatte, B.G., A new Ostrowski type inequality for double integrals, Soochow J. Math. 32 (2) (2006), 317–322.

[116] Pachpatte, B.G., New Ostrowski type inequalities involving the product of two functions, J. Inequal. Pure and Appl. Math. 7 (3) Art. 104, 2006.

[117] Pachpatte, B.G., A note on new integral inequalities of Grüss type, Demonstratio Math. 39 (2) (2006), 321–326.

[118] Pachpatte, B.G., On generalizations of Ostrowski type inequalities and some related results, An Sti. Univ. Al. I. Cuza Iaşi LII (2006), 365–375.

[119] Pachpatte, B.G., A note on Chebychev-Grüss type inequalities for differentiable functions, Tamsui Oxford J. Math. Sci. 22 (1) (2006), 29–36.

[120] Pachpatte, B.G., New generalizations of certain Ostrowski type inequalities, Tamsui Oxford J. Math. Sci., 23 (4) (2007), 393–403.

[121] Pachpatte, B.G., New Ostrowski and Grüss type inequalities, Tamsui Oxford J. Math. Sci. 22 (2) (2006), 159–166.

[122] Pachpatte, B.G., New inequalities of Čebyšev type for double integrals, Demonstratio Math. 40 (1) (2007), 43–50.

[123] Pachpatte, B.G., New Ostrowski type inequalities via mean value theorems, J. Inequal. Pure and Appl. Math. 7 (4) Art. 137, 2006.

[124] Pachpatte, B.G., Some new Ostrowski and Gtüss type inequalities, Tamkang J. Math. 38 (2) (2007), 111–120.

[125] Pachpatte, B.G., New Ostrowski and Grüss type inequalities for double integrals, Demonstratio Math. 39 (4) (2006), 783–791.

[126] Pachpatte, B.G., New Ostrowski type inequality for triple integrals, Tamkang J. Math. 38 (3) (2007), 253–259.

[127] Pachpatte, B.G., A note on Grüss type inequalities via Cauchy's mean value theorem, Math. Inequal. Appl., 11 (1) (2008), 75–80.

[128] Pachpatte, B.G., A note on new discrete Čebyšev-Grüss type inequalities, An. Sti. Univ. Al. I. Cuza Iaşi Math., LIV (2008), 117–121.

[129] Pachpatte, B.G., New Čebyšev type inequalities involving functions of two and three variables, Soochow J. Math., 33 (4) (2007), 569–577.

[130] Pachpatte, B.G., Some Grüss type inequalities for multiple integrals, Bul. Inst. Polt. Iaşi, Math. LII (LVI) (2006), 25–33.

[131] Pachpatte, B.G., New inequalities of Ostrowski-Grüss type, Face. Math. No. 38 (2007), 97–104.

[132] Pachpatte, B.G., A note on Čebyšev type inequalities, An. Sti. Univ. Al. I. Cuza Iaşi, Math. LIII (2007), 97–102.

[133] Pachpatte, B.G., New discrete Ostrowski-Grüss like inequalities, Facta Univ. (Niš) Ser. Math. Inform. 22 (1) (2007), 15–20.

[134] Pachpatte, B.G., On new inequalities of Ostrowski and Grüss type for triple integrals, Tamkang J. Math. 40 (2) (2009), 117–127.

[135] Pachpatte, B.G., New inequalities similar to Ostrowski's inequality, RGMIA Res. Rep. Coll. 12 (2) (2009), pp. 1–4.
[136] Pachpatte, B.G., On new inequalities involving convex functions, Advances in Inequalities for Special Functions, (Eds. N.S. Barnett and S.S. Dragomir), 2006, pp. 141–148.
[137] Pachpatte, B.G., New Čebyšev like integral inequalities, Advances in Inequalities from Probability Theory and Statistics, (Eds. P. Cerone and S.S. Dragomir), 2006, pp. 179–186.
[138] Pachpatte, B.G., On some discrete inequalities of Grüss type, Bul. Inst. Polt. Iaşi Math. LIII (LVIII), Fasc. 1-2 (2009), 21–28.
[139] Pachpatte, B.G., On a new generalization of Ostrowski type inequality, Tamkang J. Math., 38 (4) (2007), 335–339.
[140] Pachpatte, B.G., A note on a Trapezoid type integral inequality, Bull. Greek Math. Soc. 49 (2004), 85–90.
[141] Pachpatte, B.G., New Ostrowski type inequalities for mappings whose derivatives belong to L_p spaces, Demonstratio Math. 37 (2) (2004), 293–298.
[142] Pečarić, J.E., On the Čebyšev inequality, Bul. Sti. Tehn. Inst. Politchn Train Viua Timisora 25 (39) (1) (1980), 5–9.
[143] Pečarić, J.E. and S.S. Dragomir, A generalization of Hadamard's inequality for isotonic linear functionals, Radovi Mathematicki 7 (1991), 103–107.
[144] Pečarić, J.E., F. Prochan and Y. Tong, Convex functions, Partial orderings and Statical Applications, Academic Press, San Diego, 1992.
[145] Pompeiu, D., Sur une proposition analogue au théoreme des accrojssements finis, Mathematica (Cluj, Romaina), 22 (1946), 143–146.
[146] Rudin, W., Principles of Mathematical Analysis, McGraw-Hill Book Company Inc., 1953.
[147] Sahoo, P.K. and T. Riedel, Mean value theorems and functional equations, World Scientific, Singapore, 2000.
[148] Sofo, A., An integral approximation in three variables, J. Ineq. Pure Appl. Math. 4 (3) Art. 58, 2003.
[149] Tseng, K.-L., G.-S. Yang and S.S. Dragomir, On Quasi convex functions and Hadamard's inequality, RGMIA Res. Rep. Coll. 6 (3) (2003), 405–417.
[150] Ujević, N., Perturbations of an Ostrowski type inequality and applications, Inter. J. Math. Math. Sci. 32 (8) (2002), 491–500.
[151] Ujević, N., New bounds for the first inequality of Ostrowski-Grüss type and applications, Computers and Math. Appl. 46 (2003), 421–427.
[152] Ujević, N., Ostrowski-Grüss type inequalities in two dimensions, J. Inequal. Pure and Appl. Math. 4 (5) Art. 101, 2003.
[153] Ujević, N., A new generalization of Grüss inequality in inner product spaces, Math. Inequal. Appl. 6 (4) (2003), 617–623.
[154] Ujević, N., Sharp inequalities of Simpson type and Ostrowski type, Computers and Math. Appl. 48 (2004), 145–151.
[155] Ujević, N., A generalization of Ostrowski's inequality and applications in numerical integration, Appl. Math. Lett. 17 (2004), 133–137.
[156] Yosida, K. Functional Analysis, Springer-Verlag, Berlin, 1978.

Subject Index

FSC
www.fsc.org
MIX
Papier | Fördert gute Waldnutzung
FSC® C083411

Kolibri 360 GmbH
Ferdinand-Jühlke-Straße 7
99095 Erfurt, Deutschland
produktsicherheit@kolibri360.de